6th International Interdisciplinary Scientific Conference SOCIETY. HEALTH. WELFARE

SHS Web of Conferences Volume 40 (2018)

Riga, Latvia
23 - 25 November 2016

Editors:

Uldis Berkis
Lolita Vilka

ISBN: 978-1-5108-5727-8

Printed from e-media with permission by:

Curran Associates, Inc.
57 Morehouse Lane
Red Hook, NY 12571

Some format issues inherent in the e-media version may also appear in this print version.

Printed by Curran Associates, Inc. (2018)

For additional information, please contact EDP Sciences – Web of Conferences at the address below.

EDP Sciences – Web of Conferences
17, Avenue du Hoggar
Parc d'Activité de Courtabœuf
BP 112
F-91944 Les Ulis Cedex A
France

Phone: +33 (0) 1 69 18 75 75
Fax: +33 (0) 1 69 28 84 91

contact-edps@webofconferences.org

Additional copies of this publication are available from:

Curran Associates, Inc.
57 Morehouse Lane
Red Hook, NY 12571 USA
Phone: 845-758-0400
Fax: 845-758-2633
Email: curran@proceedings.com
Web: www.proceedings.com

TABLE OF CONTENTS

Foreword

International interdisciplinary scientific conference "*Society. Health. Welfare*", held by Rīga Stradiņš University in Latvia every second year, has proved itself over the years as a valuable platform for exchanging views and experience at international level, bringing together social, humanitarian and medical professionals, researchers and academic staff of institutions of higher education.

These collected articles are proceedings of the Part 1 of the 6th International conference. It includes **36** articles. The authors of the articles represent different countries: Latvia, Lithuania, Czech Republic, Slovakia, Russia and Kazakhstan. In general, a wide thematic diversity clearly reflects the issues which are of concern to the society as a whole and are at the focus of researchers' attention.

Following the interdisciplinary idea of the conference, the proceedings are arranged in three chapters.

The **first chapter *Society*** covers two main thematic lines: *legal and social* issues and *the rule of law and legality*.

Two articles on legal framework in health care regarding the patients' rights to high-quality health care and the responsibility of medical staff introduce the proceedings. The authors of the article *Procedural Aspects of Healthcare Quality Control in Latvia and its Effect on Legal Protection of Patients* give a comprehensive analysis of healthcare quality control mechanisms in line with the implementation of the Directive 2011/24/EU in Latvia. The authors draw attention and highlight the problems associated with the role of the Health Inspectorate and the competence of the court in assessing offence in health care. Whereas the article *Primary Law in Medicine, its Legal Nature* analyses the right to health care through the prism of economic opportunities, emphasizing the special legal status of the child.

The article *The Effectiveness of Social Work Services for Families Whose Children are in Temporary Custody* reflects the work of social workers with high-risk families. Such families are usually of multi-problem character. The authors conclude that it is not always possible to find a safe and healthy environment for children from families of social risk due to limitations of the social services system and insufficient cooperation between the various levels of decision-making bodies.

The proceedings contain several articles that analyse socially negative phenomena such as bribery, fictitious employment and corruption. The authors of the articles suggest that the development of a legal and democratic country is not at all simple.

Several articles of the **second chapter *Health*** focus on healthcare policy and modernization issues. Analysis of reform processes in the healthcare sector in the last ten years in Latvia is given in the articles *Interest Group Impact on E-health Implementation in Latvia*; *Evaluation of Healthcare Policy in Latvia* and *Latvian Health Care Competitiveness in Relation to its Infrastructure and Available Resources.* The situation in the field of health care is displayed from the comparative perspective between the Baltic and Scandinavian countries.

In today's society, the maintenance of health and personal wellbeing depends fundamentally on people's own responsibility and knowledge. Two sectors, health and education sectors are incorporated in health education. Health education and awareness of the social dimension of health are analysed in the articles *Health Literacy as a Challenge of Health Education* and *Young Adult Health Literacy on Health Promotion as Social Determinant of Health: Challenge for Health Education Practice in Lithuania*.

Complexity of the modern world and dynamic changes are today's reality. The global dimension is present in all areas of life: politics, economy, social and cultural space. Changes in the norms and values of the society affect the behaviour of the personality and often confront the aims of the personality. Therefore, adaptation to social transformations becomes a challenge. The authors of the article *The Relationship between Anomia and Maladaptive Personality Traits* analyse the reflection of general social change as "subjectively perceived anomie" on an individual level.

Several articles describe the personality health behaviour in relationship with lifestyles, as well as the environmental impact on health, such as air quality and health risk.

The role of rehabilitation is becoming increasingly important in the recovery process. The article *Physiotherapy Following Cardiac Surgery: Program Comparison* is devoted to examination and comparing the usability of two physiotherapy programmes, analysing respiratory function in patients before and after cardiac surgery.

The article *Perception of Ethical Issues by Students of Health Professions* concludes the chapter *Health*. The article focuses on education of healthcare professionals within the context of the development of cultural competence in today's multicultural society, which requires specialists to be prepared to work with patients from various social and cultural environments.

The articles included in the **third chapter *Welfare*** focus on a variety of aspects of welfare: social work in health care; the spiritual care of the dying people; adaptive behaviour as a factor in psychological health; anticipating driving behaviour in relation to socio-demographic characteristics and the system of individual values; psychological safety for a mother having children with special needs; feeling of happiness in the family; the role of physical activity in functioning; environmental accessibility and the impact of non-medical factors on patient health and quality of life.

The articles *Parenting in Low-Income Families from the Perspective of Social Work* and *Why the Share of Small Amount Pensions is so substantial in Latvia?* Conclude the proceedings. Despite the fact that the authors come from different countries and different target groups are being analysed, the two articles share a common guiding principle – '*financial vulnerability*' and the '*risk of poverty*' as important issues in the future.

The proceedings comprehensively cover the current topical issues through diversity revealing both the common and different trends in Eastern European and non-European countries. The vision and insights of the authors are able to stimulate discussion on approaches and solutions for accepting the challenges of the change in the modern society.

Prof. Margarita Teresevičienė
Institute of Education, Vytautas Magnus University, Lithuania

Statement of Peer review

In submitting conference proceedings to *Web of Conferences*, the editors of the proceedings certify to the Publisher that

1. They adhere to its **Policy on Publishing Integrity** in order to safeguard good scientific practice in publishing.
2. All articles have been subjected to peer review administered by the proceedings editors.
3. Reviews have been conducted by expert referees, who have been requested to provide unbiased and constructive comments aimed, whenever possible, at improving the work.
4. Proceedings editors have taken all reasonable steps to ensure the quality of the materials they publish and their decision to accept or reject a paper for publication has been based only on the merits of the work and the relevance to the journal.

Title, date and place of the conference

6th International interdisciplinary scientific conference SOCIETY. HEALTH. WELFARE,
23rd – 25th November, 2016

Proceedings editor(s):

Assoc. Prof. ULDIS BERKIS; Assoc. Prof. LOLITA VILKA

Date and editor's signature 15/11/2017

Procedural aspects of healthcare quality control in Latvia and its effect on legal protection of patients

A. Liepins, O. Joksts, and J. Vetra

Rīga Stradiņš University, Riga, Latvia

Abstract. According to the European Commission data, 8–12% of patients cared for in hospitals throughout the European Union have suffered accidents related to the healthcare provided for them. The Directive 2011/24/EU of the European Parliament and Council of 9th March 2011 on the application of patients' rights in cross-border healthcare provides that the Member States should ensure that patients have easily accessible and transparent appeal procedures and mechanisms that provide possibilities for legal remedies in cases of inflicted harm during medical treatment according to regulations of the respective Member State. Healthcare quality control mechanisms are intended for two major purposes: to identify accidents that have occurred during medical treatment and to prevent similar accidents from reoccurring in the future. The aim of this article is to evaluate, firstly, the procedural aspects of healthcare quality control mechanisms in Latvia and, secondly, how healthcare quality control mechanisms have been affected with the implementation of the Directive 2011/24/EU in Latvian legal order, providing for a specific legal remedy – the newly created Medical risk fund. Conclusions are made on the procedural nature of the patients' right to submit complaints and also the developments in Latvian court practice related to the healthcare quality control. The authors have also analysed regulations related to the Medical risk fund, its influence on the civil liability mechanisms of medical practitioners, as well as the patient's right to obtain just compensation for the harm inflicted to his health and the legal nature of opinions of the Health Inspectorate of Latvia.

1 Introduction

Healthcare quality is a vital matter for the society as a whole. According to the European Commission data, 8–12% of patients cared for in hospitals throughout the European Union have suffered accidents related to the healthcare provided for them. In cases when a medical practitioner has provided poor healthcare to a patient, this practitioner may be subjected to (1) criminal or (2) administrative liability and/or (3) civil liability and/or (4) disciplinary liability, and/or (5) professional liability, and/or (6) liability for the breach of a public law contract.

The European Charter of Patients' Rights, which is a doctrinal document [6] and which specifies the scope of Article 35 of the Charter of Fundamental Rights of the European

Union [3], also provides the right to healthcare quality standards, the right to safety during medical treatment as well as the right to submit complaints regarding this treatment.

The Directive 2011/24/EU of the European Parliament and Council of 9th March 2011 on the application of patients' rights in cross-border healthcare (hereinafter – the Directive) provides that the Member States have to ensure that patients have easily accessible and transparent appeal procedures and mechanisms that provide possibilities for legal remedies in cases of inflicted harm during medical treatment according to regulations of the respective Member State. As a result of this, on 25th October 2013 the Medical risk fund (*Veselības riska fonds*) concept was implemented in Latvia, which was based on the Swedish and Danish examples. Functioning of the Medical risk fund is ensured by two state institutions that are directly subordinated to the Ministry of Health of the Republic of Latvia: the Health Inspectorate and the National Health Service.

This article offers a detailed research on the following issues:

1. Administrative procedural aspects of the healthcare quality control;
2. Scope of patients' right to submit complaints in Latvia;
3. Available court control in healthcare violation cases;
4. Civil law aspects of the healthcare quality control.

In addition, it should be noted that the topicality of this article is based on the recent changes in administrative and civil liability of medical practitioners in Latvia.

2 Administrative procedures of healthcare quality control

Healthcare quality control in Latvia is carried out via administrative procedures conducted by the Health Inspectorate, which is directly subordinated to the Ministry of Health [10]. The Health Inspectorate monitors and controls the professional quality of healthcare, without any specific complaints from third parties regarding the quality of healthcare. Despite the fact that the Health Inspectorate has put forward as its priority objective to establish a united healthcare facility monitoring system, which would ensure a long-term healthcare facility monitoring program instead of examining individual cases [1], nevertheless the major importance remains with complaints submitted by patients, since these signal of a possible violation of the healthcare quality in a specific place and time. The Health Inspectorate's replies to submitted complaints, as well as its issued administrative acts and performed factual conduct are subject to a control in court, when there is a dispute of the lawfulness or validity of these replies, administrative acts or factual conduct.

On 1st March 2010 the new Law on the Rights of Patients [8] came into force, which was intended as a unified legislative act for the protection of the rights of patients, ensuring compliance with the rights and legitimate interests of patients in matters related to medical treatment. Until 1st March 2010 the regulation of patients' rights was partly provided for in the Medical Treatment Law [9].

Until 1st March 2010 Article 22 of the Medical Treatment Law provided for the right of the patient to submit complaints and to receive a healthcare quality evaluation of the provided medical treatment, however, stating that such evaluation can be received in cases specifically provisioned for in the respective legal acts of the specific Member state. Since 1st March 2010 these relevant rights of patients are enshrined in Article 18, paragraph 3 of the Law on the Rights of Patients. The right of a patient to submit complaints is considered as patient's procedural right, which also provides an opportunity for the protection of other patients' rights in cases when such rights are not ensured.

Consequently, when reviewing a patient's complaint, the Health Inspectorate examines the quality of healthcare provided to patient and – as a result of that – is authorized to make three different decisions, which have different legal consequences, that is, to acknowledge:

1) that there has been a violation of patient's healthcare and, consequently, to satisfy the submitted complaint in full. In this case the Health Inspectorate whether initiates an administrative violation procedure or acknowledges that a criminal offence has been committed. In the latter case this legal opinion is sent to the police authorities in order for them to decide whether criminal proceedings should be initiated (here it must be stressed that criminal matters are not examined within the scope of this article);
2) that there has been a violation of patient's healthcare and, consequently, to satisfy the submitted complaint in part and to initiate an administrative violation procedure, while partly acknowledging what is not to be considered as a violation of patient's healthcare. In this case the Health Inspectorate initiates an administrative violation procedure and – regarding the decision in part where there has not been a violation of patient's healthcare – it informs the submitter that the complaint has been partly rejected due to the fact that in this part there has not been a violation of patient's healthcare;
3) that there has been no violation of patient's healthcare at all. In this case the Health Inspectorate informs the submitter that there has not been a violation of patient's healthcare.

It should be noted that the various decisions of the Health Inspectorate procedurally differ from each other. The same applies to the procedures for submitting an appeal.

The Health Inspectorate decision to initiate an administrative violation procedure (investigation) is of a criminal law nature. In several cases the European Court of Human Rights has stated that administrative offence cases are similar to criminal cases ("Pfarrmeier v. Austria", "Engel and others v. the Netherlands", "Zolotukhin v. Russia", etc.) because of the fact that a person has a right to equal procedural guarantees as in criminal proceedings. In these kinds of procedures the dominant principle is the adversarial principle. If a medical practitioner is found guilty in administrative violation procedure, then the Health Inspectorate is entitled to impose an administrative sanction (quasi-criminal penalty) for committing a violation of patient's healthcare [16]. In addition it must be pointed out that from 1st July 2012 onwards an administrative offence procedure reform is being carried out in Latvia, during which the previously mentioned opinion has been maintained that administrative violation procedures have a criminal law nature and that during the investigation of a case the respective defendant is entitled to criminal procedural guarantees. At the same time it must be pointed out that the scope of the patients' rights has not been significantly affected in this reform.

The Health Inspectorate's decision that there has not been a violation of patient's healthcare is considered as a factual conduct within the scope of the Administrative Procedure Law. Until 1st March 2010 during administrative proceedings a person was entitled to submit a complaint for this factual conduct in the administrative court, which would then review the case in three instances: the court of first instance, the court of appeal and the court of Cassation. However, from 1st March 2010 during administrative proceedings the Health Inspectorate examines the complaints of healthcare violations in accordance with the Law on Submissions [7, 18]. Since this date the actions of an institution initially were controlled by two administrative court instances (the court of first instance and the court of Cassation), but after regulation amendments on March 15th 2012 (which came into force on April 18th 2012) the cases are reviewed only in the first instance court – the District Administrative Court. Consequently, during the period from March 1st

2010 until April 18th 2012 the scope of patients' rights to submit applications to the administrative courts have been significantly narrowed.

Furthermore in the annotation of the law amendments of March 15th 2012 [5] it was stated that the applications of violations of the Law on Submissions shall be reviewed in the District Administrative Court and that the rulings of this particular court will not be subject to an appeal. Such choice was made in order to reduce the large burden of cases on courts, which, however, cannot lead to a violation of person's rights, since the Law on Submissions is not intended for the protection of subjective public rights, but offers a possibility for persons to participate in the public administration.

The subjective public rights are conferred to a person by legal norms. According to the protective provisions theory, a public law norm grants the applicant subjective public rights if the norm:

1) imposes an obligation to take a certain action or refrain from it;
2) this obligation serves also for protection of individual rights;
3) the applicant is among the eligible (protected) group of persons [19].

Consequently, it must be concluded that the thesis included in the annotation of the law amendments that the Law on Submissions is not intended for the protection of subjective public rights, but instead for the participation in the public administration, is not acceptable. However the amendments in the Law on Submissions should not be considered as a mistake. The authors' point of view is that a mistake was indeed made by adopting Article 18, paragraph 3 of the Law On the Rights of Patients, which provides that the Health Inspectorate's reply to a patient's complaint must be given in accordance with the Law On Submissions, arguing that the patient's complaint belongs to the subjective public law, which serves as an individual legal remedy for the patient. The authors believe that Article 18, paragraph 3 of the Law On the Rights of Patients should be amended, excluding from this legal norm the reference that the appeal must be made in accordance with the Law On Submissions.

3 Latvian court control in healthcare violation cases

In its rulings the Supreme Court of Latvia has stated that "the court is required to examine whether the Health Inspectorate, which controls the enforcement of regulations in healthcare related matters, has ascertained all aspects of the case and made reasonable conclusions [...]. The court itself cannot evaluate the quality of healthcare provided" [20–22]. In these and many other rulings the Supreme Court has accepted the argumentation of the Health Inspectorate's decisions that were appealed before it, that evaluation of such an appeal clearly requires special knowledge in medicine.

The question whether the court itself cannot evaluate the quality of healthcare provided, is a question of debate. During an administrative violation procedure the Health Inspectorate performs a healthcare quality expertise, forms a legal opinion (which by its legal nature is considered as an evidence – an opinion of an expert which consists of the healthcare violation description within the light of medical science), prepares the protocol of an administrative violation, performs various other investigative actions and, consequently, makes a decision on the administrative liability of the respective medical practitioner. It is beyond any doubt that the judge does not have the required special knowledge in medical science, while the medical practitioners - experts of the Health Inspectorate have such special knowledge. However, a lack of special knowledge cannot be an obstacle to evaluate correlations, particularly if the party provides sufficient evidence that the legal opinion passed by the Health Inspectorate is unreasonable.

It should be pointed out that in its rulings the Supreme Court has sometimes distanced from the regular practice that the court does not evaluate the quality of healthcare provided. In these rulings the Supreme Court has stated that "the legal opinion lacks correlation between the information provided and conclusions" [24, 25]. Still, for the court to ascertain existence of factual conduct in a particular case, it would need special knowledge, in order to be able to evaluate the correlation or lack of argumentation, which the court does not have.

Thereby it is evident that the court practice in healthcare violation cases is not consistent. As a secondary consequence of this, a person cannot safely conduct their behaviour: the person must be able – if need be, by receiving advice from a specialist – to predict the consequences of her actions in a reasonable way. However, the primary problem of such inconsistent court practice is directly connected with the society's general confidence in court.

The authors believe that the court should not generally distance itself from evaluations in medical or specific scientific matters. However, that does not mean that judges should be specialists in all branches of science. Still, in unclear matters the courts should use the available procedural instruments, such as an additional expertise and/or a repeated expertise, consequently interrogating the relevant experts. Such an approach to this matter in Latvia has not been exercised yet. As a result of the procedural reform of administrative violations law, since July 1st 2012 adversarial principle is being implemented in these cases and so the courts are limited in *ex officio* appointing an expertise and inviting experts to the court hearings. However, the court may use these procedural instruments by referring to the legal clarification principle when explaining to the parties involved their procedural rights to submit relevant requests during the trial.

In cases when the Health Inspectorate has decided that there has not been a violation of healthcare quality, the court (in an administrative procedure) is not limited in *ex officio* appointing an expertise and inviting experts to the court hearings, because administrative proceedings must be carried out in accordance with the principle of objective investigation. Taking into consideration that such cases are reviewed in the court of first instance, the authors believe that it is of a particular importance to use these procedural instruments in order to determine whether there has been or has not been a violation of healthcare.

However, an additional expertise or a repeated expertise must not be appointed when a party simply raises objections against the expertise provided by the Health Inspectorate. A criteria to determine, whether an additional expertise or a repeated expertise should be appointed, is the already existing argument in court practice, which consists of a lack of correlation between the information provided and the conclusions made. It was exactly this argument, which was used in a civil case, where a patient brought an action against four medical institutions for the harm inflicted to her health. In this particular case the court – regardless of the Health Inspectorate's legal opinion which was in favour of the patient – did not find any connection between the actions of respective medical practitioners and the patient's course of illness and, consequently, her disability. In this particular case the court of appeal stated that "in the file of the case there is no evidence that would conclusively confirm that during the period of [...] in the hospital of V. there were mistakes conceded that would be in a correlation with the damage of patient's health, mutilation and the subsequent moral and material damage" [23].

4 Civil law aspects of the healthcare quality control

In most cases, violation of healthcare quality leads to inflicting harm to the patient. According to Article 2347 of the Civil Law of the Republic of Latvia [12] the patient has the right to just compensation for the inflicted bodily injury, which includes:

1) expenses for medical treatment;
2) potential loss of income (pursuant to discretion of the court);
3) remuneration (material compensation) for the suffered moral injury.

For the patient to be eligible to compensation, he must prove an existing correlation between the harm inflicted to his health and a wrongful act during the medical treatment. In Latvia such evidence may be acquired in three possible ways:

1) healthcare quality expertise and, consequently, a legal opinion of the Health Inspectorate;
2) expertise of State Centre for Forensic Medical Examination;
3) expertise of a medical commission appointed by court.

Furthermore, it must be noted that healthcare quality expertise theoretically may be performed by medical commission or a foreign forensic institution chosen by patient himself. However, to the authors' knowledge so far such situations have not occurred in Latvia.

As mentioned in the introduction of this article, on October 25th 2013 the Directive was implemented in Latvian legal system. Among other things, the Directive provisions an obligation of the Member States towards patients (including patients from other Member States) to ensure accessible procedures and mechanisms for submitting complaints, which would allow a patient to seek legal remedies in cases when the provided healthcare has inflicted harm to him. As a result of this, on October 25th 2013, on basis of the Swedish and Danish experience, there were made amendments in the Law on the Rights of Patients, providing for a specific legal remedy – the Medical risk fund.

Up until October 25th 2013, a person, who had suffered from a wrongful act during the healthcare provided, claimed the compensation from the respective medical institution or medical practitioner, which or who had provided the healthcare. In case of a dispute such actions in court were reviewed in a civil procedure and in a three level court system. Currently Latvian courts still review cases of pecuniary claims for the harm inflicted to patient's life or health until October 25th 2013. However, starting from this date the harm inflicted to patient's life or health is being solved via a specific administrative mechanism for the protection of rights of patients, the Medical risk fund.

The general principles of the Medical risk fund are enshrined in Article 17 of the Law on the Rights of Patients, but the general functioning of this fund is envisaged in the Regulation No. 1268 of the Cabinet of Ministers of the Republic of Latvia from November 5th 2013 "Functioning rules of the Medical risk fund" [17].

The institutional structure of the Medical risk fund is dual, i.e., its functioning is ensured by two institutions that are directly subordinated to the Ministry of Health: the Health Inspectorate (which provides expertise, prepares legal opinions, determines the extent of the compensation in percentage and evaluates the relation of the incurred medical expenses with the mitigation or prevention of the harm inflicted to the patient) and the National Health Service (which administers the Medical risk fund assets and on the basis of the legal opinion of the Health Inspectorate decides either to pay or refuse compensation to the patient from the assets of this fund). The decision or factual conduct of the National Health Service regarding the payment of compensation from the Medical risk fund may be challenged before the Ministry of Health and the decision of the Ministry of Health in its part may be appealed in court in accordance with the Administrative Procedure Law.

However, the patient is not entitled to challenge or appeal the legal opinion of the Health Inspectorate.

Discussing the legal nature of the Health Inspectorate's legal opinion, it can be concluded that it is neither an administrative act nor an interlocutory act in the administrative proceeding, but must be considered as the evidence – a legal opinion of an expert. Firstly, this evidence is characterized by persons who prepare it – persons who have specific knowledge and who give a specific opinion on a particular case, for whose evaluation this specific knowledge is required. Neither the Law on the Rights of Patients nor the Functioning Rules of the Medical risk fund provide for a procedure in cases when the legal opinion of the Health Inspectorate is inadmissible or improbable as the evidence (considering that it is unlikely that a disagreement could arise of the relevance of this kind of evidence). In any other type of proceedings, in the case when reasonable doubts arise it is possible to appoint an additional or repeated expertise to eliminate them.

When interviewing the responsible controlling officials (on June 26th 2015 the authors interviewed officials of the Ministry of Health who are involved in preparing the decisions regarding appeals against the decisions of the National Health Service), it was found that in cases when the legal opinion of the Health Inspectorate is not entirely clear or is incomplete and when, consequently, in regard to this legal opinion additional questions have emerged, during the appeal process the officials of the Ministry of Health take measures which are considered as a repeated expertise. Namely, if in its legal opinion the Health Inspectorate has not given assessment on some circumstances of the case, if new evidence has been submitted during the appeal procedure or if some conclusions of the opinion are controversial, the Ministry of Health requests to the Health Inspectorate to give a legal opinion to the circumstances that have not been analysed previously. Although such an action is not procedurally defined, in accordance with *ratio legis* such an action is to be qualified as a repeated expertise.

However, there is not a clear and understandable legal mechanism for appointing a repeated expertise (for comparison, such legal basis are provided in Article 182, paragraph 3 of the Administrative Procedure Law [11], Article 197 of the Criminal Procedure Law [15] and Article 125, paragraph 3 of the Civil Procedure Law [13]) in cases such as: an unreasonable legal opinion of the Health Inspectorate, significant deficiency or methodological mistakes, expert's lack of qualification, incompetence or other incompatibilities. The officials of the Ministry of Health pointed out that so far such situations have not occurred. Nevertheless, from the viewpoint of the protection of patients' rights, the lack of regulation regarding additional or repeated expertise is a step back.

With the implementation of the Directive, Paragraph 8.3 of the Rules of Functioning of the Medical risk fund provides the Health Inspectorate a right to "request to a commission (with no less than three medical practitioners and, if additional assistance is required, by inviting specialists from other medical institutions) which is formed by the head of the medical institution (for which actions the compensation is requested in the patients complaint) to provide an evaluation and opinion on the harm inflicted to the patient and the extent of it". From the authors' point of view, such a regulation, which basically provides that the violator of patient's rights evaluate its own mistakes, is nothing more than a legislative defect. In relation to this it must be pointed out that already in the Roman law there was developed the principle of *nemo iudex in causa sua* (no one should be a judge in his own case).

It should be noted that until June 27th 2015 the Health Inspectorate has not exercised this regulation yet. In the authors' opinion, the relevant regulation does not comply with the fundamental right to a fair trial (Article 92 of the Constitution of the Republic of Latvia [4, 14]. Meanwhile, until June 27th 2015 the National Health Service had received 108 submissions of compensation claims (from which 11 were refused, 1 was withdrawn and 1 was found as not filed). Regarding the submitted complaints in 8 cases the National Health Service decided to pay compensation to the patient, in 24 cases it decided to refuse the claim of compensation, from which 5 have been challenged before the Ministry of Health for the decision to refuse the claim of compensation and in 1 case challenged for the decision to refuse examination of the submitted claim [26].

At this moment, there are too few decisions of the National Health Service in which it has decided to pay compensations to patients, in order to make any assessment regarding effectiveness of the new system.

Concerning the implementation of the Medical risk fund concept, it must be noted that also here there has been a change of procedural principles. Civil procedure principles such as the adversarial principle and the principle of equality of arms have been replaced by the principle of objective investigation and other administrative procedure principles. The patient is no longer entitled to choose experts, to express an opinion on preferred candidatures of experts, to ask questions to experts or to get acquainted with the experts' qualification which in many cases is the most important issue during the stage where expertise is appointed. At the same time, according to Article 37, paragraph 2 of the Administrative Procedure Law there still remains the patient's right to apply a recusal to an expert.

The sums of compensation, to which the patient is entitled for the harm inflicted to his life and health, are determined by the court and at the discretion of the court (Article 5 of the Civil Law of the Republic of Latvia). Since October 25th 2013 this matter is regulated in accordance with the Rules of Functioning of the Medical risk fund. Determination of the severity of moral injury in percentage, which is specified in a legal norm, is a new approach to this matter in Latvia. In this case, a clear gain is the legal certainty, because on the same factual conditions there will be an equal compensation, thereby excluding the so-called human factor in the evaluation. However, at the same time here disappears the advantage of the discretion of the court, which individualised each particular case and envisaged to determine the amount of compensation according to the purpose and aim of Article 1635 of the Civil Law of the Republic of Latvia – for the physical and mental suffering which is individual for each person and not the same for everybody.

Also in Sweden and Denmark, on whose basis the Medical risk fund has been implemented in Latvia, compensations are paid for: (1) damages – loss of income, loss of pension, medical treatment expenses; (2) non-pecuniary damages – for pain and suffering (for example, in Denmark – 130 Danish kronor for each day of illness), for disability, for mutilation, for incapacity to work. The compensation for disability is paid as a lump sum. In Sweden such compensation is determined in accordance with a table approved by the Commission of Road Traffic Accidents, which is used by all Swedish insurance companies and courts [2]. This data shows the reason for excluding the discretion of the court in the administrative stage (to which public institutions also cannot be assigned) regarding the determination of the compensation to patient.

It should be noted that while performing an expertise the Health Inspectorate uses only the patient's medical documents, but does not proceed with an interview or an examination of the patient himself. Basically, during this process the patient is entitled to express himself only once, when submitting his complaint to the National Health Service. This can be regarded as an ignorance of the principle of objective investigation. It was concluded previously that so far in Latvia the court practice development in the aspect of the civil liability of medical practitioners cannot be evaluated unambiguously by the fact that in court rulings the court does not evaluate the multifactorism of failed medical treatment causalities. By transferring failed medical treatment causality evaluation from courts, that are independent bodies from the executory power of medical institutions, to a directly subordinated institution, the Health Inspectorate, and, in addition, not providing any other causal forms than only *conditio sine qua non* in the Rules of Functioning of the Medical risk fund, one cannot expect a deeper understanding in causal matters and hope for a positive development in healthcare violation cases.

Respectively, the primary aim of the Medical risk fund is to protect the patient. However, it has to be admitted that a completely different aim will be accomplished very effectively – to protect medical institutions and medical practitioners from negative pecuniary consequences in cases of a professional failure or negligence. It is going to be so, because the functioning of the Medical risk fund is based on the principle of mandatory insurance concept, as a result of which the pecuniary liability of medical institutions and practitioners according to Article 16, paragraph 2 of the Law On the Rights of Patients is limited to 142 290 euros for the harm inflicted to patient's life and health (also non-pecuniary damage) and to 28 460 euros for the patient's medical treatment expenses. However, a question arises how the patient should act in cases when s/he considers that the compensation limits provided for in Article 16, paragraph 2 of the Law on the Rights of Patients do not compensate the inflicted harm in full. In such situations the patient is entitled to file a civil claim in court. Consequently, it can be concluded that medical institutions and practitioners shall be excluded from negative pecuniary consequences in cases of a professional failure or negligence only if the patient shall be satisfied with the compensation sums paid out from the Medical risk fund.

At the same time, it must be remembered that the aim of the Medical risk fund, as defined by the Directive, is to ensure that for the patients there are available and transparent complaint submission procedures and mechanisms, which allow patients to seek remedies in accordance with the regulations of the respective Member State, in cases when the healthcare provided has inflicted harm to him. The aim of the Directive cannot be achieved if the limits of compensation provided for in Article 16, paragraph 2 of the Law on the Rights of Patients do not fully compensate inflicted harm to the patient.

Regarding this matter, the place of the disposition (or personal autonomy) principle in the Rules of Functioning of the Medical risk fund must also be discussed when implementing the concept of this fund. Annex 1 of these rules does not provide for the applicant to specify, among other information, the sum of money which they consider to be an adequate compensation for the harm inflicted to their life or health.

Returning to the matter of the primary aim of the Medical risk fund – to protect the patient – it must be noted that in cases when patient will be satisfied with the decision of the National Health Service and shall not challenge it before the Ministry of Health and/or appeal it to the court, then this aim shall be achieved. Otherwise such proceedings are simply going to be longer in comparison to civil proceedings.

5 Conclusions

The patient's right to submit complaints is considered as patient's procedural right which also provides an opportunity to protect other patient's rights in cases when such patient's rights are not ensured. During the period from March 1st 2010 until April 18th 2012 the scope of patient's right to submit applications to administrative courts was significantly narrowed.

By adopting Article 18, paragraph 3 of the Law on the Rights of Patients, a legislative mistake was made, providing in the relevant regulation that the reply of Health Inspectorate to a complaint filed by a patient must be given in accordance with the Law on Submissions, arguing that the patient's complaint belongs to the subjective public law which serves as an individual legal remedy for the patient. The authors believe that the respective regulation must be amended, by excluding the reference that the appeal must be made in accordance with the Law on Submissions.

The court practice in healthcare violation cases is not unambiguous. The primary consequence of such court practice is directly connected with the confidence of society in the actions of the court. A secondary problem is that persons cannot safely conduct their activity: the persons must be able – if need be, by receiving advice from a specialist – to predict the consequences of their actions in a reasonable way.

The courts by themselves cannot evaluate the reasonableness of medical conclusions of the Health Inspectorate without special knowledge in medical science. It should not be allowed that the courts distance themselves from evaluation of medical matters in general. However, judges should not be specialists in all branches of science. Still, in unclear matters the courts should make use of the available procedural instruments such as an additional expertise and/or a repeated expertise, consequently interrogating the relevant experts.

Considering that case in which the Health Inspectorate has acknowledged that there has not been a violation of healthcare, are reviewed in the court of first instance, it is especially important to appoint an additional expertise and/or a repeated expertise to eliminate any doubts regarding the possible violation of healthcare, if such circumstances are in dispute.

A measure to determine whether an additional expertise or a repeated expertise should be appointed can be found in the existing argument from court practice regarding the lack of correlation between the information provided and the conclusions made.

The Medical risk fund concept replaces the traditional civil liability model of medical practitioners. The Medical risk fund is both a specific insurance institute and a legal remedy for the protection of patients' rights.

Until the introduction of the Medical risk fund in Latvian legal order, which determines the extent of the compensation, there dominated the discretion of the court model, but since the introduction of this concept the severity of the moral injury in percentage is determined in accordance with legal acts. Determination of the severity of moral injury in percentage, which is specified in a legal norm, is a new approach to this matter in Latvia.

The civil law principles governing civil liability of medical practitioners have been replaced with administrative procedure principles, giving a central role to the Health Inspectorate who has the competence both to perform expertise and to determine the compensation for the moral injury in percentages.

At this moment, there are too few decisions of the National Health Service in which it has decided to pay compensations to patients, in order to make any assessment regarding effectiveness of the new system.

By transferring failed medical treatment causality evaluation from independent courts to a directly subordinated institution, the Health Inspectorate, and by not providing any other causal forms than only *conditio sine qua non* in the Rules of Functioning of the Medical risk fund, one cannot expect an understanding in causal multifactorism matters in healthcare violation cases.

From the point view of the protection of patients' rights, the absence of any regulation regarding appointment of additional or repeated expertise in relation to the expertise provided by the Health Inspectorate within the framework of the Medical risk fund concept, must be considered as a significant deficiency.

The Rules of Functioning of the Medical risk fund confer a right to the Health Inspectorate to demand for the head of the medical institution, for which actions the compensation is requested by the complaint of the patient, to appoint a commission and to provide an evaluation as well as an opinion on the harm inflicted to patient and the extent of it. Such a provision does not comply with the content of the principle of *nemo iudex in causa sua* (no one should be a judge in his own case) and the right to a fair trial.

The aim of the Medical risk fund, as defined by the Directive, is to ensure that for the patients there are available and transparent complaint submission procedures and mechanisms, which would allow patients to seek remedies in cases when the healthcare provided has inflicted harm to them. The aim of the Directive cannot be achieved if the limits set by Article 16, paragraph 2 of the Law on the Rights of Patients do not fully compensate the inflicted harm to the patient.

References

1. U. Mitenbergs, M. Taube, J. Misins, E. Mikitis, A. Martinsons, A. Rurane, W. Quentin. *Latvia: Health system review. Health Systems in Transition, 2012*, **24** (2012)
2. Cabinet of Ministers regulation, "The Medical risk fund functioning rules" the projects initial (ex-ante) impact report [http://www.vm.gov.lv/images/userfiles/phoebe/aktualitates_publiska_apspriesana_6ad08b8ff5eb59abc2257b0200464996/vmanot_210513_arf.docx (seen on 05.06.2015)]
3. Charter of fundamental rights of the European Union, 18.12.2000 Official Journal of the European Communities C **364/1** [http://www.europarl.europa.eu/charter/pdf/text_en.pdf (seen on 01.06.2015)]
4. Convention for the Protection of Human Rights and Fundamental Freedoms. The Council of Europe [http://conventions.coe.int/Treaty/en/Treaties/Html/005.htm (seen on 01.06.2015)]
5. Draft law "Amendments of the Law on Submissions" annotation [http://titania.saeima.lv/LIVS11/SaeimaLIVS11.nsf/0/850765853B51A05EC2257936004A132C?OpenDocument (seen on 02.06.2015)]
6. European Charter of Patients' Rights [http://ec.europa.eu/health/ph_overview/co_operation/mobility/docs/health_services_co108_en.pdf (seen on 01.06.2015)]
7. Law On Submissions [http://www.vvc.gov.lv/export/sites/default/docs/LRTA/Likumi/Law_On_Submissions.doc (seen on 02.06.2015)]
8. Law On the Rights of Patients [http://www.vvc.gov.lv/export/sites/default/docs/LRTA/Likumi/Law_On_the_Rights_of_Patients.doc (seen on 01.06.2015)]
9. Medical Treatment Law [http://www.vvc.gov.lv/export/sites/default/docs/LRTA/Likumi/Medical_Treatment_Law.doc (seen on 01.06.2015)]
10. Regulations of the Health Inspectorate [http://vi.gov.lv/en/start/_142/regulations-of-the-health-inspectorate (seen on 01.05.2015)]
11. The Administrative Procedure Law [http://www.vvc.gov.lv/export/sites/default/docs/LRTA/Likumi/Administrative_Procedure_Law.doc (seen on 01.06.2015)]
12. The Civil Law [http://www.vvc.gov.lv/export/sites/default/docs/LRTA/Likumi/The_Civil_Law.doc (seen on 01.06.2015)]

13. The Civil Procedure Law [http://www.vvc.gov.lv/advantagecms/docs/LRTA/Likumi/Civil_Procedure_Law.doc (seen on 01.06.2015)]
14. The Constitution of the Republic of Latvia [http://www.vvc.gov.lv/export/sites/default/docs/LRTA/Likumi/Constitution.doc (seen on 02.06.2015)]
15. The Criminal Procedure Law [http://www.vvc.gov.lv/export/sites/default/docs/LRTA/Likumi/Criminal_Procedure_Law.doc (seen on 01.06.2015)]
16. The Latvian Administrative Violations Code [http://www.vvc.gov.lv/export/sites/default/docs/LRTA/Likumi/Latvian_Administrative_Violations_Code.doc (seen on 01.06.2015)]
17. The Medical risk fund functioning rules. Cabinet of Ministers. Adopted 05.11.2013. [Came into force 23.11.2013] (Official Journal "Latvijas Vēstnesis", **228** (5034), 22.11.2013) [http://likumi.lv/doc.php?id=262102 (seen on 01.06.2015)]
18. The Supreme Court of Latvia Senate the Department of Administrative cases October 25, 2013 judgement in case No. SKA-0340-13 (A420430312) [http://www.tiesas.lv/nolemumi/pdf/130711.pdf (seen on 04.05.2015)]
19. The Supreme Court of Latvia Senate the Department of Administrative cases December 17, 2010 judgement in case No. SKA–1032/2010 (A7033910/2) [http://at.gov.lv/files/uploads/files/archive/department3/2010/ska_1032_2010.pdf (seen on 04.06.2015)]
20. The Supreme Court of Latvia Senate October 16, 2008 judgement in case No. SKA-411/2008 para. 9. [http://www.tiesas.lv/files/AL/2008/10_2008/16_10_2008/AL_1610_AT_SKA-0411_2008.pdf (seen on 04.05.2015)]
21. The Supreme Court of Latvia Senate the Department of Administrative cases October 15, 2010 judgement in case No. SKA-334/2010 para. 8. [http://www.tiesas.lv/files/AL/2010/10_2010/15_10_2010/AL_1510_AT_SKA-0334-2010.pdf (seen on 04.06.2015)]
22. The Supreme Court of Latvia Senate the Department of Administrative cases February 24, 2012 judgement in case No. SKA-64/2012 para. 11. [http://www.tiesas.lv/files/AL/2012/02_2012/24_02_2012/AL_2402_AT_SKA-0064-2012.pdf (seen on 04.06.2015)]
23. The Supreme Court of Latvia the Chamber of Civil Cases June 17, 2010 judgement in case No. PAC–0371 (C04211805) [not published]
24. The Regional Administrative Court June 10, 2010 judgement in case No. A42461207 [http://www.l2d.lv/v.php?i=16229 (seen on 05.06.2015)]
25. The District Administrative Court July 12, 2013 judgement in case No. 142251012 (court of appeal upheld the judgement) [http://www.tiesas.lv/Media/Default/Admin.tiesu%20spriedumi/Admin.raj.tiesas%20spriedumi/2013/J%C5%ABlijs/12.07.2013/AL_1207_raj_A-00849-13_26.pdf (seen on 05.06.2015)]
26. The letter from National Health Service from July 6, 2015

Primary law in medicine, its legal nature

K. Zarins

Rīgas Stradiņš University, Riga, Latvia

Abstract. The work will discuss the problems arising from the thesis that the economic opportunities are incompatible with the person's primary law – the right to life and equality of treatment. An actively maintained hypothesis claims that the country's economic opportunities and funding should not restrict or reduce a person's right to life and health. In this aspect, it will also study the role of the constitutional court.
The author points to the fact that the adoption of such, here the Supreme Court decision, successive constitutional court for a preliminary inaccurate claim and interpretation of the country's economic interests, could deny the right of people to life only after the consideration that they are of no use and financially expensive to be maintained.
Key words: Fundamental rights; survival; legal reliance; health financing; the right to life; social rights; emergency assistance; availability of resources; life dogma; primary law; rights of the child; healthy and sick persons.

1 Introduction

Where the goal is based solely and only on the recital "as they are politically correct to divide the state budget," as little as possible to pay for human survival, but the reverse is of fundamental human rights risks, expectations and breach of the principle of safety.

That would also apply to the fundamental human rights – health and the maintenance of security and legally sound rules design, uniform application and interpretation of the executive and judicial level.

2 Aim

This statement may initially look too limited and vague, but it makes one come to a conclusion when making such assumption of logical analysis, as well as the work of a broader outline. In addition, the work will discuss the fact that to look at the rights through the economic and financial prism is the same as to follow the principle – the end justifies the means.

3 Primary law in medicine, its legal nature

Viewing the matter and considering the primary rights in medicine requires a profound study of the decision made by the case conference: "Gaucher disease can be successfully treated by administering *Imiglucerase* (…) Long-term therapy (…) maintains the patient's quality of life and life expectancy." Taking this medication, the child develops normally, without any health problems, just like any of their peers, and, just like any of us, they have an opportunity to live a fulfilling life and reach a great age as an absolutely healthy individual. Gustavs may even have his own, healthy children" [1].

In the exposition of the paper, in its turn, it is emphasised that getting this medication requires annual payments beyond those that the parents can afford, as the average monthly amount, necessary for the treatment, totals to what parents are not able to earn on their own, i.e., approximately seven thousand Lats per month [2].

Pursuant to Cabinet Regulation No. 899 "Procedures for the Reimbursement of Expenditures for the Acquisition of Medicinal Products and Medicinal Devices Intended for Out-patient Medical Treatment" [3], the authority, responsible for the decisions regarding reimbursement for the expenses necessary for the treatment of a patient (a child), just like for the treatment of many other medical conditions fully covered from the state budget, is the Ministry of Health. Pursuant to the said Cabinet Regulation, the treatment of the following medical conditions is also fully covered: conditions due to use of drugs, conditions caused by alcohol abuse, conditions due to smoking, and conditions due to immoral way of life [4]. Apparently, the body, responsible for the adoption of the regulatory enactment, neither initially nor during the process of its development has included the treatment needs of a child as a special entity, even regardless of the fact that a child has a special legal status regarding their rights.

Pursuant to Cabinet Regulation No. 899, a decision regarding reimbursement for the treatment expenses necessary for a child is to be made. The State Compulsory Health Insurance Agency [5] and the Ministry of Health fully rejected the said request. For this reason, the case was brought to court.

The court joined the opinion of the Cabinet of Ministers, the Ministry of Health and the State Compulsory Health Insurance Agency that the product was not directly included in the list of reimbursed medications. Despite the fact that the texts of the Constitution and international declarations directly state that human life is one of the fundamental rights [6], this meant opposing the child's efforts to save their life and, consequently, direct interference with the assurance of the child's vital functions.

The problem issue emerges with regard to the relationship between the postulate of the state authority and the individual's relationship with their institutional design, where the state may be a non-negotiable (through legal rights institutions) in the implementation of their rights and at the same time indifferent and fundamentally restrictive to human life. (Treatment of Gaucher disease is fully financed from state budgets in practically all EU member states) [7]. This may be attributed to such an abstract concept as the dogma of financial resources (economic interests, security) and the fundamental human values, their axiological scope, as opposed to the regulatory environment.

The adjudication, issued by the Constitutional Court, undoubtedly plays a significant role in the development of the further processes. It provided an explanation that fundamental human values and rights can be interpreted, rendered and equated with fundaments of rationality so far that, due to expedience considerations these, otherwise fundamental values, may disappear entirely or, in the best case scenario, their existence is purely theoretically justified, and, for the state power, the fundamental values of every individual are something of a secondary value and reduced to financial and economic benefits.

The Constitutional Court supported the opinion and the current system that no medications are to be reimbursed for individual patients and, if they are, then only following the economic and scientific analysis [7]. The topic which needs to be studied on a wider scale would include a description of the time rights (functionality of rights), which serves to outline the economic and scientific analysis, because, while it is otherwise being reasonably carried out, the individual may have already deceased or be irreversibly crippled, thus interfering with the fundamental rights of an individual.

In Latvia, assessment of human "life", which is presumably the highest and the most precious value, is left to a public agency in the person of the State Compulsory Health Insurance Agency [8]. The interests of the society, in their turn, prevail over those of an individual. Furthermore, any individual may at any time find themselves being outside this part of the communal mass (social deviation), and recurrence of a similar precedent brings up the question – whether the principle of legal certainty is observed and assured. A precedent like this undoubtedly makes us conclude that human life can be quantified in economic, financial and scientific terms, the usefulness of its existence can be "legitimately" determined, and therefore any of the fundamental rights, to which the right to life certainly belongs, can be restricted in case of "reasonable grounds" [9].

An in-depth analysis of the adjudication issued by the Constitutional Court reveals regularities which would need to be viewed with the help of legal methodology and a subsequent linear discourse as search of elements of subsumption in them.

Clause 12 of the adjudication issued by the Constitutional Court reads [10]: "... medications are included in this list only after assessment of their therapeutic and cost efficiency."; Paragraph 12.1.1 reads "(...) The duty to assure access to a medication (...) may depend on the resources available to the state. (...) however, the extent of implementation of these rights may depend on the resources available to the state";

Paragraph 12.1.3 reads: "(...) Therefore, practically each state is to face the difficult task of dividing the resources and, thus, determining who, in what circumstances and what type of treatment would be covered. (...) Therefore, the state has a high degree of freedom of action in deciding these matters. (...) with regard to decisions regarding distribution of limited resources, the state enjoys a high degree of freedom of action. The state has the duty to assure efficient use of resources (...) achieve a fair balance (...) in the distribution of financial resources (...)";

Paragraph 12.2 reads: "Reimbursement for expenses for medications (...) is provided (...) based on availability and fairness. These [reimbursable – author's note] medications have been (...) assessed economically, (...) their costs and impact on (...) state funds are known. Their prices in Latvia are compared to the prices in other European Union states. (...) only those medications are reimbursed (...) which the competent government authority has acknowledged to be (...) cost-efficient. (...) it is calculated to what extent the costs of the new medicinal treatment exceed the previously planned costs of treatment. (...) the additional costs are calculated for one year of life achieved as result of the treatment, or the incremental cost-effectiveness ratio. (...) a medication whose incremental cost-effectiveness ratio exceeds (...) the incremental cost-effectiveness ratio for a medication already included in the list shall not be included in the list. (...) The medicinal product reimbursement system is based on (...) cost-effectiveness assessment. (...) in case of individual reimbursement of a medication, (...) therapeutic and cost efficiency are not sufficient (...)."

Paragraph 12.3 reads: "(...) the therapeutic and cost efficiency of a medication may serve as basis for different treatment. (...) the state would be entitled not to compensate the expenses for the purchase of a medication whose (...) cost-efficiency is not sufficient or has not been proven."

Paragraph 12.4.1 reads: "In the reimbursement of medications, (...) a certain budget is allocated for (...) the treatment costs per patient. (...) no separate budgets are planned for individual reimbursement (...). By (...) restricting the reimbursement volumes, (...) the state has made and attempt to achieve (...) a balance between the restrictions and the average expenses necessary for the purchase of the medication.

Paragraph 12.4.1 in the adjudication issued by the Constitutional Court further reads: "(...) a certain limit has been fixed, and it is based on the estimated mean reimbursement volume to be allocated for one individual patient. (...) the average (...) per patient expenses to be reimbursed in 2007 are 3,615 Lats."

In many situations in life and for the child in this particular case, it means early death as a juvenile, as an average of 7,000 Lats is necessary per month. However, as it is stated further in the paragraph of the adjudication issued by the Constitutional Court: "Thus it can be seen that (...) the fixed limit is based on reasonable considerations (...). The idea further unfolds in Paragraph 12.4.2: "(...) such a system has been established relying on reasonable considerations which are targeted at efficient use of the resources. (...) the state has a high degree of freedom of action in the healthcare sector; the Constitutional Court has no reason to question the reasonable balance between the interests of the society and those of separate individuals." [10]

A study of the said case leads to a conclusion that the interests of society and the diagnosis of the particular individual, which is included in the list of reimbursed medications, have been reasonably balanced and assessed, diagnoses outside these frames, which Gaucher disease is, for a child who, if timely treated, is absolutely healthy, are reasonable, and it is helpful for the society that the child is dying slowly.

This means that, not more and not less, the state determines the length of a person's life as one year, and what is behind this financial border is something that the state no longer cares for and is, in the worst-case scenario, doomed to have a lethal outcome. An increase in the numbers of people finding themselves behind this threshold, i.e., the list of reimbursed medications, brings along more negative attitudes in the society towards the restrictive nature and scope of the regulation, as it is possible that more and more people appear outside this part of the society secured by the regulatory environment. Furthermore, the approach to human life, which is a vitally important question of human existence, is, in the opinion of the Constitutional Court, based on "reasonable considerations".

Clause 5 of this adjudication issued by the Constitutional Court reads [10]: (Explanations provided by the Ministry of Health) "(...) The outcome of the price reductions in 2007 was a saving of 1.2 million Lats." Paragraph 12.3 reads: "(...) The Cabinet of Ministers informs that 66.3 million Lats have been allocated for reimbursement of medications in 2008". In this particular case, the child's share in the saving would be only 7%, or 0.1% of the funds allocated for the year.

In the adjudication, the Constitutional Court makes a division into expensive health services, which are required by a small part of the society, and cheaper health services, which would be available to a larger part of the society. At the same time, it does not deny treatment of life-threatening diseases. In such case, it would be necessary to identify who belongs to this majority in the society, and who do not fall within this definition.

A division of the part of the society with a medical condition into "those who are very sick" and "those who are little sick" but expensive, would be very much unethical and in conflict with the fundamental human rights from the legal point of view, or right would be restricted to a minimum volume of regulatory requirements [11].

The issue and the object of the study make one point to the aspect of children's rights. It should also be noted that there is a division into "very sick children" and "little sick, but expensive children". Apart from this, information on this suffering part of the society, which is subject to reimbursable drugs policy, its role in the medical law area, is very little and ambiguous. The division – those who are healthy; those who are sick, but subject to the reimbursement process; those who are sick, but are not included in such a list and finally the right to health, interaction between medical law and children's rights and quite poor description of legal research of these institutions is the basis for further discussions on the fundamental values of the meaning of constitutional law [12].

Besides, general factor destabilizing rights of patients can be a circumstance, after origination of which constantly changing wide social mass of the society can itself come behind the rights, endowed with receipt of compensated drugs and defining health limits, in the society group of sick persons separated by normative acts, which further is divided into *maintainable* and *non-maintainable*.

The most bitter and, simultaneously, the highest value of the person is life right now, since sicker persons from the point of view of the state are left with purely normative (also deviative) character. Both life and health, according to current interpretation of the norm, can only be possessed by the society; a person as an individual, especially a sick one, has very limited abilities, respectively, they have rights, but defence thereof against the political system is disproportionately difficult within the existing practice and administrative procedure.

To have a better understanding of the topic emerging from the argumentation provided by the Constitutional Court [7], it is worth taking a look at the sources used in this adjudication issued by the Constitutional Court. In its adjudication, the Constitutional Court refers to BVerfG [13].

This is indicative of a close link with the German Law in finding a solution for a problem in the Latvian legal system regardless of apparent shortcomings in the German law in the 30-ies and 40-ies, and lies in parallel with the Constitutional Court directly affection German right. For case law to be used properly and applied in practice, it is necessary to understand its sources, peculiarities and actual conditions of application. Therefore, to be able to find out whether the applied foreign case law is relevant for the particular matter, it is necessary to take a more profound look at its essence. It should also be understood that making comparisons with adjudications issued by a foreign (German) court or applying such adjudications in a particular case should involve due consideration of the peculiarities of the legal system. The adjudication would certainly contain references to the laws of the particular state, and, consequently, this would involve exploring the particular legislative system. If this is not duly implemented, it can be acknowledged that the referenced and applied source has been used superficially and is not relevant with regard to the particular matter brought before the Constitutional Court.

In the said adjudication issued by the German Constitutional Court (henceforth referred to as Adjudication), the issue is: "Whether the refusal to compensate the costs of an unregistered medication complies with the Constitution" [13].

In its adjudication, the German Constitutional Court referred to the law, code and constitution which are applicable to the particular situation. By applying this foreign legal act, the Constitutional Court believes that the provisions of the Latvian Cabinet Regulation No. 899 "Procedures for the Reimbursement of Expenditures for the Acquisition of Medicinal Products and Medicinal Devices Intended for Out-patient Medical Treatment" [3] (henceforth referred to as Cabinet Regulation) match those of "SozialgesetzbuchV", Article 12 (Paragraph 1) [14] and "Arzneimittelgesetz" Law on Sale of Medications, "AMG", Article 21, Article 25 (Paragraph 2), Article 40, Article 41 [15].

Finally, that the compliance of the Cabinet Regulation with the 1st sentence of Article 91 of the Constitution of the Republic of Latvia [16] is to be found in the compliance of the norms included in "Sozialgesetzbuch V" and "Arzneimittelgesetz" with "Grundgesetz" [17].

The purpose of the German "Arzneimittelgesetz" [18] is: "To assure due provision of people and animals with medications". Familiarisation with the said German legislative acts does not lead to such conclusions at all. This also follows from this adjudication issued by the German Constitutional Court. "The constitutional petition refers to the question about whether sickness funds should reimburse the patient for their expenses for the purchase of a medication whose authorisation has not yet been requested and which, from the regulatory point of view, must not be marketed." [13]

The petition was lodged with the German Constitutional Court by the heirs of Professor Dr W., who had obtained voluntary insurance from the sickness fund and starting taking Jomol in early 1990. The product was prohibited, and he died already on 4 September 1991. The petition was lodged exclusively regarding the fundamentally significant fact of the marketing authorisation for the medicinal product. As well as the patient's right to opt for the treatment himself. This case also involves violation of the patient's right to opt for the most suitable treatment for them, which appears to be against the provision of "Sozialgesetzbuch V" [13, 19] (according to 1st and 2nd paragraph of the justification section of the adjudication).

This matter was not heard by *the German Constitutional Court ad materiam. The constitutional petition was not accepted and was rejected. Because, as the German Constitutional Court acknowledged: "(...) the submitter (...) is not suffering losses of existential nature because, according to what they say, the insured had received the medication entirely free of charge."* [13]

The references given by the Constitutional Court to the international acts are not quite valid. This particularly follows from an analysis of the following international acts used in the adjudication issued by the Constitutional Court. Firstly, there is an emphasis on the "General Health standards" [20]. *Secondly*, is evaluated in the context of those cases the Maastricht guidelines on economic, social and cultural rights [21].

Taking a look at the said sources, it can be seen that the economic capacities of the state must in no way be linked with the primary rights of an individual (right to life and equal treatment). A statement regarding the economic capacities of the state and its finances cannot restrict or minimise any person's right to life.

By making such a statement, which is incorrect, the state might, due to economic interests, might take persons' lives just because maintaining them is inexpedient. This statement might initially seem ambiguous, but we come to such a conclusion if the assumption undergoes a systematic inductive analysis.

Viewing rights via the prism of economy and finance is the same as following the principle "the aim justifies the resources", where the justification for the aim is only "how to save", how to pay less for saving a person's life. Therefore, where we speak about economic and financial considerations, there is no place for rights. At best, they are secondary. But the fundamental values and the overall legal awareness are secured and root specifically in rights themselves, where economic considerations particularly have no place.

Work will examine the problems arising from the thesis that the economic opportunities by no means are incompatible with the person's primary law – the right to life and equality of treatment. It will be actively maintained hypothesis that claims that the country's economic opportunities and funding may not restrict or reduce a person's right to life and health. In this aspect it will also study, in addition to analyzing the role of the constitutional court. Therefore, especially the result of work, reflected credibility and foreseeability of the legal criteria included by Constitutional Court in the field of health care and its really legal character, which, as must be concluded, not always is sufficiently exhaustive.

Author binds this problem and specifies that, having adopted such Higher Court's decision and consequent Constitutional Court's verdict regarding satisfaction of such claim and legal interpretation of the state economic interests, a person may be deprived of or denied rights to the live by equality, which it is not expediently and is financially difficult to maintain such sick persons.

This assumption could initially be too limited and unclear, however, such conclusion must be made, making such logical analysis assumption, as well as research of the concept of wider matter, which invites to think that legal interests of the child and the person are not provided according to the requirements of the Constitution. Besides, the factor is disputed that with mediation of economic and financial prism cannot be accepted that the end justifies the means.

If objective is only based upon the consideration "how politically correct is division of the state budget", paying for survival (health) of the person as little as possible, and in contrast to it exist risks in the basis of human rights – i.e., violate the very human rights, legal expectations and principles and resulting safety, such considerations, as author thinks, must be seen in context with ethical and moral requirements, including in the field of rights.

It also concerns fundamental human rights – health and safety maintenance, as well as development of legally justified regulation, unified fulfilment and application and interpretation of decisions of court institutions.

4 Conclusions and proposals

Research work described problems relating to the Constitutional Court case number 2008-37-03 and the judgment of the Administrative Regional Court Case Nr. A42561808 AA43-1260-09 analysis. More specifically, the child's right to reimbursement of medicines (state-subsidized) to treat the disease diagnosed by medical genetics clinic doctors Consultation as "Gaucher disease" (Morbus Gaucher). The disease is chronic and life-threatening. In the absence of appropriate therapy, the patient experiences any damage to the bones, enlargement of the liver and spleen, decrease of white blood cells, red blood cells and platelets may cause pulmonary hypertension and other serious health problems. The disease progresses and affects the bone marrow, the bones become fragile and they can lust, it means that for a growing child it causes irreversible damage to the skeletal and bone, the child may stop growing. Moreover, if bone deformities start, they cause a severe pain; children become disabled and mutilated. If the illness is not treated, a long and painful process leads to death. So, this ruling legal analysis itself gives legal criteria and conditions under which, or the execution of which, it is possible either to ensure the child's health and life, or on the contrary, refuse, thus actually deciding children's downfall.

The Constitutional Court does not speak of the values, it creates a dangerous precedent, reflecting the state power's uncertainty and inability to defend and stand up for human core values, which in people creates fear, insecurity, as a logical reaction to the unknown law, which is rooted in the fact that without the implementation of the basic values. Thus, it cannot be expected from uniform and incorrectly applied interpretation of the norm that fundamental rights provided in the Constitution will be provided.

Describing and researching the topic and its various effects on public life in general, techniques that include peer discourse, different shots of thematic research are mostly applied and implemented. A significant place is taken by different laws and the authors of several findings and conclusions from comparison. Specific jurisprudence practice regarding perception of primary rights in medicine, its analytically practical evaluation is yet to be studied; therefore, historical, empirical and legal axiology methods should be taken into account in systematic aspect.

No less important is the intention of the legislature clarification, which runs through the interpretation methods in both the work as a whole and certain parts of it, especially when dealing with the relevant Constitutional Court case. Using system of objectives and describing methods, various norms and their compatibility are sought, therefore, search for correlation and interconnections must be stressed in the field of medical rights.

Author used for analytical distributive knowledge method that facilitates various sources and included the collection of knowledge and the main regularities of the definition of materials processing research. Therefore, author suggests using analytical right approach method, simplifying summarizing of various sources and knowledge included in them, defining main regularities resulting in clearer and more precise processing of legal research materials in medical rights.

5 Scientific capacity

Paper was developed in the field of jurisprudence, in context of legislation analysis and application, therefore, objective and scientific capacity of work can be seen in analysis of efficiency of norms and implementation thereof in further regulation. Therefore, careful scientific analysis of the subject is performed to increase scientific contribution, and conclusions obtained in result of research are consequently summarized, which author will use as a part of issues included and reviewed in doctor's dissertation.

6 Results

Looking at the sources shows that the economic opportunities by no means incompatible with the person's primary law – the right to life (European Human Rights Convention Article 2) [22] and equal treatment. Allegation of national economic opportunities and funding may not restrict or reduce a person's right to life. Assuming such incorrect statement that it may be in economic interests of the parties to deprive someone of their life, that expresses exactly in the consideration that maintenance thereof is not financially beneficial, is legally non-evaluated assumption, and, seemingly, it is invalid. This statement may even initially create fully undesired precedents, however, making such assumptions, it is considered that, systematically performing inductive analysis, we can violate fundamental human values, which is unacceptable.

See right through the economic and financial prism is the same as to follow the principle – the end justifies the means, rather, for example, by Niccolo Machiavelli's views [23, 24]. Where objective based solely and only on recital "save" as little as possible to pay for the preservation of human life. Therefore, if we speak about economical and financial considerations in rights, they have no place, at least in the healthcare field, they are secondary assumptions at the best.

But the core values and the general legal consciousness are anchored and rooted in exactly the same rights exactly where economic considerations have no place.

Basically, it is clear that the most important in the field of idea and values is the legal considerations and legal experience, not material value of the matter that reduces financial well-being, which is completely secondary value and is subjected to the main cognitive principle and legal system resulting from it, which is based upon fundamental rights and values. Consciousness is only of a person [25]. Otherwise, the economic and financial application can be justified only by a simple act of power, which has no place in the justice awareness and ethical conduct.

It is the policy of power issue of the availability of funds, how to procure and appropriate to use, but not the right question. Right there by themselves without economic conditions and their consequential of limiting conditions. Economics is the science of the material perspective [26, 27], it is the classic contradiction that the economy, which divides the only material value, cannot be divided and economic invocation cannot be the basis of human values and rationality, calculation, no matter how useful it also may seem.

Based on such unjustified assumption, Constitutional Court gives credible and objective approach, e. g., to the Ministry of Finance or the Ministry of Economics and, finally, to the Cabinet as a whole (executive institution), to evaluate the person's economic *health ability* [28], expressing it by executing and clearly defining calculations defining life expensiveness or life value, but elements of human values and rights are lost in such way.

Consequently, using this approach to society that must be divided into expensive to maintain *sick, incapable* members of society, and only stressing the largest part – healthy, to which rationally applicable legal norms would apply, would be lost the constitutional fundamental value – observing fundamental human rights to life and health. If it is clear that interests of the majority prevail over the individual, and principle exists that the strongest in rights survives, leaving sick people behind the line of life and rights, history has created such precedents that have not been successful and have been clearly rejected[a].

References

1. Spriedums Rīgā 2010.gada 11.martā. Available: http://www.globalhealthrights.org/wp-content/uploads/2013/12/No.-A42561808.pdf
2. 2009.gada 29.janvāra spriedums Nr.AA43-1260-09/7 lietā par Veselības ministrijas 2008.gada 4.aprīļa lēmuma atcelšanu un labvēlīga akta izdošanu. Available: https://www.tiesas.lv/janvaris-21
3. Ministru kabineta noteikumi Nr. 899 "Ambulatorajai ārstēšanai paredzēto zāļu un medicīnisko ierīču iegādes izdevumu kompensācijas kārtība" 31.10.2006. Available: http://likumi.lv/doc.php?id=147522
4. *1.pielikums.* Ministru kabineta noteikumi Nr. 899 "Ambulatorajai ārstēšanai paredzēto zāļu un medicīnisko ierīču iegādes izdevumu kompensācijas kārtība" 31.10.2006. Available: http://likumi.lv/doc.php?id=147522
5. Ministru kabineta noteikumi Nr.583 "Veselības obligātās apdrošināšanas valsts aģentūras nolikums" 27.12.2002. Available: http://likumi.lv/m/doc.php?id=69950&version_date=08.01.2003
6. Latvijas Republikas Satversmes tiesa. Satversmes tiesas tiesnešu Aivara Endziņa, Jura Jelāgina un Anitas Ušackas atsevišķas domas. Lieta Nr. 2000-03-01. Available: http://www.satv.tiesa.gov.lv/wp-content/uploads/2016/02/2000-03-01_Atseviskas_domas.pdf
7. Latvijas Republikas Satversmes tiesas spriedums lietā Nr.2009-12-03 "Par Ministru kabineta 2006.gada 31.oktobra noteikumu Nr.899 "Ambulatorajai ārstēšanai paredzēto zāļu un medicīnisko ierīču iegādes izdevumu kompensācijas kārtība" 92.punkta otrāteikuma vārdu "zāļu iegādes izdevumu kompensācijai piešķirto līdzekļu ietvaros", 94.punkta vārdu "izņemot šo noteikumu 100.1 punktā minēto gadījumu", 100.punkta vārdu "ne vairāk kā 10 000 latu apmērā vienam pacientam 12 mēnešu periodā" un 100.1 punkta otrā teikuma atbilstību Latvijas Republikas Satversmes 93. un 110.pantam". Available: http://likumi.lv/doc.php?id=203464
8. Vēsture. Available: http://www.vi.gov.lv/lv/sakums/_110/vesture
9. Latvijas Republikas 8. Saeimas pavasara sesijas astotā sēde, 2005. gada 26.maijā. Sēdi vada Latvijas Republikas 8.Saeimas priekšsēdētāja Ingrīda Ūdre. Available: http://www.saeima.lv/steno/2002_8/st_050526/st2605.htm
10. 2009.gada 29.janvāra spriedums Nr.AA43-1260-09/7 lietā par Veselības ministrijas 2008. gada 4.aprīļa lēmuma atcelšanu un labvēlīga akta izdošanu. Available: https://www.tiesas.lv/janvaris-21

[a] These ideas were echoed Hitler's political program, which summarizes 2.books and is called "Mein Kampf" – "Mein Kampf" (25.-1927.); further he developed this idea in National Socialist Germany 1939, September, adopting the euthanasia law – which provides all the feeble (sick, old and so on. people) complete liquidation.

11. Pārskats par Satversmes tiesas darbu 2010.gadā: Available: http://www.satv.tiesa.gov.lv/blog/2011/02/10/parskats-par-satversmes-tiesas darbu-2010-gada/
12. S. Ašnevicas – Slokenbergas zinātniskā redakcija. Medicīnas tiesības. Tiesu namu aģentūra, Rīga, 2015. P. 4.-6.
13. Beschluß vom 05.03.1997. Aktenzeichen 1 BvR 1068/96. *Vācijas Federālās konstitucionālās tiesas 1997.* gada 5. marta lēmumu lietā Nr. 1 BvR 1068/96. Available: https://dejure.org/dienste/vernetzung/rechtsprechung?Gericht=BVerfG&Datum=05.03.1997&Aktenzeichen=1%20BvR%201068/96
14. Sociālā nodrošinājuma kodekss (SGB). "Sozialgesetzbuch V" 12.paragrāfs 1.punkts. Available: https://www.gesetze-im-internet.de/sgb_5/__45.html
15. Medikamentu tirdzniecības likums. "Arzneimittelgesetz", "AMG" 21. paragrāfs, 25. paragrāfa 2. punkts, 40. paragrāfs un 41. paragrāfs. Available: https://www.gesetze-im-internet.de/amg_1976/BJNR024480976.html
16. Latvijas Republikas Satversmes 91.panta 1. teikums. Latvijas Republikas Satversme. Publicēta: "Latvijas Vēstnesis", 43, 01.07.1993., "Ziņotājs", 6, 31.03.1994., "Diena", 81, 29.04.1993. Available: http://likumi.lv/doc.php?id=57980
17. Grundgesetzfür die Bundesrepublik Deutschland, "GG". Vācijas konstitūcijas 2. pants 2. punkts 1. teikums. Available: https://www.gesetze-im-internet.de/gg/BJNR000010949.html
18. Arzneimittelgesetz, Medikamentutirdzniecībaslikums, "AMG". Available: https://www.gesetze-im-internet.de/amg1976/
19. Sociālā nodrošinājuma kodekss, "SGB", "Sozialgesetzbuch V" Available: https://www.gesetze-im-internet.de/sgb5/45.html
20. The right to the highiest attainable standard of health 28, 11/08/2000. E/C.12/2004/4. CESCR. Substantive issues arising in the Rights. General Comment Nr. 14. Available: http://www.refworld.org/pdfid/4538838d0.pdf
21. *The Maastricht Guidelines on Violations of Economic, Social and Cultural Rights, para* 6. (1997)
22. 2.pants - Tiesības uz dzīvību Available: http://at.gov.lv/lv/judikatura/ect-nolemumi/eiropas-cilvektiesibu-un-pamatbrivibu-aizsardzibas-konvencija/pec-pantiem/tiesibas-uz-dzivibu/
23. Renesanses filosofija un ētika, Available: https://kivlenieks.wordpress.com/2008/11/29/renesanses-filosofija-un-etika/
24. Deliuss K., Gatcemeiers M., Sertana D., Vunšera K., Filosofijas vēsture no antīkās pasaules līdz mūsdienām. R.: Jāņa Rozes apgāds, 2006, 11
25. Ego, apziņa Available: http://www.jungabiedriba.lv/junga-analitiska-psihologija/ego-apzina/
26. Zinātnes perspektīva civilizāciju sadursmes laikmetā, Available: http://www.lza.lv/index.php?option=com_content&task=view&id=3233&Itemid=480

27. Rima KhalafHunaidi, Abdel Latif Youssef El Hamed, Nader Fergany – Arab Human Development Report 2002 – United Nations Development Programme (UNDP), ISBN 9211261473
28. Rekomendācijas darba aizsardzības pasākumiem, strādājot ar datoru. Available: http://www.vdi.gov.lv/files/vadlinijas.pdf

The effectiveness of social work services for families whose children are in temporary custody

R. Bardauskiene and J. Pivoriene

Mykolas Romeris University, Vilnius, Lithuania

Abstract. Despite that there is an operating social support system for families, social workers are affected by factors that limit effectiveness of their activities in working with families whose children are taken into temporary custody. The article aims to uncover what hinders social worker to carry out effective work in providing social services for families whose children are in temporary custody. Qualitative research data shows that the research participants' awareness of social work effectiveness is limited to its individual components. Putting together these components one can get a broad definition of effectiveness of social work though the research participants themselves do not use such a concept. The research data reveals that micro level factors influencing effectiveness of social workers' activities working with families whose children are in temporary custody are as follows: absence of parental motivation to seek changes and unfavourable environment as well as negative community approach to social risk families. Macro level factors limiting social work effectiveness working with the families at social risk lie in the system of social services. Inadequate management of social work, limited social workers' access to resources necessary to restore family functions; too high workload for social workers are essential factors limiting social work effectiveness.
Key words: effectiveness; social worker; social services; family at social risk; temporary child custody; permanent child custody.

1 Introduction

In order to achieve positive changes in social system it is necessary to analyse effectiveness of social services. Despite the fact that the issue of effectiveness has been discussed since the beginning of the development of social work profession [1], to measure it is still a big challenge for researchers [2–4]. There is no consistent system and approved methodology for the assessment of social services quality and effectiveness in Lithuania. On the other hand it is difficult to measure an intervention on the ground of success-failure dimension. There are cases, when social workers have to make a decision which is the least harmful for a client. It is especially relevant while talking about social work with risk families, when families receive professional social services but still cannot provide safe and healthy environment for a child and he or she is taken from a family. Risk families usually have many problems and require a systematic approach from professional team.

In order to make more effective provision of social services for families at risk and to ensure children's right to grow up in family, social workers with social risk families position was established in Lithuania in 2007. It was expected to diminish the number of children who were taken out from families because of unsafe and unprotected environment. However, the Report of Ombudsmen [5] claims that the number of such cases has not diminished, social work with families is done formally only, special social services are provided insufficiently, and the causes why a child is taken from the family are not eliminated. The Report states that support system is not sufficient, it lacks services for social skills education and maintenance despite the fact that the lack of social skills is one of the main reasons why a family is enrolled in a social risk list. Psychosocial support for addicted family members has not been developed, too. Differences in infrastructure in various municipalities, insufficient number of social services providers, absence of prevention services are named as the main restrains for the effective social services provision. The legal background does not regulate assessment of social services' quality, does not define responsibilities for ineffective social services provision. In total, the restrains could be found on micro as well as on mezzo and macro levels and their elimination should be viewed in a complex way.

According to statistics [6], there were 532 647 children in Lithuania, the number has decreased by 11 000 (2.04%) since 2013. However, the number of children in risk families and number of children in custody remains the same. In the Report of State Child Rights Protection and Adoption Service under the Ministry of Social Security and Labour [7] it is said that in 2014 children in risk families composed 3.7% of all children (2012 m. – 3.80%, 2011 m. – 3.83%), the number of families whose children were in custody also slightly varied: in 2012 – 883, in 2013 – 907, and in 2014 – 817. Statistical data shows little variations in numbers which means that provided social services do not reach their targets, positive changes in families are not achieved, and social services are not effective.

Measurement of the effectiveness raises many challenges for researchers. Usually such measurement is associated with business where effectiveness is defined as a rational circulation of resources which leads to positive results, e.g. profit [8], as such usage of resources which guarantees the maximal result [9], as a state when production cost is minimal but benefit is maximal [10]. However, it is impossible to apply business vocabulary to social system and to measure effectiveness of social services with the same measures as in business. L. Zalimiene (2006) says that the difference between business and social services is that in business the biggest attention is paid to profit, but in social services there is no profit. Provided social services could bring loss in economic sense but they still are necessary. The decision not to provide uneconomical social services in order to save resources could be beneficial in short term but would cost a lot in long term if the number of services users will rise. The effectiveness of social services depends on outside factors such as policy, operations of various institutions, organizations, activities of local communities, network of clients, etc., which is difficult to predict and plan. In business loyalty of clients is crucial to effectiveness and profit while in social system clients' dependence on social services means that social services are ineffective and that one of the most important social work goal is to empower client to solve problems is not achieved [11]. J. Moriarty and J. Manthorpe (2016) define effectiveness of social services as an assessment how specific intervention or policy operates in real situation. This notion includes cost-effectiveness, impact to clients, attitudes of clients and services' providers. Authors state that effectiveness also depends on abilities of a social worker to combine various roles [12].

The aim of the article is to reveal what are restrains of effectiveness of social services for families whose child is in custody. The object is restraining factors of effectiveness.

Research questions are: how social workers define effectiveness? What are restraining factors for effective social services' provision for families?

2 Methodology

Qualitative case study methodology was used in the research. The case in the research is defined as an activity of a social worker with families whose child is in custody. Document analysis and in-depth interviews were used in the research.

Document analysis was made of five custody cases provided by Child protection office according to these criteria: child is in permanent custody, parental rights are restricted for an open-ended term. After document analysis cases were classified according to these criteria: situation in the family at the time of decision on temporary custody, situation in the family at the moment of application to the court for restriction of parental rights, duration of temporary custody, and the number of revisions of temporary custody.

For in-depth interviews informants were selected by criterion sample, selection criteria were social workers who were working with previously analysed cases from the decision on temporary custody till the application to the court for parental rights' restriction. Four social workers were interviewed as one worker had two cases. Interview included questions about social support system for families whose child is in temporary custody, social work effectiveness, factors which restrain to work effectively with families. Content analysis was used for qualitative data analysis. Research complied with ethic requirements, confidentiality and privacy was ensured during the research.

2.1 Attitude of social workers to effectiveness of activities and its restraining factors

To prove social work effectiveness is always difficult. The complex and interconnected nature of problems of families at social risk, such as alcohol, violence or poverty makes difficult to describe the effectiveness of one intervention that is highly contextual and is characterised by complex client-worker relationships from another and from the context in which it is located. For this reason, it is vital to find out what are restrains for social worker to provide effective social services to families whose children are in temporary custody.

Document analysis revealed that temporary custody and social work with family last for 8–36 months. Revision of temporary custody is done 2–3 times. In all analysed cases temporary custody was changed to permanent one, as after the revision was found that the situation in families had not changed and children could not come back to families. Interview data analysis formed three categories of the factors restraining effectiveness:

- individual (micro level);
- family and environment (mezzo level);
- social system (macro level).

Micro level factors which could influence effectiveness of social work are related with individual and professional abilities of a social worker. Qualification requirements for social workers define that a professional social worker should be able to organize effectively human, financial and other resources, while working with families at risk – assessing effectiveness of provided social services. Therefore, it is important to know social workers' understanding of their work and its effectiveness. Research data shows that informants relate effectiveness with benefit and goal achievement: *"What is effectiveness? I understand it as how useful it is (I)", "I think you can tell that it was effective when a child returns to family, when the situation in the family becomes better (I)", "it is seeking the goal (IV)".* As was mentioned earlier the effectiveness of social work is not always possible to measure by success-failure criteria. Effectiveness of intervention relates to what was done and how resources were used: *"...almost all resources were used (IV)", "the goal was not achieved, but all possibilities and resources were used (IV)", "assessing effectiveness, first of all, I would pay attention not to the result but how resources were used (IV)".* Effective social work depends on social workers' competence, e.g. ability to combine various roles, to apply knowledge: *"consultations, communication, provision of information, leading to other institutions or organizations, all these activities lead to effectiveness (II)", "social worker is a tool for achievement of results (...), it is knowledge, education, personal traits (III)", "I understand effectiveness as social worker's ability to apply knowledge, abilities, competences in practice (IV)".* Effective support should be timely: *"effective work was when we reacted immediately, the same day, we made a family plan, assigned responsibilities (II)".* It relates with clients' motivation and understanding of the problem: *"the effectiveness depends on how much a person understands his/her problems (I)"* as well as with social workers' not judgemental attitudes: *"the biggest effectiveness is when you don't have negative attitude to a person, don't have preconception (II)".*

Research data shows that informants present notion of effectiveness which is very close to the one defined by J. Moriarty, J. Manthorpe (2016) [13]. Social work effectiveness could be described as achievement of the goals, which change clients' situations and relates to usage of the resources as well as micro and macro factors, such as competence of social worker, attitudes to clients, timely support. However, research data reveals that informants do not have a clear and holistic vision what is effectiveness in social work.

Mezzo level factors are related to a risk family and its environment. The main factor which hinders effectiveness of social work is lack of clients' motivation: *"what was lacking the most, it was a person's motivation, he didn't want to change (II)", "...absence of wish to change (III)", "she wasn't motivated, even if promised, she didn't keep these promises (III)", "a client was not ready to accept the support (IV)".* It is not easy to motivate a client as motivation is influenced by inside and outside factors. It could be related with values, e.g. how important are children and how big is a wish to change for returning of children. Research data shows that for some families children are not important: *"women chose to live with a cohabitant instead if living with children but without him (I)".* On the contrary, parents for whom a child is important are motivated to change and cooperate with social workers: *"a family was motivated to get back a child; therefore they kept commitments (III)".* Other strong factors for motivation are financial resources and long-term family problems: *"his problem was long-term (II)", "the most important was alcohol problem (IV)", "a person lived without any financial resources (II)".* Another factor which hinders motivation is a denial of the problem; changes are possible only when a client understands the problem. Research data shows that people tend to deny their problems especially addiction one: *"it is denial of the problem, especially denial of addiction to alcohol (IV)".* Social workers seek to empower risk families for positive changes by educating parents parental and social skills, however, if a family is not motivated and active, activity of social worker becomes much more complicated, requires bigger time resources, higher competence, involves cooperation of various professionals.

Research data reveals a big impact of the environment, informants have told that for effective social work family environment is crucial: *"environment, neighbours, friends impeded for changes, if she would go living elsewhere, where is another environment... (III)", "there were these drinking friends, who were very important (IV)".* Lack of support from close environment and community leads to unsuccessful work as well: *"it is necessary to get understanding of the surrounding, of the society (IV)", "there is no community which would support (III)".* Negative attitude of the community prevents from effective work: *"it was impossible to overcome attitude of the community that she could change (IV)".* A community could hinder social worker's efforts: *"a community constantly interfered to social worker's job, controlled her (I)"*. The relationship between an individual and environment is explained by system theory [14–16]. It says that individuals are in constant relation with their environment and have to adjust to never ending environmental changes. Individuals change as well as the environment. If an individual is able to develop in changing environment and receives support from the environment, it means that mutual adaptation is occurring. Risk families have difficulty with adjusting to changing living conditions and challenges raised by environment. They do not receive support from the environment and this decreases their motivation and possibility for changing the way of life.

Next to micro and mezzo level factors, macro factors are present as well, they mainly include various aspects of social system, and one of them is organizing of social work. Research data reveals that such organizational limitations as high workload, paper work decrease the possibility to give enough time to work with family: *"I had twenty five families (II)", "it was not enough time for working with a family (I)", "time for family is very limited (III)", "fifty percent of my work time I spent on filling various documents and writing papers, developing plans, little time I could spent with families (III)"*.

Inter-institutional cooperation, as well as team work are important for social work effectiveness: *"it was lack of cooperation, team work (I)", "the cooperation is impeded by other institutions' negative, stigmatized attitude to a client (III)", "it is difficult to cooperate with centres for psychical health, psychiatrist, psychologist, because they don't provide any information (III)", "there is no cooperation between various institutions (IV)"*. Social workers are left alone in providing social support for families: *"it would be possible to achieve much more if social worker wouldn't stand alone (I)", "social worker in the villages is alone (II)", "when a child is in temporary custody, it is not enough with activities of a social worker, team work is needed (IV)", "if a family is in crisis, in fact social worker stays alone with a family in this crisis (IV)"*.

Research data shows the problem of social services for families accessibility: *"family lives far away, they need to travel far (I)", "activities are in the evenings, there is no transport to get, there is no transport from the village to town in the evenings (IV)", "addicted to alcohol can't get services in place (III)"*. Informants also define shortcoming of specific social services: *"there are not social housings (I)", "a client couldn't get an allowance because she is in debt for public utilities (III)"*. Data shows that for social workers it is difficult to ensure the principal of universality which is defined in The Law of Social Services [17]. It states that social services should be provided in combination with financial support, child protection, health, educational, housing and special assistance services.

Social workers while analysing their work experience with specific cases define obstacles which have hindered effective social work and children have been not returned to families. Competence of a social worker is one of the main micro factors which relates to effectiveness of social work with families. Next to micro factors, research participants name mezzo and macro factors which decrease effectiveness of social work. Such organizational factors as high workload, low accessibility of social services, limited possibilities to influence families' environment are described as negative factors. Research data shows that team work is regulated by law, however, it is minimal or absent at all, is formal, limited only to information exchange between various institutions while a child is in temporary custody. Social workers find difficulty in providing effective social services and in using resources rationally due to lack on inter-institutional cooperation. Shortcomings of social service infrastructure, inaccessibility of social services, complicating process for receiving financial support for a client, restraining factors to use various resources for social workers hinder effectiveness of social work. These challenges make impossible to achieve possible results in families' lives, to remove obstacles for successful returning child to family.

The research participants point out that their activity would be more effective if they could work in a system closely collaborating not only with the clients' environment but also with a local community, institutions, working in a team and using different and numerous resources in support process. This means that it is not enough for social workers just to intervene into micro level, but it is very important to target an intervention to mezzo and macro levels initiating changes in a community and social services system. The responsibility for social work effectiveness has to take not only social workers, but also representatives from municipalities and different ministries as well as politicians.

3 Conclusions

Social service system which has to ensure ability of risk family to function independently in society operates not effectively as number of children in custody is not diminishing, temporary custody often leads to permanent one.

Research data shows that the lack of parents' motivation for changes, not supportive family's environment and negative community's attitude to risk families causes not effective social work.

Effectiveness of social work with risk families is hindered by limitations of social service system.

Improper organization of social workers' activities, limited possibilities for social workers to use rationally available resources, too high work load causes non effective social services' provision.

Research participants have vague understanding about the notion of effectiveness of social work; they name many components of effectiveness but do not put them in one complex picture.

References

1. Cheetham, J., Fuller, R., Mclvor, G., Petech, A. *Evaluating Social Work Effectiveness*. Buckingham: Open University Press (1992)
2. Romeo, L. *Annual Report by the Chief Social Worker for Adults: one year on*. London: Department of Health. Prieiga per internetą: `https://www.gov.uk/government/uploads/system/uploads/attachment_data/file/368485/Annual_report_2014_web_final.pdf` (2014)
3. Davies, M. (Ed.) *The Blackwell Companion to Social Work*. Fourth edition. Chichester: Wiley-Blackwell (2013)
4. Adams, R., Dominelli, L., Payne, M. *Social Work: Themes, Issues and Critical Debate*. Basingstoke: Palgrave Macmillan (2009)
5. Lietuvos Respublikos *Vaikoteisių apsaugos kontrolieriaus 2012 metų veiklos ataskaita* 2013-03-29, **4-1** Vilnius (2013)
6. Statistikos departamentas. *Gyventojai ir socialinė statistika*. Oficialios statistikos portalas. Prieiga per internetą: `http://osp.stat.gov.lt/` (2016)
7. *Valstybės vaikoteisių apsaugos ir įvaikinimo tarnyba prie Socialinės apsaugos ir darbo ministerijos 2014 metų veiklos ataskaita* 2015-03-18, **5-2**. Prieiga per internetą: `http://www.vaikoteises.lt/media/file/ataskaitos/2014%20M_%20VEIKLOS%20ATASKAITA%202015-03-18.pdf` (2014)

8. Lukaševičius K., Martinkus B., Piktys R. *Verslo ekonomika*. Kaunas: Technologija (2005)
9. Mackevičius J. *Įmonių veiklos analizė. Informacijos rinkimas, sisteminimas ir vertinimas*. Vilnius: TEV (2007)
10. Newbold D. *A Brief Description of the Methods of Economic Appraisal and teh Valuation of Health States*. Joournal of Advances Nursing (2005)
11. Žalimienė L. *Socialinių paslaugų vadyba: balansavimas tarp socialinio teisingumo ir ekonominio racionalumo*. Ekonomika ir vadyba: aktualijos ir perspektyvos, 2006, (6), 183–189. ISSN 1648-9098
12. Moriatry, J., Manthorpe, J. *The Effectiveness of Social Work With Adults. A systematic Scoping Review*. King's College London. Prieiga per internetą: `https://www.kcl.ac.uk/sspp/policy-institute/scwru/pubs/2016/reports/Moriarty-&-Manthorpe-2016-Effectiveness-of-social-work-with-adults.pdf` (2016)
13. Moriatry, J., Manthorpe, J. (*The Effectiveness Of Social Work With Adults. A systematic Scoping Review*. King's College London. Prieiga per internetą: `https://www.kcl.ac.uk/sspp/policy-institute/scwru/pubs/2016/reports/Moriarty-&-Manthorpe-2016-Effectiveness-of-social-work-with-adults.pdf` (2016)
14. Pincus, A., Minahan, A. *Social Work Practice: Model and Method*. F.E. Peacock Itasca, Ill (1973)
15. Germain, C.B. *Social Work Practice: People and Environments – an Ecological Perspective*. New York: Columbia University Press (1979)
16. Andreae, D. "General Systems Theory: Contributions to Social Work Theory and Practice". In *Social work treatment: interlocking theoretical approaches*, ed. by Francis J. Turner. Oxford; New York (N.Y.): Oxford University Press. P: 242–254 (2011)
17. Lietuvos Respublikos socialinių paslaugų įstatymas. *Valstybės žinios*. 2006, **104-2367**

An overview on pedestrians involved in traffic accidents in Latvia: Years 2010–2014

A. Bukova-Zideluna[1], A. Villerusa[1], and A. Lama[2]

[1]Rīga Stradiņš University, Riga, Latvia
[2]Road Traffic Safety Directorate, Riga, Latvia

Abstract. Latvian national road accident statistics shows that for the vulnerable road users' situation is critical, since pedestrians are involved in more than a quarter of road traffic accidents. This paper gives an analysis on pedestrians involved in road traffic accidents based on the road safety accident database in Latvia for the years 2010–2014. The total number of cases does not change significantly, however there has been an increase in pedestrian fatality rates over the period. From the total number of traffic accidents with pedestrians involved 92.4% had injuries, 6.8% were lethal cases and others did not suffer from injuries. Out of 342 fatalities 37.7% occurred during the winter period, 56.1% in adverse weather (overcast, fog, rain or snow), 69.9% during twilight or darkness and 26.9% on weekends. Out of all accidents 55.3% occurred in the capital city Riga, but fatality rate was higher on main state roads. 8.1% of the total number of pedestrians involved in road traffic accidents was found to have alcohol in their blood right after the road traffic accident. Fatality rate was higher for those with exceeded BAC. Pedestrian injury risk analysis was associated with demographical and traffic-related factors, urbanization, visibility and seasonal patterns.

1 Introduction

The issue of pedestrian safety has received an increasing attention from the researchers and transport planners. Every year the lives of approximately 1.25 million people are cut short as a result of a road traffic crash. 22% of those dying on the world's roads are pedestrians [1]. These accidents consume medical resources in the emergency, hospital and rehabilitation care, the patients and their families bear a heavy physical, emotional, and financial toll.

Road infrastructure is mainly constructed with the needs of motorists in mind, although almost half of all road traffic deaths occur among the vulnerable road users, which includes pedestrians as well. In the European Union the overall long-term tendency is for pedestrian fatalities to decrease slower than other fatalities since year 2010. The number of pedestrians killed on roads in the European Union has decreased by only 11%, compared to the total fatality decrease of 18% in the period from 2010 to 2013 [2].

In 2016 European Transport Safety Council reported on Latvia having one of the highest risk numbers per head being killed as road users with more than 90 deaths per million inhabitants [3]. Data shows that 39% of all killed in road traffic crashes in Latvia are pedestrians, and the vulnerable road users are one of the top priorities defined in the national road safety program. The road safety plan proposes improvement in law and regulations, improved road user behaviour control and improved road users' awareness [4].

The objective of this research is to analyse road traffic accidents with pedestrians involved and the relations between injuries and the circumstances they occur in: the severity and types of injuries; and such causes as demographic characteristics and crash circumstances, including weather conditions and road typology.

2 Methods

This study gives analysis on road traffic accidents with pedestrians involved. In the current study the statistical database of road safety accidents and consequences in Latvia for the years 2010 to 2014, held by Road Traffic Safety Directorate, under the Ministry of Transport of Latvia, was used. The information held on the database covers close to 100% of all cases reported to the traffic police. Data is being collected from administrative protocols, which includes the protocols of road traffic accidents and fine slips from road police, and protocols and reports from other institutions of road traffic offences (for example, information on road traffic accidents without casualties is obtained from vehicle insurance companies).

Data-gathering protocol followed each case and then on-site investigation was carried out in order to identify pre-crash phenomena, such as road layout, light and weather conditions.

According to Road Traffic Safety Directorate a "death" is considered in those cases, where the victim has died on the accident site or within 30 days after. "Severe injuries" are those where the victims are hospitalized for more than one day, "minor injuries" – those hospitalized for up to 24 hours.

Regarding all pedestrians involved in road traffic accidents, the following information was gathered and analysed:

1. pedestrian-related variables: age (younger than 35, 35–54, 55 and older) and sex;
2. severity of injuries (no injuries, severe, minor injuries or death);
3. environmental variables: zone (open road, urban), time (daylight, twilight, darkness), type of day (working day or weekend), month of the year, surface and overcast weather conditions;
4. BAC (blood alcohol concentration) prior to the accident.

Data was downloaded, processed and analysed using descriptive statistics, Pearson's chi-squared and calculations of statistical reliability with SPSS for Windows 20.0. The 95% confidence interval (CI) was used to estimate the precision of the odds ratio (OR).

3 Results

From the 2010 to 2014, 5030 pedestrians were involved in reported road traffic accidents in Latvia. Fatal accident risk for pedestrians involved in traffic accidents has increased compared to all accidents with pedestrians. The annual rate of pedestrians involved in road traffic accidents is displayed in Table 1.

Table 1. The annual number of pedestrians based on severity and proportion.

Year	No injuries	Minor injuries	Serious injuries	Fatalities	Total number (% of cases in period)
2010	10 (1.0%)	786 (77.1%)	145 (14.2%)	79 (7.7%0	1020 (20.3%)
2011	6 (0.6%)	805 (80.3%)	131 (13.1%)	60 (6.0%)	1002 (19.9%)
2012	8 (0.8%)	807 (81.6%)	111 (11.2%)	62 (6.3%)	988 (19.6%)
2013	9 (0.9%)	811 (91.1%)	110 (11.0%)	70 (7.0%)	1000 (19.9%)
2014	6 (0.6%)	847 (83.0%)	96 (9.4%)	71 (7.0%)	1020 (20.3%)

3.1 Pedestrian characteristics

There were 2157 male (42.9%) and 1758 female (35.0%) pedestrians; but in 22.2% of cases sex was not recorded in the police protocol. Most pedestrians (46.5%) were under 35, but rate of serious injuries was higher for the 35–54 year old group. Fatality rate was higher for those 55 years and older. The characteristics of pedestrians involved in road traffic accidents according to the age are displayed in Table 2.

Table 2. The number of accidents and severity of injuries based on age.

Age	No injuries	Minor injuries	Serious injuries	Fatalities	Total number (% of cases in period)
Younger than 35	28 (1.2%)	2007 (85.8%)	236 (10.1%)	68 (2.9%)	2339 (46.5%)
35–54	6 (0.5%)	891 (76.7%)	158 (13.6%)	107 (9.2%)	1162 (23.1%)
55 and older	5 (0.3%)	1158 (75.7%)	199 (13.0%)	167 (10.9%)	1529 (30.4%)
Total number (%) of injury by severity in all ages	39 (0.8%)	4056 (80.6%)	593 (11.8%)	342 (6.8%)	5030 (100%)

Differences between sex and injury/fatality rate (Pearson's chi-squared test $\chi^2 = 80.334$; p-value < 0.0001) are statistically significant. Differences between age groups and injury/fatality rate (Pearson's chi-squared test $\chi^2 = 136.54$; p-value < 0.0001) are statistically significant.

3.2 Characteristics of accident conditions

A total of 40.4% of pedestrians were involved in road traffic accidents during twilight or dark, and 42.6% of accidents occurred during overcast, fog and rain or snow. The light and weather conditions are shown in Table 3.

Police protocols gave incomplete reports on accident conditions, which mean there was some missing data on the light and weather conditions in total numbers. The differences between light condition and injury/fatality rate (Pearson's chi-squared test $\chi^2 = 163.384$; p-value < 0.0001) are statistically significant, fatality rate is higher for twilight and dark conditions (odd ratio OR = 3.750; confidence interval CI 2.953–4.761). Differences between weather condition and injury/fatality rate (Pearson's chi-squared test $\chi^2 = 22.717$;

p-value < 0.0001) are statistically significant, fatality rate is higher for adverse weather (OR = 1.792; CI 1.436–2.236).

Table 3. Characteristics of pedestrians involved in accidents based on the light and weather conditions.

Conditions		No injuries, minor or serious injuries	Fatality	Total number (%) of all cases by accident condition
Light condition	Daylight	2796 (96.4%)	103 (3.6%)	2899 (59.6%)
	Twilight, dark	1730 (87.9%)	239 (12.1%)	1969 (40.4%)
	Total number (%) of injury by severity	4526 (93.0%)	342 (7.0%)	4868 (100%)
Weather condition	Not adverse weather	2735 (94.8%)	150 (5.2%)	2885 (57.4%)
	Adverse weather (overcast, fog, rain or snow)	1953 (91.1%)	192 (9.0%)	2145 (42.6%)
	Total number (%) of injury by severity	4688 (93.2%)	342 (6.8%)	5030 (100%)

3.3 Seasonal and weekday characteristics

Looking at the time of a year, more pedestrians suffered in accidents during autumn (September, October, and November) and winter (December, January, and February). Comparing the severity of injuries across the seasons, data shows that fatality percentage was higher in autumn, but serious injury percentage was higher in winter and spring. The correlation between the season and injury/fatality rate is statistically significant (Pearson's chi-squared test $\chi^2 = 31.899$; p-value < 0.0001).

There were four times more pedestrians involved in traffic accidents on weekdays than on weekends, but rates for fatality and serious injuries were higher on weekends than on weekdays. The analysis of injuries on weekdays compared to weekends shows statistically significant differences (Pearson's chi-squared test $\chi^2 = 21.591$; p-value < 0.0001). Fatality rate is higher for weekends (OR = 1.467; CI 1.143–1.883). Characteristics of pedestrians involved in road traffic accidents according to the seasons and the day of week are shown in Table 4.

Table 4. Characteristics of pedestrians involved in accidents and severity of injuries based on the seasons and the day of week.

Season/ weekday	No injuries	Minor injuries	Serious injuries	Fatality	Total number (%) of all cases
Winter	7 (0.5%)	1090 (79.8%)	171 (12.5%)	98 (7.2%)	1366 (27.2%)
Spring	10 (0.9%)	864 (80.4%)	135 (12.6%)	66 (6.1%)	1075 (21.4%)
Summer	14 (1.2%)	962 (84.0%)	120 (10.5%)	49 (4.3%)	1145 (22.8%)
Autumn	8 (0.6%)	1140 (78.9%)	167 (11.6%)	129 (8.9%)	1444 (27.7%)
Weekdays	29 (0.7%)	3276 (81.9%)	443 (11.1%)	250 (6.3%)	3998 (79.5%)
Weekend	10 (1.0%)	780 (75.6%)	150 (14.5%)	92 (8.9%)	1032 (20.5%)
Total	39 (0.8%)	4056 (80.6%)	593 (11.8%)	342 (6.8%)	5030 (100%)

3.4 Characteristics of pedestrians involved in road accidents based on the time of accident

During the day time there have been mostly minor injuries (12:00–15:59), while serious injuries and fatalities occurred more during the night time (00:00–03:59). In total number of all cases more pedestrians suffered in traffic accidents in late afternoon (16:00–19:59). Table 5 presents the overall percentages of pedestrians injured according to the time of accident.

Table 5. Characteristics of injured pedestrians based on the time of accident.

Time	No injuries	Minor injuries	Serious injuries	Fatality	Total number (%) of all cases by time
00:00–03:59	1 (0.6%)	126 (70.8%)	27 (15.2%)	24 (13.5%)	178 (4.6%)
04:00–07:59	0	187 (79.6%)	27 (11.5%)	21 (8.9%)	235 (6.1%)
08:00–11:59	4 (0.6%)	518 (83.5%)	72 (11.6%)	26 (4.2%)	620 (16.2%)
12:00–15:59	11 (1.3%)	704 (85.4%)	75 (9.1%)	34 (4.1%)	824 (21.5%)
16:00–19:59	8 (0.6%)	1038 (79.9%)	147 (11.0%)	113 (8.5%)	1336 (34.8%)
20:00–23:59	2 (0.3%)	475 (73.6%)	98 (15.2%)	70 (10.9%)	645 (16.8%)

3.5 Characteristics of pedestrians injured based on the place of accident

55.3% of pedestrian road accidents occurred in the capital city, Riga, 30.8% happened in other big cities and 14.8% on the main roads, regional and local roads. 28.6% of the serious injuries occurred on local roads, and 28.2% of fatalities occurred on main state roads. Differences between the accident place and injury/fatality rate are what statistically significant (Pearson's chi-squared test χ^2 = 631.210; p-value <0.0001). Table 6 presents characteristics of pedestrians according to the place of accident.

Table 6. Characteristics of pedestrians based on the place of accident.

Place	No injuries	Minor injuries	Serious injuries	Fatality	Total number (%) of all cases by place
Main state roads	2 (0.5%)	221 (52.7%)	78 (18.6%)	118 (28.2%)	419 (9.4%)
Regional roads	0	91 (46.4%)	53 (27.0%)	52 (26.5%)	196 (3.9%)
Local roads	0	49 (63.6%)	22 (28.6%)	6 (7.8%)	77 (1.5%)
Capital city – Riga	24 (0.9%)	2392 (86.7%)	254 (9.2%)	88 (3.2%)	2758 (55.3%)
Other city	13 (0.8%)	1263 (82.2%)	185 (12.0%)	75 (4.9%)	1536 (30.8%)

3.6 Blood alcohol concentration

During the five years of under study, 407 (8.1%) pedestrians were found to have alcohol in their blood right after the road traffic accident. Fatality rate is higher for those with exceeded BAC (OR = 1.167; CI 0.794–1.714). Table 7 shows overall tendencies in correlation between the pedestrians involved in traffic accidents and BAC.

Table 7. Blood alcohol concentration among pedestrians involved in traffic accidents.

Alcohol presence	No injuries	Minor injuries	Serious injuries	Fatality	Total number (%) of all cases by BAC
No alcohol	38 (0.8%)	3740 (80.9%)	540 (11.7%)	305 (6.6%)	4623 (91.9%)
Positive BAC	1 (0.2%)	316 (77.6%)	53 (13.0%)	31 (9.7%)	407 (8.1%)

4 Discussion

Through the analysis of traffic police data, this study systematically examined the causes of injuries among pedestrians involved in traffic accidents in Latvia. In the period from 2010 to 2014, 5030 pedestrians were involved in road traffic accidents. The total number of cases does not change significantly, however there has been an increase in pedestrian fatality rate over the period.

Both – injuries and fatalities – in pedestrian accidents were correlated to male gender. In comparison a larger percentage of female than male pedestrian deaths occur in other European Union countries [5], while Abay showed that one of the factors that increased the risk of a fatal injury for pedestrians was the male gender [6]. Elderly pedestrians are particularly at risk – the proportion of pedestrian fatalities increases for the age group of 35–44 as compared to those under 35, and this increases even more for the group aged above 55, which is relevant to other findings. European Commission showed that around 44% of all killed pedestrians were 65 years old or older [2]. The increased fatality rate can be explained by physical vulnerability as older people are less able to endure injuries.

The role of light circumstances is important since results of the study show that twilight and darkness are related to a high proportion of fatal pedestrian accidents. This study shows that adverse weather conditions (overcast, fog, rain or snow) are related to pedestrian fatalities. Most fatalities occurred during the autumn, the percentage of serious injuries was higher in winter and spring suggesting that the greatest proportion of pedestrian traffic accidents happened due to bad visibility. The analysis showed that injuries mostly occurred in a late afternoon, suggesting that the greatest proportion of traffic accidents were related to traffic after working hours. However, the biggest part of serious injuries and fatalities occurred at late evening or night time. This indicates that a high proportion of fatal traffic accidents are related to visibility. These results are in line with the findings of other research studies on the fundamental limitations of vision [7, 8]. In these cases, use of reflective materials on the pedestrian could likely reduce fatality risk while walking in dark conditions.

This study shows that pedestrians face a large risk on urban roads. However, the largest percentage of serious injuries occurred on local and regional roads, and the largest percentage of fatalities occurred on the main state roads. The traffic mix outside cities makes the pedestrians share the road with vehicles, forcing them into dangerous situations in fast-moving traffic.

The current study also shows the relation between fatalities and pedestrian intoxication, showing that drunk walking is related to injuries and fatalities in road traffic accidents. Öström showed positive blood alcohol in 19% of the fatalities in Sweden [9], while in the United States nearly four out of ten pedestrians killed had had some level of alcohol involvement at the time of the crash [10]. Alcohol use among pedestrians has been a relatively neglected issue when compared with alcohol use among drivers. Educational programs can be designed to inform pedestrians about the dangers posed by alcohol consumption.

5 Conclusions

Pedestrian injuries and death arc rclated to preventable risk factors. In this study pedestrian fatalities are analysed based on demographical factors (males, older adults), blood alcohol concentration and accident circumstances (adverse weather, dark light conditions, night time, main state roads and weekends).

Making walking safer is critical for reducing the number of road traffic deaths. A key strategy for achieving a safe traffic system for pedestrians is to separate different kinds of road use, eliminating conflicts between high-speed vehicles and vulnerable road users. It is clear that both environmental and behavioural measures aimed at reducing collisions will be important in reducing the overall incidence of pedestrian injuries.

This study has some limitations. Protocols on accident conditions were reported incompletely, which means there was some missing data in the total numbers. The results of the analyses allow an overall assessment of pedestrian safety level in Latvia. Certainly, the effort of data collection is an on-going challenge and the analysis presented in this study in the future could be combined with more detailed analysis of additional data to emphasize the problem.

References

1. World Health Organization, *Global status report on road safety*, 2–8 (2015)
2. European Commission, *Road safety in European Union: trends, statistics and main challenges* **13** (2015)
3. European Transport Safety Council, *Ranking EU progress on road safety: 10th Road Safety Performance Index Report* **17** (2016)
4. Latvijas Republikas Ministru Kabinets, *Ceļu satiksmes drošības plāna 2014.-2016.gadam kopsavilkum* (2014)
5. European Transport Safety Council, *Risk on the roads – A Male Problem* **25** (2013)
6. K.A. Abay, Res in Transp Econ *Examining pedestrian-injury severity using alternative disaggregate models* **43**, 123–136 (2013)
7. M. Sivak, J. Luoma, M.J. Flannagan et al., J Safety Res *Traffic safety in the U.S.: Re-examining major opportunities* **38,** 337–355 (2007)
8. J.M. Sullivan, K.J. Flannagan, Accid Anal Prev *The role of ambient light level in fatal crashes: inferences from daylight saving time transitions* **34**, 487–498 (2002)
9. M. Öström, A. Eriksson, Accid Anal Prev *Pedestrian fatalities and alcohol* **33,** 173–180 (2001)
10. Department of Transportation, National Highway Traffic Safety Administration *National Pedestrian Crash Report* **26** (2008)

Fictitious employment contracts in loan recovery processes in Latvia

R. Neilands

Rīga Stradiņš University, Riga, Latvia

Abstract. Latvia from 2008 till 2013 experienced serious financial crisis. The results of the crisis were significant financial loss in banking area, plenty of foreclosures, plenty of corporate and private bankruptcy processes. One of issues Latvia faced in the foreclosure proceedings was the reaction of debtors – some of them used all possible means (including illegal) to obstruct foreclosure. One of such means is a conclusion of fictitious employment contracts on the mortgaged real estate maintenance and later obtaining a court judgment on a wage recovery.
The aim of the paper is to research how fictitious employment contracts are used and to propose a solution for the fictitious labour agreements issue.
Methods of qualitative research were employed in the paper – comparative method, analytic method, inductive method, and deductive method.
Key words: loan recovery; fictitious employment contracts.

1 Fictitious employment contracts

One of the ways how legitimate interests of mortgagees are infringed is entering into fictitious employment contracts for maintenance of the mortgaged real estate and obtaining of a court judgment on recovery of a salary therefore. Pursuant to sub-paragraph 1 of paragraph one of Article 628 of the Civil Procedure Act, first the judgment execution costs related to sale of the real estate are to be covered from the funds received for mortgaged real estate that is sold, and subsequently claims of those employees shall be satisfied who are involved in maintenance of the real estate along with social insurance contributions associated with their salaries. The claims secured by mortgages come only fourth in the ranking (after the salaries paid for maintenance of the mortgaged real estate, taxes and real burdens related to the real estate, which are due and payable).

In practice, the said provision of the Civil Procedure Act is widely abused. A debtor (mortgagor), fully aware of a pending recovery in respect of the mortgaged real estate owned by him caused by his failure to perform his commitments towards the creditor, enters into a fictitious employment contract with some friend or acquaintance for maintenance of the mortgaged real estate. No salary is actually paid. Instead, judicial proceedings are used as a way to legalize the claim for salary. The fictitious employee brings an action in court against the employer (owner of the mortgaged real estate) for recovery of the salary due for maintenance of the real estate. During the judicial proceedings the fictitious employer acknowledges the claim brought against him resulting in the court passing a ruling to fully satisfy the fictitious employee's claim for the salary. Later on, a writ of execution is issued based on such judgment, which entitles the fictitious employee to take precedence over the mortgagee and receive the funds, which, in essence, are due to the mortgagee. If such malicious scheme succeeds, the mortgagor and the fictitious employee share the monies received among them.

The category of fictitious salary claims also has an inherent aspect that the salary established for such employee is usually incommensurately high in comparison with work remuneration for a similar job in the relevant period and in the particular region; it is brought for a period of several years; since the beginning of employment the employee usually has not once received any salary at all, and no social insurance contributions or the personal income tax has been paid for him; the employer has not reported to the State Revenue Service on hiring the employee and on the fact that such employment relationship exists under the procedure set forth by law [1].

2 Case law in matters of fictitious employment contracts

The Department of Civil Cases of the Supreme Court has evaluated legality of one scheme of fictitious salary in its judgment of 9 November 2015 in case No. SKC-184/2015. On 26 November 2010, JSC "Rietumu banka" had brought a claim against a mortgagor and a fictitious employee for recognition of the said transaction null and void stating that it has become aware of a claim of the fictitious employee against the mortgagor being satisfied by the judgment of the Riga City North District Court of 7 June 2010, where recovery of a salary in the amount of LVL 4230 was claimed on the basis of an employment contract, under which the mortgagor had hired the fictitious employee as a manager of an apartment. In the opinion of JSC "Rietumu banka", the employment contract is fictitious, entered with retroactive effect and for the purpose to reduce the compensation due to the bank from the sale of the real estate [2].

By the judgment of 13 March 2012 of the Riga City North District Court, the claim of JSC "Rietumu banka" was dismissed; and, upon adjudicating the matter in the court of appeals, the Civil Court Panel of the Riga Regional Court also resolved to dismiss the claim on 28 December 2012. However, the Department of Civil Cases of the Supreme Court recognized that the judgment of the Riga Regional Court is to be dismissed based on the following reasoning:

1. Article 1415 of the Civil Act stipulates that unauthorized and indecent act, the purpose of which is contrary to religion, law, or moral principles, or which is intended to circumvent the law, may not be a subject-matter of a lawful transaction; consequently, such a transaction is void. Even though the abovementioned legal provision is incorporated in the section of the Act concerning the subject-matter, it actually regulates the issues concerning objective and motive (causa) of the transaction, which is an autonomous and independent element of the transaction and should be assessed in isolation from the subject-matter of the transaction. Consequently, even if a transaction is lawful in appearance, upon evaluating underlying motives, there is a possibility to recognize such a transaction null and void. So, in order to establish elements of the legal contents of Article 1415 of the Civil Act, establishing of circumstances, in which the transaction was closed, and actual motives of intentions of the parties to the transaction is necessary [2].

2. JSC "Rietumu banka" has submitted evidence, which confirms that: (1) the mortgagor was notified about withdrawal from the facility agreement and possible aiming of the recovery of the debt at the mortgaged apartment on 18 February 2009; (2) the employment contract between the mortgagor and the employee hiring the latter as the manager of the apartment is dated 10 March 2009; (3) the mortgagor's title to the real estate was corroborated with the Land Register on 18 April 2007; (4) as at 1 April 2009 the mortgagor had a debt accrued in the amount of LVL 1498.49 towards the manager of the respective residential building – LLC "Realko" – for the services rendered. Arising out of

the foregoing, the mortgaged real estate was already in freehold of the mortgagor before the contract for management thereof was concluded; however, there is no information in the case file how and whether at all the real estate owned by the mortgagor was managed before conclusion of the contested employment contract [2].

Furthermore, arising out of the evidence filed, on 10 March 2009 the mortgagor had undertaken a new obligation to pay LVL 470 per month to the employee as the apartment manager in a situation, where, first of all, he already had a debt of almost one and a half thousand Latvian lats accrued towards building manager LLC "Realko" (a fact that casts doubt on his solvency and financial capacities in general); and, secondly, he undertook to pay for a service (apartment manager), which he did not need before. Furthermore, there is no information in the database of the State Revenue Service about income gained by the employee and his work place during the period between 1 March 2009 and 31 May 2009, and the mortgagor has not been registered as an employer during the same period; as well as, irrespective of the absence of a dispute between litigants in respect of this aspect (which is confirmed by the fact that the mortgagor has, without reservation, acknowledged the employee's claim for recovery of the salary) and the salary not being received since 1 June 2009 (for almost a year he allegedly had worked without remuneration), the employee has brought an action in court to protect his interests only on 9 April 2010, when a judgment was passed on recovery of the loan facility in favour of the bank [2].

In the opinion of the Department of Civil Cases, the aggregate of all abovementioned aspects infers interconnection, which suggests that the employment contract was not entered into for a purpose to establish employment relationship. In view of the procedure established in Article 628 of the Civil Procedure Act for division of the funds received from sale of a mortgaged real estate, satisfaction of the claim arising out of the employment contract unambiguously infringes the right of JSC "Rietumu banka" to receive the funds obtained from enforcement of the mortgage in full. The Department of Civil Cases considered that the circumstances outlined by the claimant definitely shed doubts about legitimacy of the employment contract. However, the Riga Regional Court has assessed the established facts in isolation and individually, thus failing to gain correct idea about circumstances of the dispute conformant to the objective reality. To this end, the legal doctrine stipulates that facts are to be evaluated in their entirety and not in isolation for the purpose to clarify actual circumstances of the case. Selection of individual facts and erroneous interpretation thereof precludes arriving at the truth. Individual facts, isolated from their causal relationship, forfeit their effect as evidence [2].

Based on the foregoing reasons as well as other reasons, the Department of Civil Cases of the Supreme Court revoked the judgment of the Civil Court Panel of the Riga Regional Court and forwarded the case for repeated adjudication in the court of appeals.

In another case, a complaint of an employee about actions of a bailiff and a calculation made omitting the employee's claim therein was dismissed by the decision of the Aizkraukle District Court of 28 February 2012. In the reasoning section the court, among other, pointed out a necessity to tie the salary with the period of recovery. By the decision of Zemgale Regional Court of 10 April 2012 the decision of the Aizkraukle District Court was left unmodified [1], and the decision has come into effect.

3 Prospective solution in counteracting fictitious employment contracts

Priority of the salaries paid for maintenance of the mortgaged real estate enjoyed over satisfaction of mortgage claims is no modern-day invention. Historically, the salary for maintenance of mortgaged real estate had to be satisfied before covering the mortgage claims either.

Chronologically, the Civil Procedure Statute (*Civilprocesa nolikums* in Latvian) promulgated in 1914 in the Russian language, the Civil Procedure Statute in Latvian passed by the Codification Department of Saeima (Parliament) in 1932, and the Civil Procedure Act passed in 1938, were in force during the first period of independence in Latvia. All these statutory laws had priority set for salaries to be compensated before mortgages.

Professor Bukovskis has noted in 1933 when speaking about priority of salaries that the new law has returned to the principle stipulated in paragraph 4 of Article 1890 of the Civil Procedure Statute, according to which there has to be definite relation between claims of employees and the real estate sold. In the legislative body's opinion, work of these persons employed at the real property is aimed at maintenance of the real property in good order during such a period, when its owner, in view of the pending public sale of the property, has taken poor care of it (neglected) [3]. Professor Bukovskis has stated the foregoing in response to comments of A. Hasmanis and Baron A. Nolkens, officials of the Ministry of Justice of the Russian Empire, on Article 153 of the Statute for Transformation of the Judicial System in Baltic Provinces [4]. It should be noted that paragraph 4 of Article 153 of this Statute prescribed that the salary for property management or for performance of work at the property shall be satisfied for one year before the date of auction.

Professor Bukovskis has also noted that it is impossible to extend priority of the salary to claims of a schoolteacher for delivering lessons to children of the debtor in the houses on sale; whereas, claims of a gardener or another worker, if these persons have performed any work at this real property or outside it, while, nevertheless, for this real property, such work being aimed at preservation or maintenance of this property, shall be satisfied under the procedure set out by the Civil Procedure Statute [3].

Through this historical retrospect a solution can be found for the issue with fictitious salaries. Since initially the objective of prioritizing the salary over mortgages was to ensure maintenance of the mortgaged property, because its owner/proprietor ceased to take care of the property, this objective should be reinstated, thus precluding possibility of abuse of this priority in bad faith.

Article 603 of the Civil Procedure Act prescribes a procedure for attachment of the real estate. Pursuant to paragraph one of this article a bailiff attaches the real estate upon a request of a debt collector.

Conversely, Article 605 of the Civil Procedure Act stipulates:

1) The attached real estate shall remain under management of the former possessor or manager until conveyance thereof to the new owner.
2) The possessor or manager shall preserve the attached real estate in the same condition as it was at the time of attachment and together with the same movable assets.
3) If the possessor or manager of the real estate is not known, the bailiff may appoint a manager of the real estate at his/her own discretion. The real estate manager appointed by the bailiff bears the same liability as the custodian of the movable assets referred to in this act.
4) The possessor and the manager of the real estate shall provide accounts to the bailiff for the period of management of the attached real estate. The revenues gained from the real estate shall be transferred to the bailiff and added to the amount received from sale of this real estate.

Attachment and management of the real estate is aimed at preservation of the real estate in the condition as it was at the time of attachment. Consequently, the salaries associated with maintenance of the real estate should be specifically tied to management of the real estate after attachment thereof. In such a way priority of the salaries would be ensured in line with its historical objective (to ensure maintenance of the property in proper order until auction), as well as the possibility to make use of paragraph one of Article 628 of the Civil Procedure Act for the purpose to procure fictitious salaries would be precluded. In order to comprehensively prevent dishonest use of this provision, it should be stipulated that only an account of a real estate manager (reports on work performed and expenses on maintenance of the property) approved by the bailiff can serve as basis for disbursement of the manager's salary.

4 Conclusion

Taking into account the foregoing considerations, sub-paragraph 1 of paragraph one of Article 628 of the Civil Procedure Act should be restated in a new wording:

"Article 628. Division of funds received from sale of mortgaged real estate

(1) First expenses on enforcement of a judgment related to sale of the real estate shall be covered from the funds received for such mortgaged real estate sold, and then claims shall be satisfied in the following order of priority:

1) A claim of the real estate manager for payment of a salary approved by the bailiff, including social insurance contributions associated with the manager's salary. The claim referred to in this paragraph shall be applicable only to the period between the date of attachment of the real estate and the date of conveyance of the real estate to the new owner."

References

1. Stepanovs J. Četrpadsmitā sadaļa. Piespiedu izpildes līdzekļu piemērošana. 75. Nodaļa. Piedzīto summu sadalīšana starp piedzinējiem. No: Torgāna K. zin. red. Civilprocesa likuma komentāri. III daļa (61.-86. nodaļa). Rīga: Tiesu namu aģentūra **546** (2014)
2. Judgment of the Department of Civil Cases of the Supreme Court of the Republic of Latvia of 9 November 2015 in case No SKC-184/2015. Available on: https://manas.tiesas.lv/eTiesasMvc/nolemumi/pdf/238166.pdf
3. Bukovskis V. *Civīlprocesa mācības grāmata*. Rīga: Autora izdevums **656** (1933)
4. Гасманъ А., Нолькенъ А. Положенія о преобразованіи судебной части и крестьянскихъ присутственныхъ мѣестъ въ Прибалтійских губерніях и правила о приведеніи означенныхъ положеній въ дѣйствіе. 2-ое пересмотрѣнное и дополненное изданіе Министерства Юстиціи. С.-Петербугъ Типографія правительствующаго сената **137** (1890)

The motivational peculiarities of bribe-takers

A. Krivins

Daugavpils University, Daugavpils, Latvia

Abstract. Criminology (as the scientific study of criminal behaviour on the individual and social level) and psychology (as the science, which seeks to understand individuals and groups) pay special attention to motivation – a theoretical construction, used to explain behaviour. In this article, the motivation of recipients of bribery is widely investigated. In addition to such well-established motives as greediness, selfishness, prevalence of personal interest, the author analyses also other criminal behaviour determinants. The works, which are used in this research are of the following authors: John Broadus Watson, Edward Lee Thorndike (Behavioural); Sigmund Freud, Erich Fromm (Psychoanalysis); Carl Rogers, Abraham Maslow (Existential-humanistic theories); William Isaac Thomas (Thomas theorem); Florian Witold Znaniecki (sociological theory); Erik Homburger Erikson (theory on psychosocial development of human beings); Jean William Fritz Piaget (theory of cognitive development); Herbert Lefcourt, Ronald Smith (*Locus of Control*); etc. As the result of this research the author has concluded that bribe-takers' behaviour is determined by such psychological constructs as the desire for power, fear, envy, mental deflections, mania, feeling of exclusion, the need for extreme, oppressed creativity.
Key words: bribe-takers; bribery; motives; motivation.

1 Introduction

Motivation represents the reasons for people's actions, desires, and needs, what causes a person to want to repeat behaviour and *vice versa* [1].

Studying of bribe-takers' motivation permits to define causes which determine the illegitimate behaviour of bribe-takers and permits to develop mechanisms for more efficient prevention of bribe taking. In order to ensure more efficient fighting with corruption, it is of vital importance to determine what urges an official to become a bribe-taker. Notwithstanding the importance of the problem, this sector still presents lack of quality-targeted studies. Nowadays there is still no scientifically based psychological conception for researching the bribe-taking problem.

In psychology, as opposed to law, the term "motive" is used in various meanings (for instance, instincts, ideals, needs, emotions, etc.).

The formal constituent elements of bribe taking determine also its subjective side – it is a deliberate offence characterized by a direct intent: a state official acknowledges the harm of their action – bribe taking – and performs it deliberately [2].

Yet, when returning to the issue of distinguishing bribe taking and using an official position in bad faith for purposes of acquiring property that is afforded by a state official, unfortunately it must be stated that this issue has not been duly addressed in the theory of criminal law, only a reference can be found that, from the subjective point of view, the crime envisaged in Article 320 of the Criminal Law may be committed only with direct intent [3] and that, when regarding motives of bribe taking, it is of no importance whether a state official takes a bribe with the intention to acquire wealth or due to any other motive [4].

The motive of a criminal offence is the inner encouragement, drive, propensity of a person that directs the will of the guilty party towards a criminal offence.

Within the frames of the article, the author understands the term "motive" as inner force related to person's needs and that motivates this particular person to perform definite actions or, in particular cases, ensures her failure to act.

The article is structured on the basis of Maslow's hierarchy of needs [5]: 1) Physiological Needs; 2) Security Needs; 3) "Belongingness" Needs; 4) Esteem Needs; 5) Self-Actualization Needs; 6) Self-transcendence Needs. Maslow's theory suggests that the most basic level of needs must be met before the individual will strongly desire (or focus motivation upon) the secondary or higher level needs [6].

2 Materials and methods

The methods of the present research were: literature review, statistical data, discourse analysis and teleological reduction.

3 Results

3.1 Accepting bribes as physiological needs

Physiological needs are the physical requirements for human survival. Air, water, and food are metabolic requirements for survival in all animals, including humans. Clothing and shelter provide necessary protection from the elements.

Trubnikov, a specialist in philosophical anthropology, has noted that all person's thoughts and actions are aimed at defeating death. At first, a person wishes to delay death by satisfying their physiological needs (firewood, food, clothes), but, having understood that the physical life is passing, a person tries to extend it by using more durable forms, such as literature, art, etc. [7].

According to Carl Jung, the will to live is the most important motivator of a person's behaviour. Also Ernest Bekker believed that the main driver of a person's actions is fear of death [8, 9]. This opinion relatively coincides with Schopenhauer's thoughts – Schopenhauer puts a central emphasis on will and in particular has a concept of the "will to live".

As one of examples when bribes are taken in order to ensure physical survival (ensure for oneself the minimal level of food, dwelling, etc.) it is possible to mention the lower civil servants of the Russian Empire in the late 19th century.

On the basis of that-time sources, professor Golosenko concludes that a part of the Russian civil servants had salaries insufficient for surviving in the late 19th and early 20th century [10].

Namely, N. Rubakin's study regarding income of Russian civil servants in the late 19th century proves that there existed the so-called "slaves of state service" – officials with salary of only 7 roubles per month. At the same time price for food at that period of time was quite high – for instance, 10 eggs cost round 40 kopecks, but unsalted butter – 75 kopecks for 410 grams [11].

According to the research, conducted by N. Benediktov, regarding dwelling issue, the average civil servants quite often rented damp, dark rooms with no utilities, i.e. premises that were poorly suited for living due to hygienic conditions, due to the lack of money [12].

It is quite difficult to find such examples in the 21st century. Thus it must be concluded that nowadays this motive of bribe taking cannot be regarded as highly relevant.

3.2 Accepting bribes as safety needs

Once a person's physical safety needs are relatively satisfied, their safety needs take precedence and dominate behaviour. Safety and Security needs include: personal security, financial security, health and well-being, safety net against accidents/illness and their adverse impacts.

As it is noted by Erich Fromm, a philosopher, psychologist and sociologist, in society where the main goal is to possess it seems that exactly possessing is the point of existence and a person per se is nil if she does not own anything.

In Fromms' view, neither possession, nor power, nor sensuous satisfaction, can fulfil man's desire for meaning in his life; he remains in all this separate from the whole, hence unhappy. Only in being productively active can man make sense of his life, and while he thus enjoys life, he is not greedily holding on to it. He has given up the greed for having, and is fulfilled by being; he is filled because he is empty; he is much, because he has little [13, 14].

Like Edward Lee Thorndike ("we study animal behaviour, not animal consciousness, for the ultimate purpose of controlling behaviour") [15], John Broadus Watson put the emphasis on external behaviour of people and their reactions on given situations, rather than the internal, mental state of those people. Watson pushed for psychology to no longer be considered the science of the "mind". Instead, he stated that psychology should focus on the "behaviour" of the individual, not their consciousness [16].

Nevertheless, in Rogers' view the human being as: "capable of evaluating the outer and inner situation, understanding herself in its context, making constructive choices as to the next steps in life, and acting on those choices" [17, 18].

This short philosophical review makes to think whether a bribe-taker may achieve the desired when looking for safety. The following part of the article will present references to the fact that large sums of money provide for feeling of safety, yet this safety is fictitious.

3.3 The will to power

When examining criminal cases related to bribe taking, quite often it is possible to state that officials obtain remarkably huge sums of money in their possession as the result of criminal activities [19–21].

Those sums considerably exceed personal needs. During search it can be seen that in some cases the huge bribes are not even spent, but are kept in cash [22].

Certainly it is difficult to spend the illegally acquired means so that it remains unnoticed, yet any bribe-taker can acknowledge this beforehand. However, notwithstanding the saved amount of the illegally obtained money, officials usually continue to take bribes till the moment when criminal proceedings are commenced against them. What are the causes of such an ill-considered action? What can money give that is not used (or just a small part of which is spent)? If money in this aspect is not intended for spending, then what function does it perform?

We also need to pay attention to such an aspect as sense of power in addition to the first element that we have examined before (sense of safety). In consumer society only money gives weight and is a measure of achievements. The preserved money means potential opportunities. In this way the money permits to feel strong and powerful.

In order to better understand this phenomenon, let us pay attention to what happens in wild nature. Every animal occupies a particular place in its pack – a rank that they fight for among themselves. As the result hierarchy is formed that shows the social standing of animals. For example, dominance in wolves typically is determined through strength and skill [23]. However, in modern human society the status is defined by money (money is the potential opportunity to gain power and influence).

Friedrich Nietzsche in his work "Beyond Good and Evil" [24] equates a concept "Will to power" (German: der Wille zur Macht) to life *per se*: "Even the body within which individuals treat each other as equals, as suggested before – and this happens in every healthy aristocracy – if it is a living and not a dying body, has to do to other bodies what the individuals within it refrain from doing to each other: it will have to be an incarnate will to power, it will strive to grow, spread, seize, become predominant – not from any morality or immorality but because it is living and because life simply is will to power."

On the one hand, Friedrich Nietzsche attributed the concept "will to power" to living beings as such, without distinguishing activity of these living beings (humans among them). If the power is mostly related to brutal force in wild nature, then in the present-day society the rights and possibility to rule, to determine and subject (i.e. power) to a great extent depends also on such constructs as functions and authorities.

Functions and authorities regarding administration, supervision and decision-taking of a public resource considerably exceed possibilities of an average citizen that creates real risk of corruption per se. In addition, acquisition of a status of a public official is a quite easy way to obtain power. At the same time it can be concluded that quite often it is not enough to receive a position in order to feel complete sense of power.

On the other hand, the feeling of power in itself affects the person's thinking process (see: Lammers – power, like alcohol, brings about a kind of moral Myopia – provokes behaviour that is described as a hyper-self-confident or hyper-assertive) [25].

It can be concluded from the aforementioned that a person who has received a position of an official with the aim to obtain sense of power (in order to mitigate the inferiority complex developed during childhood) will engage in bribe taking with high probability. It is also confirmed by conclusions made by Alfred Adler, who states: "The Ruling or Dominant type strive for power and are willing to manipulate situations and people, anything to get their way. People of this type are also prone to anti-social behaviour." [26].

In this context, the Adler's interpretation seems interesting: to convert feelings of inferiority to superiority [27]. If the corrective factors were disregarded and the individual overcompensated, then an inferiority complex would occur, fostering the danger of the individual becoming egocentric, power-hungry and aggressive or worse [28].

When regarding the phenomenon of "will to power" in the frames of bribe taking, we need to be very careful. Under no circumstances can we automatically relate the fact of power acquisition with willingness to take bribes. We should pay attention not to the fact that a person wishes to obtain a status of a public official, but to the motivation of this particular person, for instance, a person who has received the position not in order to acquire the sense of power, but for other motives (to improve public conditions, to help the working poor, etc.) will doubtfully engage in corruption.

3.4 Accepting bribes as "Belongingness" needs

According to Maslow, humans need to feel a sense of belonging and acceptance among their social groups, regardless whether these groups are large or small.

This need for belonging may overcome the physiological and security needs, depending on the strength of the peer pressure. It is well-known that in countries where social institutions are efficient, bribe taking becomes a marginal phenomenon [29]. Nevertheless in countries, where bribe taking can be regarded as an admissible phenomenon, one of the motives to take a bribe is fear of breaking "rules of the game", fear of becoming a "black sheep".

In a world of hunchbacks, a fine figure becomes a monstrosity [30] and it is of extreme difficulty for an official to act honestly when starting one's career in an institution where bribe taking is accepted as a standard.

On the basis of conclusions of the social identity theory and self-categorization theory, it can be concluded that adaptation in a collective body envisages acknowledgment and acceptance of standards effective in that particular collective body. In case the conflict between an individual's core values and collective body's core values is not eliminated it leads to disintegration – isolation of an individual and exclusion from a particular group.

Thus, on the one hand, an official is influenced by actual conditions (the bribe taking subculture existing in the society). On the other hand, it is convenient for a person to agree with standards existing in a society and a collective body as this is the direction where "belongingness". This desire is so universal that the need to belong is found across all cultures and different types of people [31].

Belongingness to a group enables a person to feel their value [32]. It is of particular significance in a work place.

Employees want to fit in at work as much as students want to fit in at school. They seek the approval of their leaders, bosses, and other employees in order to be accepted. A sense of belongingness increases a person's willingness to assist others in the group by the group rules [33].

Sometimes this sense of belongingness is achieved regardless of inconvenience, awaiting the very individual: "With the social animal controlled by herd instinct, it is not the actual deed which is instinctively done, but the order to do it which is instinctively obeyed. The deed, being ordained from without, may actually be unpleasant, and so be resisted from the individual side, and yet be forced instinctively into execution" [34].

Thus in countries where bribe taking is a common phenomenon, belongingness prepares good soil for corruption to take roots. While in countries, where corruption is actually condemnable, belongingness helps to ensure efficient fight against bribe taking. This phenomenon is reinforced by fear of standing out, the herd instinct ("power of crowds") [35].

Although Jean William Fritz Piaget believed that a person forms her own moral views, even when a child [36]. Piaget believed that children made moral judgments based on their own observations of the world.), yet in this aspect we have to agree with the thesis expressed by Gustave Le Bon – "... the individualities in the crowd who might possess a personality sufficiently strong to resist the suggestion are too few in number to struggle against the current. /.../ the individual in a crowd differs essentially from himself. Even before he has entirely lost his independence, his ideas and feelings have undergone a transformation, and the transformation is so profound as to change the miser into a spendthrift, the sceptic into a believer, the honest man into a criminal, and the coward into a hero" [35].

Erik Homburger Erikson (1902–1994) showed how personality development in certain key individuals can induce widespread cultural changes.

Perhaps more intriguing, and troubling, is that some individuals' personalities and moral dispositions may change when they become part of a group. "Social identity theory" suggests that a person can think an act is immoral before becoming a part of the group, and then their personality changes to conform to the predominant identity of the group. Research suggests that this social identity theory may indeed explain why corruption can persist in a system long after the original corrupt actors leave [37].

In countries with ancient corruptive traditions a person is often regarded as a biological construct of little significance, as an insignificant part of a state mechanism. One of the most real possibilities to improve the situation could be rise of a value of each individual personality in society.

3.5 Accepting bribes as esteem needs

All humans have a need to feel respected; this includes the need to have self-esteem and self-respect. Maslow noted two versions of esteem needs: a "lower" version and a "higher" version. The "lower" version of esteem is the need for respect from others. This may include a need for status, recognition, fame, prestige, and attention. The "higher" version manifests itself as the need for self-respect. For example, the person may have a need for strength, competence, mastery, self-confidence, independence, and freedom.

Striving for esteem in the modern world is closely related to the perception of society regarding what really is to be respected. In the modern consumer society it is usually money, power, and success.

Alfred Adler [38] noted that one of the most important phenomena in the human nature is an individual's drive for superiority and success.

Individuals who have a high external locus of control tend to have higher levels of psychological and physical problems. These people are also more vulnerable to external influences [39, 40].

According to William Isaac Thomas (Thomas theorem), if men define situations as real, they are real in their consequences [41].

When a person feels underestimated, unsafe or inferior, they will always want to obtain higher position in order to receive compensation and sense of content in that way. Often sense of lack of compensation and fear of not achieving the desired goal cause additional problems. Namely, in case the desire for superiority becomes extremely strongly expressed, it could cause a sense of envy in parallel.

As Helmut Schoeck puts it [42], a person, who has a passive envy of something, can very easily turn into an aggressive criminal (in our case – a bribe-taker).

3.6 Accepting bribes as self-actualization

This level of need refers to what person's full potential is and the realization of that potential. Individuals may perceive or focus on this need very specifically. Maslow believed that to understand this level of need, the person must not only achieve the previous needs, but master them.

According to Maslow, a personality who self-actualizes lives fuller life and with greater sense, than an average individual who is mainly motivated by satisfaction of their needs [5]. When characterising people regarding self-actualization, Maslow indicates the following peculiarities of such people: more effective perception of reality and more comfortable relationship with her; adoption (self, others, and nature); spontaneity, simplicity and naturalness; the centrality of the task; some thedi-lennosti and need for privacy, autonomy, independence from the culture and environment; permanent freshness assessment; mysticism and the experience of higher States; the sense of belonging, of being one with the other; the deeper interpersonal relations; democratic character structure; differentiating the means and the objectives of good and evil; philosophical, non-hostile sense of humour; savakta-lysed creativity; the resistance of acculturation.

Self-actualization is a difficult and painful process involving continuous challenges to one's capabilities [43] wrote, "It involves the courage to be. It means launching oneself fully into the stream of life."

Rogers offered [44] five specific characteristics of the fully functioning person: Openness to Experience; Existential Living; A Trust in One's Own Organism; A Sense of

Freedom; Creativity (Creative individuals are not known for conformity or passive adjustment to social rules; due to their lack of defensiveness, they are not concerned with approval from others for their behaviour. The fully functioning person is thought to be more capable of adapting to and surviving drastic life changes due to the creative and spontaneous element).

In the following part of the article we need to answer a question whether these five specific characteristics of the fully functioning person can be attributed also to a bribe-taker.

Does the bribe-taker possess openness to experience, existential living, and a sense of freedom? Can the bribe-taker be creative? In other words, can the bribe-taker experience self-actualization?

It must be concluded that self-actualization may be varied, i.e., self-actualization can be expressed not only in forms approved by the society. The humanistic approach [ccc] envisages that the inner harmony is possible when a person is a self-actualizing personality. Nonetheless finding an occupation that is satisfactory for the person should be settled for each person individually. It is not excluded that this harmony is achieved also by individual people who have decided to become bribe-takers.

The Personal Orientation Inventory (POI) by Everett L. Shostrom. According to the POI testing, the level of self-actualization amongst criminals was higher than among college students [45]. In this aspect particular attention should be paid to creativity. Sigmund Freud has searched for pathological causes of a genius; Florian Witold Znaniecki has indicated that a person possesses also the irrational next to the rational [46].

This irrational is to great extent related to creative activity as creative activity differs from a natural process exactly in a way that it provides for a new, previously non-existing element [46]. The aforementioned confirms that both creativity and crime may have common features. For instance, just like a criminal offence, also creativity is closely related to violation of standards (perceptions) existing in society, both creativity and crime are not possible without breaking the previously set limit, without reaching another dimension. Both a criminal and a creative person destroy barriers with their activity.

It is not possible to assert that any criminal is more creative than an average person, yet it is significant that, according to the results of Den Arieli's research, greater success in manipulations is achieved by creative personalities with non-standard approach to problem solving and with developed imagination [47].

In this relation it is of significance to pay attention to the assumption that bribe taking may also be the creative potential used in a wrong (destructive) way.

The role of a bribe-taker, especially in the high level bribe taking, requires a person to possess particular characteristics. These people are usually better adapted to changes, they possess plasticity, adaptability. Typical bribe-takers feel the state of affairs and perfectly understand hierarchy within the system, understand nuances of social interaction, are socially active and communicative.

Gilinsky [48], a criminologist, believes that a person's activity can be manifested both in a positive (creative activities) and a negative way (crimes). In this plane, if an official's creativity is artificially suppressed, bribe taking can be interpreted as a protest against routine, social pressure, and lack of creativity.

3.7 Self-transcendence

In his later years, Maslow explored a further dimension of needs, while criticizing his own vision on self-actualization [49]. The self only finds its actualization in giving itself to some higher goal outside oneself, in altruism and spirituality [50–52].

Self-transcendence as assessed in the Temperament and Character Inventory originally had three subscales [53, 54].

A study relating the Temperament and Character Inventory to the Five factor model personality traits (1. Self-forgetful vs. self-conscious experience; 2. Transpersonal identification vs. self-isolation; 3. Spiritual acceptance vs. rational materialism; 4. Enlightened vs. Objective; 5. Idealistic vs. Practical) found that self-transcendence was most strongly related to openness to experience and to a lesser extent extraversion [55].

The level of self-transcendence illustrates which level of Maslow's hierarchy of needs bribe-takers nevertheless do not reach.

4 Conclusions

1. When examining the fact of bribe taking, we primarily need to understand that any action is affected by various elements; in addition, all these elements are interrelated. For instance, it would be wrong to indicate that the bribe-taker's motive is greediness as such, without examining causes of appearance of greediness in the personality of the particular bribe-taker. Largely, it is possible to confine oneself to the motive of greediness only in case if a bribe-taker tries to satisfy the primary (physical) needs. Yet, if money serves to satisfy secondary – psychological – needs, we should search for deeper motives of bribe taking.
2. Exaggerated drive for material benefits in the bribe-taker's personality may form under influence of several causes – sublimation of childhood fear and poverty, search for safety, willingness to overcome the sense of weakness, feeling of being unprotected, of being inferior, will for power, for self-actualization, fear of becoming a "black sheep", etc.
3. The bribe-taker's motivation is to great extent not a problem of sociology or economics, but a problem of philosophy and psychology. Bribe taking by itself indicates that a person has problems (unsatisfied needs). Nowadays the issue of physical survival due to lack of financing (according to Maslow – physiological needs) currently is not important for officials. Thus, we need to analyse official's psychological needs. In this aspect it must be concluded that in any case the fact of bribe taking is a proof (confirmation) that the person's life philosophy has failed. If we reduce the Fromm's conclusions to the problem of corruption, then bribe-takers live with a willingness to convert everything surrounding into their property, they want to control. Whereas the real life envisages freedom, independence, autonomy and critical thinking, inherent activity (to use one's potential, talents, to develop, hold an interest, achieve particular goals, and share with others).
4. Accepting bribes as self-actualization is a problem that the society will have to face even more often. If bribe taking takes place on the level of self-actualization, then it is not really possible to talk about an individual's discontent per se. Bribe taking as self-actualization is a deviation that does not comply with social norms.

References

1. A.J. Ellliot, M. Covington, Approach and Avoidance Motivation, Educational Psychology Review **13**, 2 (2001)
2. `www.at.gov.lv/files/uploads/files/6_Judikatura/Tiesu.../AT_petijums_kukulosana.doc` (2015)
3. U. Krastiņš, V. Liholaja, A. Niedre, Krimināltiesības. Sevišķā daļa. Trešais papildinātais izdevums. Rīga : Tiesu namu aģentūra, 766.lpp. (2009)

4. D. Mežulis, Korupcijas ierobežošanas krimināltiesiskie aspekti. Rīga : Latvijas Vēstnesis, 79. lpp. (2003)
5. A. Maslow, Motivation and Personality, N.Y. (1954)
6. F. Goble, The third force: The psychology of Abraham Maslow. Richmond, CA: Maurice Bassett Publishing, pp. 62 (1970)
7. Н.Н. Трубников, О смысле жизни и смерти, М., С. 84 (1996)
8. E. Becker, The denial of death. N.Y., Free Press (1973)
9. E. Becker, The Birth and Death of Meaning. N.Y., Basic Books (1971)
10. http://www.jourssa.ru/sites/all/files/volumes/1999_3/Golosenko_1999_3.pdf (1999)
11. Н. А. Рубакин, К статистике русского чиновничества, Спутник Чиновника, №№3,4.; № 4, с. 1-2 (1912)
12. Н. Бенедиктов, Жилищная нужда чиновников, Спутник Чиновника, №4. с. 21-22 (1911)
13. E. Fromm, Marx's Philosophy of Man (1961)
14. E. Fromm, Man for Himself, Rinehart & Co., New York (1947)
15. http://www.muskingum.edu/~psych/psycweb/history/thorndike.htm
16. http://www.newworldencyclopedia.org/entry/John_B._Watson
17. C. Rogers, Carl Rogers on personal power. N.Y., Delacorte Press, p.15 (1977)
18. An Analysis of Carl Rogers' Theory of Personality by Dagmar Pescitelli. http://pandc.ca/?cat=carl_rogers&page=rogerian_theory (1996)
19. Head of Russia's anti-corruption agency is found with $122m in CASH in his house. http://www.dailymail.co.uk/news/article-3784186/Head-Russia-s-anti-corruption-agency-122m-CASH-house.html#ixzz4V0Kunm9v (2016)
20. 11 Ashmiany Customs Officials' High-Profile Trial Started In Minsk https://charter97.org/en/news/2016/11/18/231524/ (2016)
21. Pictured: Inside former Ukraine general prosecutor's luxury home that was "built on corruption". http://www.mirror.co.uk/news/world-news/pictured-ukraines-former-general-prosecutor-3181937 (2014)
22. Senior Russian anti-corruption official arrested after police find $123 million in cash at his home. http://www.independent.co.uk/news/world/europe/senior-russian-anti-corruption-official-arrested-after-police-find-123-million-in-cash-at-his-home-a7239041.html (2016)
23. Wolves of the World. http://wolph.onbile.com/packhierarchy
24. F. Nietzsche, Beyond Good and Evil (1886)
25. J. Lammers, A.D. Galinsky, E.H. Gordijn, S. Otten, Power increases social distance, Social Psychological and Personality Science **3**, 282–290 (2012)
26. H. H. Mosak, M. Maniacci, A Primer of Adlerian Psychology, 64–65 (1999)
27. A. Adler, The Neurotic Character (1912)
28. A. Adler, Understanding Human Nature, Chapter 11, Aggressive Character Traits (1927)
29. http://ipulsar.net/newsnew-727.html
30. H. de Balzac, La Comedie Humaine (1851)
31. R.F. Baumeister, M.R. Leary, The need to belong: Desire for interpersonal attachments as a fundamental human motivation, Psychological Bulletin **117**(3), 497–529 (1995)
32. W. McDougall, An Introduction to Social Psychology, Boston, Luce (1912)

33. D.N. Den Hartog, A.B. de Hoogh, A.E. Keegan, The interactive effects of belongingness and charisma on helping and compliance, Journal of Applied Psychology **92**(4), 1131–1139 (2007)
34. W. Trotter, Instincts of the Herd in Peace and War, T. F. Unwin Ltd, London. p. 48 (1916)
35. G. le Bon, The Crowd: A Study of the Popular Mind (1895)
36. https://en.wikipedia.org/wiki/Jean_Piaget
37. E.H. Erikson, http://www.encyclopedia.com/people/medicine/psychology-and-psychiatry-biographies/erik-erikson
38. A. Adler, The Education of Children. Gateway Editions, Ltd South Bend Indiana (1978)
39 H.M. Lefcourt, Locus of Control: Current Trends in Theory and Research. NJ: Lawrence Erlbaum Associates (1976)
40. R.E. Smith, Effects of coping skills training on generalized self-efficacy and locus of control, Journal of Personality and Social Psychology **56**(2), 228–33 (1989)
41. W.I. Thomas, D.S. Thomas, The child in America: Behavior problems and programs. New York, Knopf, 571–572 (1928)
42. H. Schoeck, Envy: a theory of social behaviour, Liberty Fund, Indianapolis (1969)
43. http://www.texcpe.com/html/pdf/fl/FLHP.pdf
44. C. Rogers, On Becoming a Person: A Therapist's View of Personality. Boston, Houghton Mifflin (1961)
45. E.L. Shostrom, Man, the manipulator: The inner journey from manipulation to actualization. Nashville, TN: Abingdon (1967)
46. F. Znaniecki, Wstęp do socjologii. Poznan **116** (1922)
47. http://www.learningfirst.org/predictably-irrational-conversation-best-selling-author-dan-ariely
48. Я. И. Гилинский, Н. А. Исаев, Ю. А. Клейберг и др.,Творчество как позитивная девиантность. Российский государственный педагогический университет им. А. И. Герцена, Санкт-Петербургский Центр девиантологии, СПб., Алеф-Пресс, 279 с. (2015)
49. A.H. Maslow, Critique of self-actualization theory, in: E. Hoffman (Ed.), Future visions: The unpublished papers of Abraham Maslow (Thousand Oaks, CA: Sage), pp. 26–32 (1996)
50. A.H. Maslow, The farther reaches of human nature, Journal of Transpersonal Psychology **1**, 1–9 (1969)
51. A.H. Maslow, The farther reaches of human nature, New York: The Viking Press (1971)
52. M.E. Koltko-Rivera, Rediscovering the Later Version of Maslow's Hierarchy of Needs: Self-Transcendence and Opportunities for Theory, Research, and Unification, Review of General Psychology **10**(4), 302–317 (2006)
53. C.R. Cloninger, DM. Svrakic, T.R. Przybeck, A psychobiological model of temperament and character, Archives of General Psychiatry **50**(12), 975–90 (1993)
54. D.A. MacDonald, D. Holland, Examination of the psychometric properties of the temperament and character inventory self-transcendence dimension. Personality and Individual Differences **32**(6), 1013–1027 (2002)
55. F. de Fruyt, L. Van De Wiele, C. Van Heeringen, Cloninger's Psychobiological Model of Temperament and Character and the Five-Factor Model of Personality. Personality and Individual Differences **29**(3), 441–452 (2000)

SHS Web of Conferences **40**, 01007 (2018) https://doi.org/10.1051/shsconf/20184001007
Int. Conf. SOCIETY. HEALTH. WELFARE. 2016

Legal analysis of systemic investment protection regulation in the European Union's financial sector

L. Bocs

Rīga Stradiņš University, Riga, Latvia

Abstract. After the Treaty of Lisbon the European Union has an exclusive and uniform competence regarding investment agreements within its common commercial policy. Yet the political events in 2016 showed that there are still many regional differences politically and economically, especially after the so-called Brexit and negotiations with the United States of America in relation to transatlantic trade and investment. Therefore, the aim of the research is to determine the legal framework and related problems for unified investment protection within the European Union. Using descriptive, logical and deductive methodology the paper establishes a juristic base consensus for trade and investment policies, concludes that so far those policies have been systemically neglected due to regional differences in economic development and accordingly suggests to unify and protect the common investment policies by using already existing regional judicial mechanisms of member states within a unified code of conduct.

1 Introduction

The European Union (EU) has a uniform competence for investment agreements within the common commercial policy. Political congruency faces increased difficulties because of interdependencies, procedures within the so-called Banking Union and unified policy arrangements for sustained investment. Several *ex ante* considerations have already been demonstrated by the author relating to the international scientific conference "Society. Health. Welfare" in 2014 and have perilous implications for unified policies.[a] Considering the events in 2016 relating to the British referendum to leave the EU, also known as "Brexit", as well as the wide media coverage and criticism for lack of transparency within EU negotiations with the United States of America for the Transatlantic Trade and Investment Partnership (TTIP), the EU wide legal base for unified investment protections needs to be determined by general consensus.

The aim of the research is to determine the normative regulations and associated problems for the enforcement of EU law within the EU's financial sector for unified

---

[a] See also : L Bočs. The difficulties with the European Union's financial sector's legal regulation. Available at: `http://www.shs-conferences.org/articles/shsconf/pdf/2016/09/shsconf_shw2016_00019.pdf` [visited 31.01.2017].

investment protection and to give an assessment of possible solutions, especially in relation to subsidiary investment protection capabilities within the EU member states that have differently developed economies. The published paper will be addressed in a video commentary by the author, and also recorded and made available as an audio article for listening[b] with a creative commons license[c].

The research area is EU public law, the subject of the paper being investment law. Using descriptive, logical and deductive methodology within an *ex ante* doctrinal analysis, the normative supervision of EU investment protection is evaluated in relation to the hypothesis that the centralization of investment protection centralizes volatility into a systemic risk for the whole EU. The results show that unified investment legal protection can be achieved within a comprehensive framework according to existing member state judicial procedures for investment protections since it relates to prior established bilateral treaties and protection of property rights in general. Therefore the merit for legal analysis lies in the common commercial policy interests for unified investment protection and the author suggests to implement national law attributes for investment protection within a decentralized, nevertheless systemically congruent, legal regulation for coherent investment protection.

2 Investment policy of the EU

Article 3 of the Treaty on the Functioning of the EU sets out areas where the EU has exclusive competence and commercial policy is one of them. Under it the EU has sole responsibility for *inter alia* trade and investment issues after the so-called Treaty of Lisbon came into force in 2009, consisting of foreign direct investment (capital for commercial activity purposes) and also portfolio investment (capital for asset purchases and passive income) in EU territory [1]. Therefore EU centralized supervision of investment policies is a fairly recent endeavour without established best practices.

Investment policies in the EU have been national matters through bilateral agreements for most of the EU's existence and the exclusive competence of the EU has faced criticisms regarding proportionality and fair dispute resolution negotiations [2]. Such arguments arise out of general concerns regarding the balancing of state and public interests and private investors for the mutual benefit and proportional protection of legal interests [3]. By foreign direct investment agreements the EU plans wide opening and synchronizing of economies to mostly foreign corporations for added capital value within the EU through integrated policies [4]. This means that regional differences in the EU have to be addressed in order to be integrated in a truly unified commercial policy.

By way of EU regulation the EU has granted protection for bilateral agreements until the EU has made separate ones [5]. Separate EU agreements will be put in place since trade and investment are to be continued, yet in 2016, for instance, the negotiations with the United States of America for the Transatlantic Trade and Investment Partnership (TTIP) Agreement came under severe scrutiny *inter alia* because of private arbitration clauses which would limit public authorities to fully engage in democratic supervision [6]. So in order to actually establish centralized investment policies in the EU, the democratic needs of all member states have to be identified and respected since regional oppositions to EU policies can lead to events like the so-called Brexit, where the United Kingdom decided in a referendum to abolish its membership in the EU.

---

[b] List of public files by Lauris Bočs. Available at: `http://files.fm/laurisbocs` [visited 31.01.2017].

[c] Creative Commons license CC BY 4.0. Available at: `http://creativecommons.org/licenses/by/4.0/` [visited 31.01.2017].

3 Investment issues under EU commercial policy

The need for the EU to have a centralized trade and investment policy is related to the single market concept which comprises also single currency, capital flow and regional congruency aspects [7]. That is a logical system of integrated components, yet the specific details of investment policies are related to essential economic factors like, for instance, legal certainty, asset protection and dispute resolution on several political levels [8]. So in order to fully establish a single commercial policy, a common approach for these matters should be established since, for instance, dispute resolution has a connection to established legal jurisprudence.

The biggest issue of investment in the EU is the arbitration of possible disputes, yet within legal research this can be considered a purely political decision with several options available since the EU does not have prior experiences within the common commercial policy principle [9]. Since this paper is related to the international scientific conference "Society. Health. Welfare" in 2016 and its context of integrated social sciences, the author points out interdisciplinary research as a fundamental cornerstone relating to the efficacy of arbitration of investment disputes between states and investors, which has not yet been addressed in a coherent legal framework.

Established historical and economic research suggests that European relations for foreign trade and investment are more stable for foreign direct investment, so for national economies there is a reduced volatility and fluctuation of financial flows [10]. So the legal democratic interest for economic stability and growth leads to the possible conflict of a too broad state's forthcoming to investors by establishing investment law practices with compromised democratic justifications [11]. As recent statistics indicates (Fig. 1), the clearly leading countries in the EU for foreign direct investment, for instance, are those with a very favourable taxing system for foreign corporations, meaning that international corporations can use them as legal tax havens. So the essential need for countries is the economic development, but a private investor has too much risk for public interests can have long-standing consequences, such as lack of transparency, unfavourable arbitration in disputes and an overwhelming dependency on private capital sources for public officials without the public *fiscus* actually benefiting much. So considering the regional tax laws as well as common commercial policy needs, a unified approach would be to implement simple EU directive measures for ensuring congruent external investment options under a centralized investment framework that is yet to be developed under EU law and poses a separate challenge because of aforementioned issues.

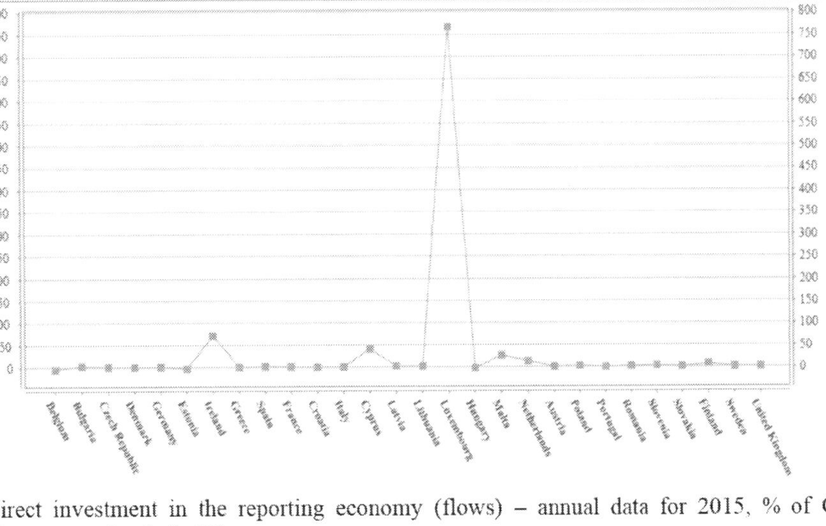

Fig. 1. Direct investment in the reporting economy (flows) – annual data for 2015, % of GDP (Source: Eurostat, code: tipsbp90).

4 Conclusions

According to the aforementioned legal evaluation, the author concludes the following:

1. The Treaty on the Functioning of the EU has created a central, but legally unsophisticated investment protection mechanism with a somewhat new common commercial policy approach under which the EU acts as a representative entity for its member states for *inter alia* investment issues, so the need is to establish EU wide political congruence for that.
2. The EU faces criticism whenever it challenges democratic principles of public transparency for the sake of private interests for capital allocation, so the lobbying and sustainable development criteria should be as big of an issue within the common commercial policy as trade and investment issues.
3. A common commercial policy can only be achieved under a unified code of conduct, because congruent and predictable economic decisions are a key factor for private capital allocation, so EU policy should reflect the proportional needs of society for the creation of long term wealth which could prevent separate member states to forsake the legal interests of their residents for short-term economic gain.
4. While regional economic differences relating to matters such as economic growth, employment and tax rates are still an issue in the EU, the arbitration for investment dispute resolution is overwhelmingly planned to be a unified approach so the risks for regional indifference for economic matters become increased and should be hedged against at least by national authorities.
5. It is clearly questionable issue whether or not all EU member states are willing to risk their position as legal tax havens, so an EU directive for minimum aims within investment policies could relate to a more diversified spectrum of investment locations in the EU since exceptional tax havens are just another form inter-EU competition which is contrary to a single market principle.
6. Conditions for investment *de lege ferenda*, i.e., future EU investment regulations are yet to be developed in a consistent framework, so the author points out the economic sector of public procurements as a possible role model for EU wide implementation of investment matters, since tender operations do take into account regional differences and operate with the intent of public-private cooperation for the public good with the practical effectiveness of regional member state courts.

References

1. C.M. Brown, Changes in the Common Commercial Policy of the European Union After Entry into Force of the Treaty of Lisbon: A Practitioner's Perspective. In: *European Yearbook of International Economic Law. Common Commercial Policy after Lisbon*. M. Bungenberg, C. Hermann, eds. (Springer, Berlin, 2013, 163–186, P.175)
2. B.A.M. Araujo, *The EU Deep Trade Agenda* (Oxford University Press, Oxford, 2016, P.115)
3. L. Markert, The Crucial Question of Future Investment Treaties: Balancing Investors' Rights and Regulatory Interests of Host States. In: *International Investment Law and EU Law*. M. Bungenberg, J. Griebel, S. Hindelang, eds. (Springer, Berlin, 2011, pp. 145–172, P.170)
4. P. Strik, *Shaping the Single European Market in the Field of Foreign Direct Investment* (Hart, Oxford, 2014, P.9)

5. EU Regulation Nr.1219/2012 of 12 December 2012 establishing transitional arrangements for bilateral investment agreements between member states and third countries, Article 3
6. D. Collins, *An Introduction to International Investment Law* (Cambridge University Press, Cambridge, 2017, P.44)
7. A. Ezrachi, *EU Competition Law. An Analytical Guide to the Leading Cases*. Fifth Edition (Hart, Portland, 2016, P.44)
8. S. Meilinger, *The New EU Competence for Foreign Direct Investment. Legal Questions of its Implementation* (GRIN, München, 2014, P.3)
9. M.A. Gwynn, *Power in the International Investment Framework* (London, Macmillan, 2016, P.211)
10. Economics Department (EIB), *Investment and Investment Finance in Europe* (European Investment Bank, Luxembourg, 2016, P.158)
11. D. Schneiderman, Compensating for Democracy's "Defects": The Case of International Investment Law. In: *The European Crisis and the Transformation of Transnational Governance: Authoritarian Managerialism versus Democratic Governance*. C. Joerges & C. Glinski, eds. (Hart, Oregon, 2014, pp. 47–70, P.48)

Conditions for the lawful exercise of the right of self-defence in international law

V. Upeniece

Rīga Stradiņš University, Riga, Latvia

Abstract. The Charter of the United Nations was thought to establish a normative order, maintain international peace and security. According to the Article 51 of the Charter of the United Nations "Nothing in the present Charter shall impair the inherent right of individual or collective self-defence if an armed attack occurs" [1]. However the Article 51 does not propose a legal definition of the conduct which is considered as an armed attack or the commencement of such an attack. It does not propose strict criterions for the use of force for self-defence. As a result different interpretations of this norm have been arising and continuing to change in response to new situations and threats.
Key words: self-defence; armed attack.

1 Introduction

In the General Treaty for Renunciation of War as an instrument of National Policy (Kellogg–Briand pact) the Contracting Parties condemned a recourse to a war for the solution of international controversies, renounced it as an instrument of the national policy in their relations with one another and agreed that the settlement or solution of all disputes or conflicts, which might arise among them, would never be solved except by pacific means [2]. According to the Preamble of the Treaty, if the Contracting Party seeks to promote its national interests by resort to a war, it will be denied the benefits furnished by this Treaty [3]. Thus, the commencement of hostilities against the Contracting Party which started the war was lawful and the war of self-defence was not prohibited by the Treaty [4].

2 Discussion

There are no debates about the statements that the right for self-defence, which is embodied in the Article 51 of the Charter of the United Nations, is the temporary right until the Security Council takes measures necessary to maintain international peace and security. The Article 51 of the Charter of the United Nations states that the force can be used in self-defence only in relation to an armed attack [5]. The International Court of Justice in the case *Nicaragua v. United States* (1986) determined that an armed attack must raise to certain "scale and effect", therefore not all attacks can be qualified as armed attacks [6]. The international Court of Justice explained the meaning of the "armed attack" as "the most grave form of the use of force" [7] without going deeper into details and explanations of the meaning of the words "grave form of the use of force".

In the case *Islamic Republic of Iran v. United States of America* (2003) the International Court of Justice pointed out that the burden of proof of the fact of showing the existence of such attack rested on the State which justified its own use of force as self-defence [8]. It is clear that the use of force is lawful when the force is used as a response to the armed attack which has taken place. Obviously, the existence of an armed attack prevents any uncertainty. For this reason there is also a reasonable point of view that the lawful self-defence is restricted to the response to the armed attack [9]. But if an actual attack has not yet occurred, it is necessary to understand whether the state can lawfully use the force for self-defence against a threat. Many scholars are against all forms of preventive actions. The others recognize the use of force for self-defence in response to imminent attack as lawful one.

The scholars often mention imminence as a criterion that justifies the use of force for self-defence. In the UN Secretary-General's High-Level Panel on Threats, Challenges and Changes it was mentioned, that the threatened state can take military action as long as the threatened attack is imminent [10]. The criterion of imminence correlates with the Caroline formula (1842), according to which the self-defence can only be exercised in the situations where the need to respond is instant, leaving no moment for deliberation [11]. There is no legal definition of imminence [12]. As examples of events that may indicate the presence of imminent threat of an attack mostly are mentioned: the enlargement of armed forces, procurement of weapons and belligerent declarations of intention to start a war, the other threatening activities and the other evidences that prove that an attack will take place in the nearest future [13].

Some scholars argue that imminence means the clear and present danger of an attack of an adversary, which is highly likely to occur. The others claim that the reference to the imminence of an attack is more the attempt to predict the future than constitution of the certain fact, because in different situations there may be a certain degree of uncertainty about the perfection of received information about the adversary's plans, intentions, and motivations [14]. According to Dinstein the armed attack of the adversary must be evident to the victim state, namely, the victim state can lawfully intercept the armed attack when it is in process of being started [15].

The criterion of imminence was also discussed in the Security Council after the Israeli attack on the Iraqi Osiraq nuclear reactor in 1981. Israel declared that "in the light of Iraqi declarations and deeds, and Iraq's refusal even to sign an armistice agreement with Israel, Israel had full legal justification to exercise its inherent right of self-defence to abort the Iraqi nuclear threat to Israel" [16]. The United Kingdom and Sierra Leone used the Caroline formula to explain the illegality of Israeli actions. The representative of Sierra Leone mentioned that "it has long been accepted that, for it to be invoked or justified, the necessity for action must be instant, overwhelming and leaving no choice of means and no moment for deliberation." The above-mentioned countries condemned the Israeli attack generally because the Israeli action was based on the long preparations and planning. Almost all the other states also condemned Israeli action generally because the threat was not imminent [17].

There is an opinion that the armed response is lawful only if it conforms to all conditions of necessity, proportionality and immediacy. The principle of necessity and immediacy, as expressed in the Carolina doctrine, is considered to be a part of the customary international law. According to the international customary law the self-defence is lawful if the response is necessary, proportional and immediate. The International Court of Justice made references to the principles of the international customary law in such cases as *Nicaragua v. United States, Islamic Republic of Iran v. United States of America,*

Democratic Republic of the Congo v. Uganda and in the other cases. In *Nicaragua* case the Court recognized, that the Charter of the United Nations "does not contain any specific rule whereby self-defence would warrant only measures which are proportional to the armed attack and necessary to respond to it, a rule well established in customary international law" [18]. At the same time the Court evaluated the principle of immediacy, by stressing that "the reaction of the United States in the context of what it regarded as self-defence was continued long after the period in which any presumed armed attack by Nicaragua could reasonably be contemplated" [19]. In *Oil Platform* case the Court referred to the principle of necessity and proportionality and pointed out that "the United States must also show that its actions were necessary and proportional to the armed attack made on it and that the platforms were a legitimate military target open to attack in the exercise of self-defence" [20]. At the same time in the case of *Armed activities in Congo* the Court indicated that "since the preconditions for the exercise of self-defence do not exist in the circumstances of the present case, the Court has no need to enquire whether such an entitlement to self-defence was in fact exercised in circumstances of necessity and in a manner that was proportionate" [21]. However in these cases the Court did not examine the question of the use of force for self-defence in the situation of existence of the threat of the use of armed force from the part of adversary.

The principles of necessity and proportionality are recognized as requirements of outstanding legal and practical importance for the right of self-defence [22]. The necessity requires the verification that the conflict cannot be solved by peaceful measures. There is an opinion, that the situation of absolute certainty that no other measures could prevent enemy attack also justifies the necessity, for example, in the situations of the direct threat to the state survival when the other options, except the use of force, cannot change the situation in a positive way [23]. The condition of the necessity requires the presence of conclusions, which are based on the proved facts, that an armed attack is imminent and requires the response [24]. Thus the use of force for self-defence should be a response to the real threat to the survival of a state. At the same time this principle contains the valid obligation to verify whether there are no other peaceful settlements of the conflict [25].

The principle of proportionality means that the use of force must be limited to the neutralization or abolition of the attack against which a state is defending itself [26]. According to Cassese the principle of proportionality requires not only the insurance of the balance between the injury caused by the wrongdoing state and the countermeasures, or at least the insurance that the countermeasures do not seriously exceed the injury created by the wrongful act, but also follows the aim to force the offender to discontinue its wrongful conduct [27]. The principle of proportionality does not require that the weapons used for self-defence must be the same as those, which are used for the attack. That is why the general prohibition of the first use of nuclear weapons is not derived from the principle of proportionality [28]. At the same time the International Court of justice could not reach a definitive conclusion about the legality or illegality of the use of nuclear weapons by a state in an extreme circumstance of self-defence, where its very survival would be at stake [29]. Consequently, the interpretation of this principle as one of the previously mentioned principle of necessity depend on the circumstances of each certain case and that is why is flexible.

The other criterion which requires an examination is the principle of immediacy. This criterion is also like a controversial issue. There is a point of view that the action in self-defence must immediately follow the start of an attack. As Dinstein explains the isolated armed attack may not be the reason for starting a war for self-defence. At the same time he agrees that the principle of immediacy cannot be explained directly, because of different bureaucratic procedures the state faces with the beginning of hostilities by the adversary – the certain period of time passes after the state officials make a decision to act in response and give the instruction to the armed forces. Therefore if the interval between an armed attack and a war of self-defence is long, a war may still be lawful if the delay is objectively justified [30]. The other scholars link the criterion of immediacy with the criteria of necessity and also deny that the beginning of an armed attack must be sufficiently close to the action for self-defence. Again the interpretation of closeness in time between these two actions is dependent on the context of each situation.

And lastly, the main disputable and controversial question is when or in which point of time the actions for self-defence become lawful. As it was mentioned above there is an opinion that the anticipatory self-defence is admissible if it complies with the requirements, which are set by Webster in the *Caroline Case*, namely, if the necessity of self-defence is instant, overwhelming and leaving no choice of means, and no moment for deliberation [31]. Still there is no unified attitude among the states to problem of lawfulness of the use of force for self-defence in the situation when the missile is already launched in the direction of the victim state and in the situation when a radar guiding the missile is locked on readiness to fire in the direction of the victim state [32]. The opponents of the anticipatory self-defence argue that *Caroline* formula does not broaden the right of self-defence incorporated in the Article 51 of the Charter of the United Nations, because the Article 51 replaces the customary right of self-defence and prevails over it as a *lex specialis*. The main limitation of the use of force for self-defence derives from the phrase "if an armed attack occurs" which should be read in connection with the main purpose of the Charter of the United Nations to limit the unilateral use of force in international relations. According to this point of view the use of force for self-defence is lawful only after the armed attack had already been launched [33].

The International Court of Justice did not examine the question if the meaning of the phrase "if an armed attack occurs" actuates the Article 51. The International Court of Justice cites the Article 51 in the case *Nicaragua v. United States* and specifies that "the inherent right (or "droit naturel") which any State possesses in the event of an armed attack covers both collective and individual self-defence" and that in the examined case "reliance is placed by the Parties only on the right of self-defence in the case of an armed attack which has already occurred, and the issue of the lawfulness of a response to the imminent threat of armed attack has not been raised" [34]. In case the *Armed activities in the Congo* the Court also concluded that the force had not been used against an anticipated attack [35].

Some scholars recognize the legality of the interceptive self-defence. According to Dinstein the interceptive self-defence is lawful under the Article 51 of the Charter of the United Nations because it takes place in the situation when an armed attack is currently in progress. In such a situation he recognizes an armed attack as imminent and practically unavoidable. At the same time Dinstein stresses that the use of force should not be based only on the assumptions, expectations or fear [36].

The other opinion was expressed by Israel before the Security Council anent its actions against the Iraqi Osiraq nuclear reactor and is based on the idea that it would be cruelly to allow the adversary to deliver the first and perhaps the fatal blow in a nuclear age and epoch of developed technologies of destruction, which make it possible to destroy a state with great speed from long distances. The different interpretation of the Article 51 of the Charter of the United Nations would force the victim state passively except its fate and protect the aggressor right to the first strike [37].

The state's position in the issue of Iraqi Osiraq nuclear reactors acknowledges that mostly all the states recognize that the use of force in the period before an armed attack had occurred is not admissible if the force is used against an attack which is not clearly imminent. Although according to Lubell the states are playing the game of probabilities and trying to settle on the notion of "certain enough" that an attack is imminent [38]. There is an opinion among another group of scholars, that the use of force for self-defence is not restricted by the criterion of imminence in the situation of conviction that the attack will take place the nearest future.

3 Conclusions

The ways and means to ensure peace are diverse. At the same time the right to act for self-defence is reserved in the international law. Different points of view concerning the problem of the use of force for self-defence in response to the future attack which is imminent indicates its controversial nature. However the rapid changes in the modern world pose new challenges which need appropriate solutions.

Adopted in 1945 with the purpose to maintain the international peace and security the Charter of the United Nations is still the main source of the rules for evaluating the legality of the use of armed forces. The Charter of the United Nations contains both the prohibition on the use of armed force and the reference to the collective security system which is created to guarantee obedience for such a prohibition and is the first exception from the prohibition of the use of force expressed in the Article 2 (4) of the Charter of the United Nations. The second exception is embodied in the Article 51 and provides the legal bases for the use of force for self-defence.

The world is still facing the situations where an established rule – the prohibition of the use of force, continues to be challenged. Obviously, the purpose of the Article 51 of the Charter of the United Nations is dual, because from one side it poses a restriction to uncontrollable unilateral use of force, and from the other side it provides the states with the instrument which the states would use for their survival.

There is no definition, which would be applicable to all situations, including the situation when there is a threat of the use of force. There are debates about the right of the state to use force for self-defence against a threat of the use of force for the purpose to avoid being physically destroyed. So the question of the moment in time which is considered to be the point of commencement of an armed attack is also open. The scholars generally speak about criterions and preconditions of the use of force with precaution and avoid from giving precise definitions. Such situation may be explained with the reasonable disinclination to open the door for uncontrollable and unilateral abuse of this right. And at the same time, the shortage of precise definition opens the door for existence of different positions and interpretations.

References

1. Charter of the United Nations, available at: http://www.un.org/en/sections/un-charter/chapter-vii/index.html
2. General Treaty for Renunciation of War as an Instrument of National Policy (Kellogg-Briand Pact), 1928, available at: http://www.yale.edu/lawweb/avalon/imt/kbpact.htm
3. See ibid.
4. Dinstein Y. War, Aggression and Self-Defence, Cambridge University Press, 2005, p. 83–84
5. Charter of the United Nations, available at http://www.un.org/en/sections/un-charter/chapter-vii/
6. Schmitt M.N. Cyber operations and the jus ad bellum revisited. *Villanova Law Review*, 2011, **56**, p. 569–606

7. *Nicaragua v. United States*, judgement of International Court of Justice, 1986, para. 191, available at: http://www.icj-cij.org/docket/files/70/6503.pdf
8. Case Concerning Oil Platforms, 2003, 42 ILM 1334, 1356 (2003) (Islamic Republic of Iran v. United States of America), Judgement of 6 November 2003, para. 64 and 72, available at: http://www.icj-cij.org/docket/files/90/9715.pdf
9. Simma B., The Carter of the United Nations, commentary, second edition, Volume I, 2002, New York: Oxford University Press, 805p.
10. Report of the Secretary-General's High-Level Panel on Threats, Challenges and Change. A More Secure World: Our Shared Responsibility, A/59/565/, 2 December 2004, p. 62
11. The United States Secretary of State Daniel Webster dated 24 Apr. 1841, in *Caroline Case,* 29 British and Foreign State Papers (1841) 1137–1138, available at: http://avalon.law.yale.edu/19th_century/br-1842d.asp
12. Lubell N., The Oxford Handbook of The use of force in international law, edited by M. Weller, Oxford University Press, 2015, p. 702
13. Anad R., Self-Defence in International Relations, Palgrave Macmillan UK, 2009, 5p.
14. Muller K.P., Catillo J.J., Morgan F.E., Pegahi N., Rosen B., Striking first. Preemptive and Preventive Attack in the U.S. National Security Policy. RAND Corporation, 2006, XIV
15. Dinstein Y. War, Aggression and Self-Defence, Cambridge University Press, 2005, p. 187
16. United Nations Security Council document S/PV.2283 of 15 June 1981, para. 79, available at: http://repository.un.org/bitstream/handle/11176/67170/S_PV.2288-EN.pdf?sequence=17&isAllowed=y
17. United Nations Security Council document S/PV.2283 of 15 June 1981, available at: http://repository.un.org/bitstream/handle/11176/67170/S_PV.2288-EN.pdf?sequence=17&isAllowed=y
18. *Nicaragua v. United States*, judgement of International Court of Justice, 1986, para. 176, available at: http://www.icj-cij.org/docket/files/70/6503.pdf
19. *Ibid,* para. 237
20. *Islamic Republic of Iran v. United States of America,* judgement of International Court of Justice, 2003, para. 51, available at: http://www.icj-cij.org/docket/files/90/9715.pdf
21. *Democratic Republic of the Congo v. Uganda,* judgement of International Court of Justice, 2005, para. 147, available at: http://www.icj-cij.org/docket/files/116/10455.pdf
22. Simma B., The Carter of the United Nations a commentary, second edition, Volume I, 2002, New York: Oxford University Press, 805p.
23. Green J.A. The ratione temporis elements of self-defence, *Journal on the Use of Force and International law*, 2015, **2**, No.1, p. 101
24. Shaw M.N., International law, Cambridge University Press, 2003, 1031p.
25. Dinstein Y., War, aggression and self-defence, Fourth edition, Cambridge University Press, 2005, 237p.
26. Gutman R., Rieff D., Dworkin A., Crimes of War: What the public should Know, W.W. Norton & Company, 2007, 341p.
27. Cassese A., International Law, Oxford University Press, 2005, 306p.

28. Simma B., The Carter of the United Nations a commentary, second edition, Volume I, 2002, New York: Oxford University Press, 805p.
29. Legality of the threat of use of nuclear weapons, Summaries of Judgments, Advisory Opinions and Orders of the International Court of Justice Not an official document, 1p., available at: http://www.icj-cij.org/docket/files/95/7497.pdf
30. Dinstein Y. War, Aggression and Self-Defence, fourth edition, Cambridge University press, 242–243p.
31. The United States Secretary of State Daniel Webster dated 24 Apr. 1841, in *Caroline Case*, 29 British and Foreign State Papers (1841) 1137–1138, available at: http://avalon.law.yale.edu/19th_century/br-1842d.asp
32. Gray C. International Law and the Use of Force, Oxford University press, New York, 2000, 96p.
33. Simma B. The Charter of the United Nations a commentary, second edition, Volume I, 2002, New York: Oxford University Press, pp. 803–806
34. *Nicaragua v. United States*, judgement of International Court of Justice, 1986, 103p., para. 193, 194, available at: http://www.icj-cij.org/docket/files/70/6503.pdf
35. *Democratic Republic of the Congo v. Uganda,* judgement of International Court of Justice, 2005, para. 143, available at: http://www.icj-ij.org/docket/files/116/10455.pdf
36. Dinstein Y., War, Aggression and Self-Defence, fourth edition, Cambridge University press, 2005, p. 191
37. United Nations Security Council document S/PV.2283 of 15 June 1981, para. 81–85, available at: http://repository.un.org/bitstream/handle/11176/67170/S_PV.2288-EN.pdf?sequence=17&isAllowed=y
38. Lubell N. The Oxford Handbook of The use of force in international law, edited by M. Weller, Oxford University Press, 2015, p. 714

Lapse of rights connected with assignment of legal claims, problematic issues and their solution

V. Nemenova

The University of Economics and Culture, Riga, Latvia

Abstract. Impact of transfer of legal claim on the destiny of rights security and connected claims may differ. The aim of the article is to analyze the transfer of rights connected with transferred legal claims, to survey the problematic issues and to suggest methods of their solution. The author analyzes legal norms, judicial practice of Germany, Switzerland, Russia and Latvia, as well as opinions of law scientists about transfer of guarantee, lien, penalty, interest, losses compensation and agreement of court of arbitration in connection with assigned legal claims. The author offers her own point of view regarding the issue and suggestions on amendments in the legal acts.
Key words: cession; guarantee; pledge; penalty; interest.

1 Introduction

Consequences of cession (assignment) are always limited with lapse of right to legal claims.

The scope of rights under transfer according to the Civil Law of Latvia is defined in Article 1806:

"The Assignee does not acquire more or bigger rights with cession than the assignor used to have before, but the claim itself goes to him with all rights belonging to it and existing at the moment of cession, even if they are justified basing on the personal benevolence towards the assignor, so far as they have no directly defined exception from this condition.

Unpaid claim interest if it is not directly contracted also goes to the assignee.

The assignor shall transfer to the assignee everything that suits as claim evidence or that may facilitate its collection, as well as everything that he has received from the debtor already after cession."

A. Tuhrs [1] states that transfer of rights connected with legal claims to the assignee is the cession agreement's *legal concomitant consequences*, which commence also in case when the agreement has nothing mentioned about it. True, their commencement may be excluded by agreement between the assignor and the assignee or agreement between the creditor and the person providing security (for instance, creditor and guarantor agree that the claim against the guarantor will not be transferred to other person transferring the main legal claims).

The impact of transfer of legal claims on the destiny of rights security may be different.

Aim: to analyze the transfer of rights connected with the transferred legal claims, to study the issues and offer their ways of solution.

2 Discussion and results

Ensuring the claim under transfer ***with security***, the new creditor receives both the conditioned and the actual claim towards the guarantor. Many authors [2–5] believe that in case when the cession to the guarantor is excluded, it is terminated. However, the issue on the destiny of claim towards the guarantor in case of its exclusion from the secured rights is closely connected with the issue on the possibility of isolated transfer of this claim. Legal scholars of Germany and Switzerland [1, 6, 7] believe that this issue shall be resolved affirmatively, therefore it is necessary to allow it as possible for an assignor to keep the claim against the guarantor in case of isolated transfer of the main claim.

Guarantee is the duty undertaken by the agreement to bear responsibility towards the creditor for the debt of the third person, however, without releasing the latter from his debt [8]. The guarantee, like the mortgage rights, is the type of rights reinforcement. Still, unlike the mortgage agreement, in case of guarantee a separate agreement with the third person on claim security is concluded.

Direct participants of guarantee establishment are warrantor and creditor. The main debtor stays aside. It arises from here that the main debtor has not only no necessity to participate in guarantee establishment, but also to know about it at all [9].

Although the guarantee is closely connected with the debt liability, its establishment, scope and termination, it is a separate contractual liability, which in individual cases may exist without the basic debt [8], as well as the guarantee claim term and fulfillment may not coincide with the term of execution of debt defined in the agreement. Herewith the issue is disputable regarding the guarantee transfer in case of cession, as with cession only the claim is transferred against the debtor, and not the whole contractual relation and agreements, which have been concluded with the third persons. Likewise problematic issue may be noticed in implementation of cession execution, as, for example, the assignor must duly inform the debtor about the cession [8]; in this case the guarantor may fulfill the liability for the assignor, without knowing about the claim transfer and independently from the debtor.

A. Grūtups states in the comments of the Civil Law that there is no law, which would require that the guarantee must be in force only against strictly defined creditor. As the guarantee is an accompanying liability, which (usually) does not exist permanently, but accompanies the (main) debt liability, but the latter may be transferred to other person, then there is no grounds to think that in this case the guarantee would be discharged, even more because usually the guarantor sees no difference about any specific creditor he bears responsibility as a guarantor [10]. The author believes that one shall agree with the given comment, as when the guarantee does not go with debt claim in case of cession, the scope and security of the claim subject would be considerably changed.

The guarantor's rights arising from the guarantee contract may be the legal claim transfer according to the Article 1793 of the Civil Law, because according to the Article 1704 of the Civil Law, in what amount the guarantor satisfies the creditor, in that amount he receives claims of the latter towards the main debtor. In this case automatic transfer of legal claims is fulfilled from the creditor to the guarantor, without will expression of the creditor, and the guarantor owns regress rights towards the debtor.

Alongside with other secured rights, also the *lien* goes to the assignee.

One more accompanying related right, which goes to the assignor, is legal transfer of lien in case of claim cession basing on the Article 1278 of the Civil Law, which defines that the lien is such a right to foreign object, on the grounds of which this object ensures to the creditor his claim in such order, that he may get from it payment of this claim.

The lien being the real right allows the pledge using the pledged property as the debt liability security because he may receive debt payment from the pledged thing itself or money received for its sale [11]. The lien shall be recognized as related right; and as such in case of cession follows the main claim. As the mortgage being the lien is entered into the Land Book, then, unlike other cases, for example, in case of renewal, to enter cession into the Land Book, no consent of mortgager is needed. The assignee may request, depending on the conditions of pledge agreement, incontestable compulsory execution or real estate sale at free auction through court." [11]

The Commercial Pledge Law [12] regulates rules on transfer of commercial pledge in case of secured claim cession. First part of Article 35 of the Commercial Pledge Law defines that assigning the secured claim, the assignee receives also all rights of the person taking commercial pledge, unless the assignor and the assignee have not agreed otherwise, and the second part of the article of the mentioned law defines that regarding conclusion of cession agreement the person holding commercial pledge register makes a note on the grounds of a written notification of the person taking commercial pledge.

Penalty is one of the most often used types of reinforcement of law of obligation for related guarantee, pledge (mortgage) and earnest money. Penalty regulation in the Article 716-1724.[1] of the Civil Law gives wide opportunities to define it in different ways and volumes, and exactly the penalty scope issues become the basic dispute subject of many cases under consideration at the court. There should be a large number of cases of application of accumulated "expressed unjust penalties" to start considering, what in reality are the penalty functions, what is the penalty application in good faith and whether the fact that the law does not define the rule on penalty decrease means that any exceptions are banned. The searches for a new approach were facilitated also by looking-through of documents of the European Union and the European Council, civil law unification projects, laws and practice of other countries. Judicature turn took place, which initially was marked in the resolution of Senate from 2004 in the Case No. SKC-441, resolution from 2005 in the Case No. SKC-48 and resolution from 2007 in the Case No. SPC-43 and explicitly justified by the resolution of the Senate from May 14, 2008 in the Case No. SKC-212. In relation to the court's possibility to fight against excessive penalties, the practice and Senate's judicature has developed, overtaking the amendments in the Civil Law (Article 1724[1]), which generally gives rights to decrease the penalty if it is excessive [13].

The lawmaker has registered the rights of the court to decrease the penalty up to reasonable amount, if it is excessive, also in the Article 1724.[1] of the Civil Law, as amended on June 4, 2009, which have come into effect on July 1, 2009.

By amendments from June 20, 2013, which came into effect on January 1, 2014, the Article 1716 of the Civil Law defines that the penalty for failure to timely (within the term) fulfill liabilities, may be increasing, however, without exceeding 10 percent of the property value of failure to fulfill the delayed liabilities.

The most important thing is that the penalty institute must correspond to the civil law task and the idea of justice, like the European Union's policy and attitude towards the penalty sanctions in private law. While developing the amendments, the lawmaker pointed that the current legal regulation facilitates the social stratification tendencies, which leads to a part of society finding itself under yoke of debts when riches of other people multiply.

Civil Law amendments, which have been made by the law of Saeima from June 20, 2013 and have come into effect on January 1, 2014, shall be considered one of the most considerable amendments in the part of Law of Obligation since renewal of the Civil Law performance. They will sharply change the meaning of penalty in civil circulation and shall be considered a step of lawmaker in civil law modernization and even getting closer to the system of Anglo-Saxon law [14].

In Great Britain and other Anglo-Saxon law system countries, like Ireland, the USA, Australia and Canada, there are mutually equal rules in relation to the penalty, namely, the penalty, which actually performs the function of penalty fee (*penalty*), is forbidden. Only penalty is allowed, which performs the function of loss compensation. Currently, in the American law, two criteria are formulated, which shall come true for the penalty to not be admitted the penalty fee (*penalty*), namely, first of all, it cannot be disproportionate in relation to potential or actual losses, which have arisen due to failure to fulfill liabilities. Second-of-all, potential or actual losses must be clearly provable [15]. Similar approach was used in the case *Dunlop Pneumatic Tyre Co Ltd v New Garage and Motor Co Ltd. [16]*, which is one of the significant court judgments in cases on penalties in Great Britain. In the mentioned case, the House of Lords defined the principles how to define, whether the penalty's clause is actually a punishment and thus it shall be acknowledged inapplicable. Australia's Supreme Court has also referred to this case within the case *Ringrow Pty Ltd v BP Australia Pty Ltd. [17], as well as Irelands Supreme Court in* the case *O"Donnell v Truck and Machinery Sales Limited [18]*. Canadian Supreme Court has used similar approach in the case *Elsley v. J.G. Collins Ins Agencies [19], in which it rejected higher compensation for failure to observe the agreement provisions than the losses actually caused by failure to fulfill liabilities* [20].

Payment of penalty correlates with creditor's losses.

The penalty is the purest self-enrichment at the expense of the debtor, if it is defined in the amount, which considerably exceeds the loss of profit caused by non-fulfillment [21].

Civil Case Department of the Supreme Court's Senate of the Republic of Latvia has stated [22] that the penalty has the function of punishment and loss compensation. In cases when it is concluded from the case that the penalty by its amount, in comparison to the possible losses caused by violation, perform mostly the function of punishment, its application shall be evaluated according to the general principles of application of legal means with nature of punishment, including principles of punishment proportionality and justice. The general law principles are generally accepted guidelines arising from the legal state principle, main regulations, which the respective law system or sector bases on. The general law principles are unwritten primary rights of the European Union. By cancellation of agreement, the grounds for further application of penalty provided in it is lost, still, it does not exclude the rights to demand already accumulated penalty.

The performance provided by penalty is encouragement of the debtor to fulfill liabilities, threatening to impose additional duties on him. That is why the security function is performed by it only until commencement of *condicio juris*. As the rights, which ensure liability fulfillment and which go to the assignee in case of main claim transfer, only the *conditioned claim for penalty payment* is put forward. Already existing claim on penalty payment in case of transfer of main legal claims does not go to the assignee *ipso jure*, like some civil law scientists believe [6, 7], but it is kept by the assignor [1, 23–26].

The Civil Case Department of the Senate of Superior Court of the Republic of Latvia has stated [27] that if the contracted rights of contract termination before term are used, when the contract is terminated before term, the legal grounds disappears also regarding calculation and collection of contracted interest and penalty according to the contract provisions, because there is no legal grounds for their application after contract termination. The contracted interest and penalty in cases, when the party legally withdraws from the contract, shall be compensated till the moment, when the contract ceases to exist.

The claim on ***losses compensation*** has *permanent nature* in relation to the rights, from the violation of which it has arisen. This claim continues to exist even if the violated rights have been terminated, and in case of their alienation is not transferred *ipso jure* to the new rights owner, but is alienated with a separate cession agreement [28]. The German legal scholars [5, 29, 30] state that the claim for losses compensation, which have arisen before cession debtor term delay, in case of transfer of the delayed legal claims *is not transferred* to the assignee as the rights related to it, but is kept by the assignor.

Analogous approach shall be used defining the destiny of claims arising at the moment of assignment regarding payment of penalty and interest for use of foreign money, which refers to the claim on loss compensation, like the claim on compensation payment.

It is necessary to pay special attention to the instruction given in Part 2 of Article 1806 of the Civil Law, that the *unpaid* ***claim interest****, even if it is not directly contracted, is also transferred to the assignee.*

Such a definition causes many questions. First of all, it is not explained, what interest we speak about – interest on late payments or interest that must be paid for use of money, which is granted according to the loan agreement, credit agreement, as commercial credit etc. Such differentiation in relation to Part 2 of Article 1806 of the Civil Law should be performed, because it is necessary to define the destiny of claim on unpaid interest like it is done regarding the claims on caused losses, when the debtor does not follow the payment term, as well as for payable penalty – if there is no separate agreement, they all are kept by the assignor. Second of all, hardly any connection between the right to unpaid interest and the transferred legal claims can be seen, which would request their joint transfer to the assignee. If BGB in this issue has no special instructions, the German civil law scientists [5, 24] usually state that the interest are due to the previous creditor with the commenced term.

In relation to the issue on mutual interconnection of interest and losses – the professor K. Torgāns states that the answer shall be sought in Article 1788 of the Civil Law, in which it is clearly stated that the interest by its economical nature is the compensation for uncollected profit, i.e., a component of losses. It is represented also in the Regulation 2000/35/EEK (submission 17), which states that the steps to be taken shall be implemented, without affecting the regulations of countries, according to which the state judge may grant to the creditor additional remuneration for losses caused by late payments of the debtor, taking into account that these costs may already be compensated with interest of late payments. Especially one should mark out that the interest serves for compensation of costs, i.e., losses. It means that the interest payments serve for compensation of losses, and thus false interpretation of Article 1722 of the Civil Law must be stopped, which sometimes appears in the statements of claim and other legal documents [20].

The perspectives of claim circulation largely depend on the assignee's real possibilities to implement his rights in relation to the acquired claim, including also by force. In this case, it is required to separately examine the transfer of jurisdiction agreement rights to the assignee. In law literature, different points of view regarding this issue may be found. So, V.A. Belovs, analyzing the legal acts of the Russian Federation, assumes that the proviso made by the court of appeals (court of arbitration) is not either condition of transferred rights or condition of their implementation. It forms the condition of protection of violated rights that is why it is not affected by the Article 384 of the Civil Code of the Russian Federation, which refers only to rights in their normal status. As the assignor bears no responsibility for excessive acts during rights implementation, it is unjust to impose to the assignee a burden to observe the protection conditions, which the assignor has agreed on. That is why the rights according to the arbitrage proviso go to the assignee only in case if it is directly provided for by the assignment agreement [31].

Opposite opinion in this issue belongs to L.A. Novoselova, who believes that the forced defense possibility is an integral element of subjective rights. That is why, acquiring the rights to demand, the assignee hereby, according to general conditions, undertakes a duty to follow the condition of forced implementation of rights [32]. R.S. Bevzenko also assumes that: "Consequent application of the opinion to the right of filing a claim as to subjective material civil law element must inevitably lead to a conclusion that in case of rights transfer proviso performance remains regarding disputes' contractual jurisdiction connected with execution of transferable claim" [33].

Many Russian law scientists see this position as a more consequent, as discussing the issue on the destiny of arbitrage proviso, in the first place it is necessary to consider the consequences of one or other decision from the point of view of the debtor's status constancy provision in case of transfer. From these positions, exclusion of the assignee from the field of activity of the arbitrage proviso considerably affects the debtor's status, who has agreed with the defined conditions regarding claim set forth to him. And vice versa – the assignee, concluding an assignment agreement, can look through the agreement's conditions, from which the transferable claim has arisen. If he acquires these rights for his possession, it means that he considers the rules of its implementation and protection acceptable that is why he does not need any protection in this situation.

It is necessary to note that in the Russian Federation this position on following of arbitrage disclaimer the rights under transfer (claim) in practical aspect is not disputed by anyone.

A more different approach, which is connected with the arbitrage court clause in case of cession, is used in the judicial practice in Latvia.

In Latvia's practice, courts of arbitration often satisfied the claims of collection of assignees' debts, where the assignee has not been a member of agreement of court of arbitration, and later it was asked to the district court to issue a writ of execution for forced execution of the decision, which quite often was satisfied. The main argument of the Senate, cancelling the district court's judgments on issue of writs of execution for forced execution of resolutions made by courts of arbitration, is the Article 1800 of the Civil Law, basing it by the fact that transition of the regulation of court of arbitration (clause of court of arbitration) is impossible, because the assignee is not the party of the agreement (from which the liability has arisen), as well as he is not the members of agreement of court of arbitration.

AT Civil Cases Senate's opinion has gained support in law doctrine. Senate's opinion has been supported by such civil law scientists like E. Moris [34] and L. Lapsa [35].

Opposite opinions are also met in the Latvian legal literature. Contradictory opinions on the border of assignment being a specific claim in the legal transaction state that this issue is not unambiguous. The authors conclude that in case if the assignment agreement includes a clause of court of arbitration, then, taking the principles of legal transaction as a free agreement included into the Civil Law into account, the assignee gets also the rights to resolve a dispute, which is connected with the assigned claim, at the court of arbitration. Besides the authors believe that this condition will not worsen the debtor's condition, because the agreement on dispute resolution by the court of arbitration has been achieved already between the assignor and the debtor. Analyzing the procedure of court of arbitration, A. Miķelsone has noted that assigning rights, the liabilities previously undertaken by the creditor (assignor) and arising from the respective transaction are binding to the assignee. If the assignor has agreed on transfer of the possible dispute to the court of arbitration, these regulations are in force and they shall be executed also by the assignee [36].

On September 11, 2014, a new Arbitrage Court Law was adopted, which came into effect on January 1, 2015, which gained wide resonance among the Latvian law scientists during development and adoption of draft law [37–43]. The author positively evaluated the fact that the lawmaker has included the long-expected legal norm into the Part 4 of Article 13 of the Arbitrage Court Law: "Assigning the claim, the assignee receives the legal claims, but not the clause of arbitrage court included into the contract on consideration of civil dispute by the arbitrage court".

3 Conclusion

Securing the claim under transfer *with a guarantee*, the new creditor receives both the conditions and the actual claim against the guarantor. Together with other secured rights the assignee receives also the *lien*.

As the rights ensuring liability fulfillment and going to the assignee in case of transfer of the main claim, only the *conditioned claim for penalty payment* is brought forward. The author offers to refer the penalty to non-assignable claims and respectively to change the legal norms of the Civil Law.

The claim on losses compensation has *permanent nature* in relation to the rights, from violation of which it has arisen. This claim continues to exist, even if the violated rights have been terminated, and in case of their alienation does not go *ipso jure* to the new rights owner, but is alienated with a separate cession agreement.

The interest by its economical nature is the compensation for uncollected profit, i.e., a component of losses. The author suggests expressing the second sentence of Article 1806 of the Civil Law in the following wording: The unpaid claim interest, which has payment term already commenced, if the contract does not define otherwise, is kept by the assignor.

The Arbitrage Court Law adopted on September 11, 2014, which came into effect on January 1, 2015, shall be also positively evaluated. Its Part 4 of Article 13 contains a long-expected legal norm: "Assigning the claim, the assignee receives the legal claims, but not the clause of arbitrage court included into the contract on consideration of civil dispute by the arbitrage court".

References

1. A. Tuhr (1925). Allgemeiner Teil des schweizerischen Oblihationenrechts. Halbbd.2. Tübingen
2. O. Gierke (1917). Deutsches Privatrecht. Bd.3.München und Berlin
3. H.G. Lambsdorff, B. Skova (1994). Handbuch des Bürgschaftsrechts. München
4. G. Lüke (1995). Grundfragen des Zessionrechts//Juristische Schulung
5. O. Palandt (2004). Bürgerliches Gesetzbuch. Kurzkommentar. 63 Aufl. München
6. L. Enneccerus, Th. Kipp, M. Wolff (1927) Lehrbuch des bürgerlichen Rechts. Marburg
7. F. Leonhard (1929). Allgemeines Schuldrecht des BGB. München und Leipzig
8. Civillikums. Ceturtā daļa. Saistību tiesības. / Latvijas Republikas likums. Pieņemts Latvijas Republikas Saeimā 28.01.1937.; stājies spēkā 01.03.1993. Ziņotājs Nr.4,, 1992.gada 30.janvāris
9. Сводъ гражданских узаконений губерний Прибалтийских т. 2. (1914). Сост. В. Буковский
10. Autoru kolektīvs prof. K. Torgāna red.(2000). Latvijas Republikas Civillikuma komentāri. Saistību tiesības. Rīga: Mans īpašums
11. G. Višņakova, K. Balodis (1998). Latvijas Republikas Civillikuma komentāri. Lietas. Valdījums. Tiesības uz svešu lietu. Rīga: Mans īpašums
12. Komercķīlas likums. 21.10.1998 / Latvijas Republikas likums. Pieņemts Latvijas Republikas Saeimā 21.10.1998.; stājies spēkā 01.03.1999. Latvijas Vēstnesis Nr.337/338. 11.11.1998
13. Tiesu prakse līgumsodu piemērošanā, http://www.at.gov.lv/.../Apkopoj_ligumsodi_5.11.2013.do
14. K. Freimanis, R. Niedra. Līgumsodu griestu noteikšana Eiropā: salīdzinošs pētījums, http://www.juristavards.lv/doc.php?id=263853

15. Social Transformations in Contemporary Societi clauses within different legal systems, 2013. http://www.stics.mruni.eu/wp-content/uploads/2013/06/153-162.pdf
16. Dumlop Pneumatic Tyre Co Ltd v New Garage and Motor Co Ltd, (1915) AC79. http://www.bailii.org/uk/cases/UKHL/1914/1.html
17. Ringrow Rty Ltd v BP Australia Pty Ltd (2005) HCA 71
18. O"Donell v Truck and Machinery Sales Limited 1998 4 IR 191
19. Elsley v. J.G. Collins Ins Agencies, (1978) 2 S.C.R. http://www.scc-csc.lexum.com/decisia-scc-csc/scc-csc/en/item/2597/index.do
20. K. Freimanis, R. Niedra Līgumsodu griestu noteikšana Eiropā: salīdzinošs pētījums . Pieejams: http://www.juristavards.lv/doc.php?id=263853
21. K. Torgāns Līgumsoda anomālijas Direktīvas 2011/7/ES normu kontekstā. Pieejams: http://www.juristavards.lv/doc.php?id=252065&show=true
22. Latvijas Republikas Augstākās tiesas Senāta Civillietu departamenta 2008. gada 14.maija Spriedums Lietā Nr. SKC – 212/2008
23. P. Oertmann (1906). Das Recht der Schuldverhältnisse. 2 Aufl. Berlin
24. K.W. Nörr, R. Scheyhing, W. Pöggeler (1999). Sukzessionen: Forderungszession, Vertragsübernahme, Schuldübernahme. 2. Aufl. Tübingen
25. Е. А. Крашенинников (1999). Основные вопросы уступки требования // Очерки по торговому праву: Сб. науч. тр. / Под ред. Е. А. Крашенинникова. Вып. 6. Ярославль
26. В. В. Грачев (2006). Притязание на неустойку // Очерки по торговому праву. Ярославль, Вып. 13
27. Latvijas Republikas Augstākās tiesas Senāta Civillietu departamenta 2013.gada 30.janvāra Spriedums lietā Nr. SKC-644/2013. http://at.gov.lv/files/uploads/files/archive/department1/2013/644-skc-2013.doc
28. М.М. Агарков (2002). Избранные труды по гражданскому праву. в 2-х т. Москва
29. J. Gernhuber (1974). Synallagma und Zession// Festschrift für Ludwig Raiser. Tübingen
30. J. Esser, E. Schmidt (1993). Schuldrecht: ein Lehrbuch. Bd. 1. Teilbd. 7 Aufl. Heidelberg
31. В. А. Белов (2000). Сингулярное правопреемство в обязательстве. М.: ЮрИнфоР
32. Л. А. Новоселова (2003). Сделки уступки права (требования) в коммерческой практике. Факторинг. Москва. Статут
33. Р. С. Бевзенко (2009). Некоторые вопросы судебной защиты права при уступке требований // Закон. № 10
34. E. Moris (2005). Cedētas tiek prasījuma tiesības, bet nevis strīda izskatīšanas kārtība. Jurista Vārds, 12.04.2005, Nr.368
35. J. Lapsa. Par jauno šķīrējtiesu likumu. Pieejams: http://vpstiesa.lv/page/lv/par-jauno-skirejtiesu-likumu
36. A. Miķelsone (2005). Par izmaiņām šķīrējtiesas procesā. Jurista Vārds. 15.03.2005/NR.10
37. G. Litvins (2014). Šķīrējtiesas kompetenci jāļauj apstrīdēt vispārējās jurisdikcijas tiesā. Jurista Vārds, 9.12.2014/Nr.48 (850)
38. T. Krūmiņš (2014). Šķīrējtiesas izņēmums Briseles režīmā. Jurista Vārds 7.10.2014/NR. 39 (841)

39. I. Kucina (2014). Par šķīrējtiesu regulējumu un tā pilnveidi. Jurista vārds, 3.06.2014/NR. 22 (824)
40. S. Bērtaitis (2014). Valsts nedrīkst pieļaut prettiesisku praksi strīdu alternatīvajā risināšanā. Jurista Vārds, 3.06.2014/NR. 22 (824)
41. J. Lapsa (2014). Šķīrējtiesu regulējuma novitātes. Jurista Vārds. 3.06.2014/NR. 22 (824)
42. M. Tipaine (2014). Šķīrējtiesu reputāciju grauj neitralitātes trūkums. Jurista Vārds, 3.06.2014/NR. 22 (824)
43. A. Fillers. Līgumiskās cesijas un šķīrējtiesas līguma mijiedarbība// Pieejams: http://www.juristavards.lv/doc.php?id=265042&show=true

Issues in separation of criminal procedural functions

S. Kaija

Rīga Stradiņš University, Riga, Latvia

Abstract. The term "criminal procedural function" has historically been used by scientists. In order to identify primary issues in procedural functions, the competition principle was analyzed. This strengthened such important issues as separation of prosecution and court functions, parties' equality, independence of courts, etc. Therefore, the concept of three concept functions – prosecution, defence and adjudication – was developed in criminal proceedings. In the context of the Criminal Procedure Law having taking effect, this term was included into the law, including the principle of separation of criminal procedural functions. The aim of this paper is to offer the insight into interpretation of the concept of the criminal procedural function with the special focus on specific issues in function separation. Section 17 of Criminal Procedure Law determines the function of control of restrictions of human rights in a pre-trail. The functions of prosecution, defence and adjudication do not determine their priority over all other criminal procedural functions. These functions are claimed to be separate and therefore cannot be applied to the same person. This paper examines the equivalence of procedural functions as one of the most fair of the court elements. In the end, key conclusions are summarized.
Key words: criminal procedural function; defence; prosecution; court adjudication.

1 Introduction

The term "criminal procedural function" has historically been used by scientists [1, 2]. In the context of the Criminal Procedure Law (henceforth CPL) having taking effect, this term was included into the law, including the principle of separation of criminal procedural functions [3]. Thus, this term has undergone transformation from a theoretical to an applied concept that has received the legal status and that is used in CPL today.

Aim. The aim of this paper is to offer the insight into interpretation of the concept of the criminal procedural function with the special focus on specific issues in function separation.

Materials and methods. In order to attain the aim, the paper examines legal regulations and analyzes relevant legal literature and court cases. In the end, key conclusions are summarized. The following research methods were applied: 1. Historical method (the analysis of historical development of criminal procedural functions), 2. Analytical method (the overview of legal literature, periodicals and other sources), 3. Descriptive method (detailed analysis of the research subject).

2 Results and discussion

The division of the criminal procedure into particular stages characterizes its structure, succession of implementation of procedural tasks, however, it does not offer the opportunity to spot and distinguish the nature of procedural activities, the so-called functions [1]. **A function is not only a particular criminal procedural activity but also the entirety of activities and decisions that serve the same aim.**

Different views are expressed in the literature regarding the concept of the criminal procedural function. Some lawyers interpret it as directions of activities that are implemented during the court proceedings [3]. According to others, the division of procedural activities into functions represents particular types (elements) of such activities with their own aims that are realized through special roles of the parties involved in the activities [2].

The Supreme Court points to actions that are viewed as criminal procedural actions, which, in turn, is determined by the Criminal Procedure Law. In order words, criminal procedural actions are the actions that are regulated by the Criminal Procedure Law [4]. Specifically, criminal procedural activities are the activities that are regulated by the Criminal Procedure Law. Therefore, if a function is provided for in the sources of rights in the criminal proceedings, it is done so in order to be consistent with the features of public administration functions. It is a criminal procedural function that is administered according to the Criminal Procedure Law and other sources of criminal procedure rights.

It is feasible to determine functions of the persons involved in criminal proceedings by assigning rights and duties. Mostly this depends on whether a particular person acts in defence of private or public interests. For persons that have personal interests in the outcome of criminal proceedings (the suspect, accused, victim, etc.) the functions mostly derive from rights that have been assigned to them. Conversely, the functions for officials that deliver criminal proceedings are assigned based on their duties.

In order to identify primary issues in procedural functions, the competition principle was analyzed. This strengthened such important issues as separation of prosecution and court functions, parties' equality, independence of courts, etc. Therefore, the concept of three concept functions – prosecution, defence and adjudication – was developed in criminal proceedings. This view of the concept is also supported by the University of Cambridge professor J.R. Spenser, who indicated that traditionally a prosecutor prosecutes, a defender defends and the court adjudicates [5].

However, this concept touched neither on the functions triggered in the pre-trial investigation nor the functions of witnesses, experts, etc. According to Professor A. Liede, the following key functions could be identified within the criminal court proceedings: an investigation function, incrimination and defence functions, adjudication function as well as the function of the supervision of the Office of Prosecutor [1]. A few authors suggest that criminal procedural functions be divided into the main or basic functions, subsidiary functions and supplementary functions. However, this classification of functions might not be necessary. Figuratively speaking, criminal procedural functions could be compared to human blood-vessels. All blood-vessels are essential for healthy life of a human being and therefore all of them are equal. The only difference lies in that some of them ensure greater blood flows than others, yet overall, their integrity secures healthy functioning of a human body [6]. This comparison could easily be applied to criminal proceedings: any minimization of the value of any criminal procedural function leads to less attention to attaining the goals of criminal proceedings, which results in the inability of criminal procedural activities to fully fulfil the assigned duties. Having completed the analysis of the Criminal Procedure Law norms, it is plausible to conclude that the legislator does not provide any grounds for the division of criminal procedural functions into categories based on their significance.

Section 17 of CPL determines the function of control of restrictions of human rights in a pre-trail. The functions of prosecution, defence and adjudication do not determine their priority over all other criminal procedural functions. These functions are claimed to be separate and therefore cannot be applied to the same person. Precisely, the legislator has determined the functions that can be separated in criminal proceedings, thus, prohibiting the realization of two or more particular functions by one person in one case of criminal proceedings.

The legislator does not provide any reason to question the nature of another criminal procedural function that is not determined in Section 17 of CPL, for example, Section 36 of CPL points to that the public prosecutor in criminal proceedings performs the following functions: supervision of investigation, investigation, criminal prosecution, sustaining of state incrimination and other functions identified in the Criminal Procedure Law.

A specific person involved in criminal proceedings can simultaneously realise multiple functions except for the functions that are specifically separated.

Constructing Section 17 of CPL, the legislator has specified four procedural functions and most importantly has stressed that these functions can be separated. This separation means that relevant spheres must be functionally and institutionally separable.

In the absolute majority of cases, this is how it works, specifically, **at the pre-trial stage the control over the restrictions of human rights is implemented by the investigating judge, the accusation is delivered by the prosecutor, the defence is conducted by the person who has the rights of defence, his/her representative and his/her attorney, in turn, administration of justice is delivered by a judge or court.** One circumstance does not fit into the scheme, that is, in accord with the subsequent formulation of the text in the Criminal Procedure Law, the investigating judge controls the restriction of human rights only on the limited scale, in other words, only within the framework of specific warrant. Other controlling activities are mostly assigned to the public prosecutor [7]. The investigation judge realizes the control function of the restriction of human rights not only in the pre-trial but also in specific cases determined by the CPL in the Court of First Instance until the beginning of the case adjudication.

The second part of Section 41 of the CPL claims that from the Court of First Instance to the commencement of trial of a case, an investigating judge has a duty to decide on the following: (1) the application of an accused in relation to the amending or revocation of security measures; (2) the proposal of a public prosecutor in relation to the selection or amendment of a security measure; (3) the acquaintance of a person involved in criminal proceedings, who has the right to get acquainted with the materials of a criminal case, with special investigative actions that are not attached to the criminal case (primary documents).

It is important to stress that an investigating judge shall not be permitted to replace a person directing the proceedings and a supervising public prosecutor in pre-trial criminal proceedings by giving instructions regarding the direction of an investigation and the performance of investigative actions.

The impartiality of the court is viewed based on the procedure of the court composition, and the fact that in criminal proceedings a judge participates who has previously been involved in the proceedings indicates that the impartiality requirement is breached. In order to avoid it, Section 52 of the CPL claims that he investigating judge shall not be the person who has been the person directing the proceedings or supervising public prosecutor in the same criminal proceedings. A judge shall not participate in the adjudication of a case if he or she: (1) has participated in the criminal proceedings in any status; (2) is in kinship to the third degree, affinity to the second degree, or married to another judge involved in the adjudication, the maintainer of prosecution, or the public prosecutor who has transferred the criminal case for trial, or if he or she has a common household with the referred to judge, maintainer of prosecution, or public prosecutor.

Today the quality of criminal proceedings largely depends on the rights to fair, open and timely case hearing in the independent and impartial court determined by the law. According to Chapter 8 of the Constitution of the Republic of Latvia, a fundamental human right is state recognition and defence of fundamental human rights consistently with the Constitution, laws and binding international agreements [8]. Section 92 of the Constitution also guarantees that anyone can defend his/her rights and legal interests in a fair court.

This guarantee is consistent with the most important international legal acts, such as Section 10 of ANO General Human Rights Declaration, Section 6 of the European Convention of Human Rights and Protection of Fundamental Freedom and Section 14 of ANO International Pact of Civil and Political Rights.

According to Part 1 of Section 6 of the Convention of Human Rights and Protection of Fundamental Freedom that defines civil rights and duties as well as the validity of incrimination in criminal proceedings, anyone has rights to fair and open timely hearing in an impartial court determined by law. The court decision is to be publicly announced, however, the presence of press and public in the proceedings can be fully or partially prohibited for ethical reasons or in the interests of public order and state security in the democratic society that demands the protection of the interests of the minor or private life of the participants of the proceedings, or else to the extent that the court considers absolutely essential in special circumstances when the publicity endangers the interests of justice.

Section 48 of the CPL claims that considering the criminal case the court has the warrant of the person directing the proceedings in administration of criminal proceedings and securing the procedural order, as well as the exclusive rights to administer justice. Consistently with Section 23 of the Criminal Procedure Law, the court considers the validity of incrimination in acquitting innocent persons or pleading persons guilty of committing a criminal offence and determining the mandatory adjustment of criminal legal relations for state institutes and persons. In turn, the accusation function in criminal proceedings is delivered by the public prosecutor. Although the concept of the state prosecution is mentioned in Section 42 of the Criminal Procedure Law, the classical definition is not included. Part 1 of Section 42 of the Criminal Procedure Law states that "Public prosecution in the first instance is maintained by the public prosecutor who has submitted the criminal case to a court. A senior public prosecutor can assign the case to another public prosecutor." Yet, Part 2 of the same section states that "As far as possible, instances of public prosecution appeal in court are maintained by the same prosecutor who presented the case in the Court of the First Instance. A senior public prosecutor can assign the case to another public prosecutor."

Currently criminal proceedings are formed in such a way that the same public prosecutor can be involved throughout criminal proceedings. In criminal proceedings the public prosecutor fulfils the supervision of investigation, investigation, prosecution, maintenance of state prosecution and other functions outlined in the law (Sect. 36 of the CPL). Initially, the public prosecutor is the public prosecutor supervising investigation (Sect. 37 of the CPL). The supervising public prosecutor acquires the status of the person directing the proceedings from the moment he/she takes over the administration of criminal proceedings and decides about the initiation of criminal prosecution (Sect. 38 of the CPL). Public prosecution in the Court of the First Instance is delivered by the public prosecutor who has submitted the criminal case to court (Sect. 42 of the CPL), etc. Each public prosecutor is controlled by senior public prosecutor who, following the law, controls the way the public prosecutor uses his/her warrant [9].

Incrimination restrictions and content are determined by the public prosecutor; it is the exclusive function of the public prosecutor to decide if a person is to be held criminally liable and if yes, then for what type of an offence. In turn, **incrimination of the public prosecutor determines the restrictions of court hearing.** Without intervening into incrimination maintenance and delivery of defence, by fulfilling its function, the court participates in the examination of evidence.

In order not to intervene into realization of separate functions in criminal proceedings, the law puts forward a number of restrictions to the court and judge. For example, the court and the judge:

1. do not make the decision on the initiation of criminal proceedings. Part 2 of Article 476 of the CPL is an exception example that states that the court makes a decision on the criminal procedure separation, which is simultaneously considered as the decision about the initiation of new criminal proceedings;
2. cannot send the criminal case for supplementary investigation;
3. cannot continue adjudication if the public prosecutor has declined incrimination in the order consistent with Article 476 of the CPL;
4. following own initiative, cannot acquire evidence or vary it in court hearing, except for the cases when the defendant conducts his/her own defence and the court has justified doubts about the defendant's accountability or possible guilt in the incrimination (Part 2 of Article 455 of the CPL);
5. the court rights to state other circumstances that are different from the incrimination are restricted.

According to Part 3 of Article 455 of the Criminal Procedure Law, the court can recognize as proved the factual circumstances different from incrimination for offence only if these circumstances do not worsen the condition of the defendant's and the defendant's rights for defence are not breached [10]. This condition includes not only the court's rights to replace factual circumstances that have not been proved in the offence with evidence but also exclude from incrimination factual circumstances that have not been proved and if there are legal grounds to reclassify the incriminated criminal activities if this does not worsen the defendant's condition and his/her rights for defence are not breached [11].

According to the European Convention of Human Rights and Basic Freedom Protection Act, there exist main guarantee categories for fair case examination: (a) organic guarantees that are aimed at ensuring that such case examination is possible, which assigns state institutions a number of duties (form a legal court, impartiality, publicity of the case hearing, etc.); (b) functional guarantees that are aimed at ensuring equality at all stages of the procedure [12]. Article 6 of the Convention requires that procedural equality insurance is attributed to the functional guarantees.

The concept of fair court includes the principle of equal opportunities for all parties involved, which provides an opportunity for all parties involved to present case circumstances and which forbids any party to be assigned more significant advantages in comparison to the opponent [13].

Persons involved in the criminal procedure have warrants (rights and duties) that ensure tasks that are determined for them in legislative acts and provides for equal realization of guaranteed rights. The principle of equivalence of procedural warrants applies to entire criminal proceedings. Procedural equivalence of warrants **does not mean that the rights and duties of the persons involved in criminal proceedings must be absolutely identical.**

It is important to stress that this principle is regulated generally, and "equivalent realization" obviously represents a value concept that is filled in each specific case. Similarly, it must be noted that this principle is not identical to the equality principle. While the equality principle asserts equal attitude to the persons in the same status and same circumstances, the **procedural warrant equivalence principle can be attributed to various process participants who on a particular occasion in the process were assigned the same duties, rights,** etc. This principle means that relevant process participants have to ensure equivalent (but not necessarily the same) opportunities to use them, etc. Adjusting various criminal procedure norms to practice, directly practicing employees must be the ones who thoroughly observe the realization of equivalence ideas [7].

The quality of arms principle is both a significant element of the rights to fair court and is special manifestation of legal equality principle that refers to all types of cases subjected to court. This determines that in case hearing the rights of the participants involved in the process have to be fairly balanced, in other words, **each participant has to be ensured with adequate opportunities to use procedural tools and must not be subjected to groundless unfavourable conditions in comparison to other process participants.**

Therefore, the principle of procedural equality leads to the state duty to form such case hearing regulation that would ensure it. This, however, **does not mean that in some cases (or specific category cases) such advantages could not be groundless [14].**

The defence that is conducted by the person who has the rights of defence, his/her representative and defender, in the CPL is regarded as "weaker" than the incrimination that represents state interests. Therefore, the defence receives additional procedural advantages (**defence privileges** (*favor defensionis*)), for example, the presumption of innocence (praesumptio innocentiae), doubt interpretation in favour of the person who has the rights of defence (in dubio pro reo), mandatory defence in cases determined by law, rights (not duties) to provide testimony, etc.

The legislator has included the principle of rights of defence in Article 20 in the Criminal Procedure Law. This article asserts that each person regarding whom an assumption or allegation has been expressed that such person has committed a criminal offence has the right to assistance of a defence counsel, that is, the right to know what offence such person is suspected of committing or is being accused of committing, and to choose his or her position of defence.

The rights of defence as the essential component of the rights to fair court are constitutionally guaranteed fundamental human rights. Article 92 of the Constitution of the Republic of Latvia asserts that anyone has the right to the defender's help.

Legal profession is an essential component of the state legal judicial system. Only having the legal profession is it possible to converse about the actions of the state legal judicial system, courts and state investigation institutions in a state that functions based on law [15].

The rights of defence include a number of aspects:

1. Assignment of procedural rights to the person that has the rights of defence. This person has the rights to know incriminated suspicions, to defend own rights using any possible legal means, etc.
2. The person that has the rights of defence can conduct own defence or has to have real opportunities to use the service of a defender.
3. The duty of the person directing the proceedings is to ensure that the person who has the rights of defence is provided with an opportunity to defend his/her case using all legal means. The duty of the person directing the proceedings is not only to ensure the

application of such rights but also to defend such rights and not to obstruct their application. The person directing the proceedings must understand that only by observing such personal procedural guarantees in the criminal procedure is it possible to ensure the effective application of the Criminal Law norms and fair regulation of criminal legal relations in specific criminal proceedings and to overcome the protest of the person who has the rights of defence.

However, the person directing the proceedings:

- does not make a deal regarding the defence and does not invite a specific lawyer to act as a defender; however, this person has to provide the person involved in the case with all necessary information and provide with an opportunity to use communication devices to invite a defender (Article 80 of the CPL);
- in cases when the person who has the rights of defence has not made a deal about the defence and yet the defender's participation is mandatory or the person wishes to have a defender, the person directing the proceedings has to inform a senior certified attorney in the relevant court territorial jurisdiction about the necessity to ensure a defender's participation in criminal proceedings (Article 80 of the CPL);
- in cases specified in Article 81 of the Criminal Procedure (in specified procedural activities), the person directing the proceedings has to invite a defender in accord with the defence schedule of the relevant senior attorney in specific procedural activities.

The University of Zurich Criminal Law Professor Stefan Trechsel claims that the arguments that justify the offer of legal help to the person that has the right of defence could be viewed from technical, psychological and humanistic perspectives [16].

The decisions of the European Court of Human Rights have included such aspects. For example, in the case of *Salduz v. Turkey* the court drew attention to the fact that the person who has the rights of defence at the pre-trial stage is particularly vulnerable because that person does not have specialized knowledge about criminal proceedings, his/her status, interrogation, evidence acquisition procedures whose knowledge is very important, particularly at the initial stage of investigation. It is exactly the evidence obtained at the initial investigation stage that often sets incrimination restrictions. In most cases such vulnerability can be overcome by the involvement of a defender in the defendant's criminal proceedings [17]. The decision of this Court has impacted on legal regulations of many states [18].

The European Court of Human Rights further asserted the necessity of engaging a defender in the criminal proceedings in the case *Melnikov v. Russia* in which the submitter complained about the absence of a defender in confrontational situations. The Court highlighted that due to absence of the defender, the submitter could not have understood the confrontation procedure and therefore could not have effectively used his rights of defence [19].

The humanistic aspect has been approved by the European Human Rights Committee in its report on the case *Can v. Austria*. Although the provision of emotional support is not included into the duties of a person who offers legal help, this aspect can be used as an additional argument when claiming the necessity of good legal support in criminal proceedings [20].

As a topical matter it is important to mention the European Parliament and Council Directive (EU **2016/343 (09.03.2016)) regarding the necessity to strengthen certain aspects of the presumption of innocence and rights to participate in case hearing in person in criminal proceedings [21], whose goal is to ensure the rights to fair court in criminal proceedings, determining minimal limits of the presumption of innocence and the right to participate in case hearing in court in person.**

The Directive refers to the: (1) requirement that the suspect or the accused shall be considered innocent while his/her guilty has not been proved in accord with the law; (2) regulations about the defence of the suspect or the accused against the authorities announcing the guilt prior to the judgement of conviction; (3) regulations pertaining to the fact that the burden of proof lies with the person directing the proceedings and the fact that the reasonable doubt about the guilt of the accused results in this person's pleaded not guilty; (4) rights of the suspect or the accused not to give testimony against self and not to cooperate with the investigation; (5) rights to remain silent and (6) rights participate in the case hearing in court in person.

The Stockholm programme has already focused on strengthening of the person's rights in criminal proceedings. In Point 2.4 of the programme the European Council invites the Committee to submit a proposal strengthening the rights of the suspect or the accused determining the rights to common minimal standards of the fair court. Three actions have already been passed:

1) European Parliament and Council Directive (2010/64/EU (20.10.2010)) on the rights to oral and written translation in criminal proceedings;
2) European Parliament and Council Directive (2012/13/EU (22.05.2012)) on the rights to information in criminal proceedings;
3) European Parliament and Council Directive (2013/48/EU (22.10.2013)) on the rights to the defender's help in criminal proceedings and European arrest warrant process, on the rights to inform the third person upon restriction of freedom and on the rights to communicate with the third persons or consulting institutions during the period of restriction of freedom [22].

The Directive will supplement the legal regulation that has been created by the European Convention of Human Rights and Charter of Fundamental Rights. It will strengthen mutual trust amongst the institutions of member states and will ease the process of mutual recognition of judgements in criminal cases.

3 Conclusions

1. A function is not only a specific criminal procedural action; it is the entirety of actions and decisions that are advanced to attain a specific aim.
2. It is possible to determine the functions of persons involved in criminal proceedings by assigning rights and duties.
3. The analysis of CPL norms has revealed the fact that the legislator does not provide grounds for separation of criminal procedural functions according to their distinct features.
4. The legislator has determined functions that are separate in criminal proceedings and that forbid the application of two or more separate functions to the same person within the scope of one case of criminal proceedings.
5. A person involved in specific criminal proceedings can simultaneously be applied multiple functions, except for separate functions. The equivalence principle of procedural warrant applies to entire criminal proceedings and has a significant role in ensuring the separation of procedural functions.
6. The equivalence principle of procedural warrant does not mean that all rights and duties of the persons involved in criminal proceedings must be absolutely identical.
7. The concept of fair court includes the principle of opportunity equivalence of the parties involved, which includes the provision of an opportunity to lay out the case circumstances to all parties involved in the proceedings and forbids offering significant advantages to any party as compared to the opponent.

8. However, this does not mean that in some cases (or in some category cases) such advantages cannot be justified. For example, to a certain extent the defence might be regarded "weaker" than the prosecution that stands for state interests. Therefore, the defence receives supplementary procedural advantages (defence privileges).

References

1. Liede A. Latvijas PSR kriminālprocess. Vispārīgā daļa. Tiesu pierādījumi. Rīga, Zvaigzne, 13.lpp. (2010)
2. Ларин А.М. Расследование по уголовному делу, процесуальные функции. М., Юрид. лит., с. 4 (1986)
3. Kriminālprocesa likums. Spēkā no 01.10.2005. Latvijas Vēstnesis, 11.05.2005. **74**., Ziņotājs, 09.06.2005. **11** (2005)
4. Latvijas Republikas Augstākās tiesas Senāta Administratīvo lietu departamenta 2010.gada 21.maija lēmums, lieta nr. SKA – 528/2010 (2010)
5. Mireille Delmas-Marty, J.R. Spencer. European Criminal Procedures. Cambridge University Press, ISBN 0521591104, 9780521591102, p.160 (2002)
6. Зинатуллин З.З., Зинатуллин Т.З. Еще раз об уголовно-процессуальных функциях. //http://kalinovsky-k.narod.ru/b/ufa20042/zinatullin.htm
7. Meikališa Ā., Strada-Rozenberga K. Kriminālprocesa veidi, uzbūves raksturojums un pamatprincipi// Kriminālprocess. Raksti 2005-2010. Rīga: Latvijas Vēstnesis, 63.lpp. (2010)
8. Latvijas Republikas Satversme. Spēkā no 07.11.1922. Latvijas Vēstnesis, 01.07.1993.Nr.**43.** (1922).
9. Kaija S. Prokurora loma kriminālprocesa atjaunošanā sakarā ar jaunatklātiem apstākļiem. Rīgas Stradiņa universitātes Juridiskās fakultātes elektroniskais juridisko zinātnisko rakstu žurnāls Socrates, **2**(5). http://www.rsu.lv/images/stories/dokumenti/publikacijas/socrates_5_2016.pdf (2016)
10. Latvijas Republikas Augstākās tiesas 2014.gada 26.marta lēmums krimināllietā Nr. 11354005611, Nr. SKK-0019-14 (2014)
11. Latvijas Republikas Augstākās tiesas 2014. gada 30. janvāra lēmums krimināllietā Nr. 11370003509, Nr. SKK – 4/2014 (2014)
12. Комментарий к Конвенции о защите прав человека и основных свобод и практике ее применения.// Pieejams – http://www.hri.ru/docs/?content=doc&id=371
13. Law and practice of the European Convention on Human Rights and the European Social Charter, Council of Europe, 1996, p.172. Citēts pēc. Satversmes tiesas 2003.gada 27. jūnija sprieduma lietā Nr. 2003-03-01., 6.p. (2003)
14. Latvijas Republikas Satversmes komentāri. VIII nodaļa. Cilvēka pamattiesības. Autoru kolektīvs prof. R.Baloža zinātniskā vadībā. Rīga: Latvijas Vēstnesis, 134.-135. lpp. (2011)
15. Niedre A. Latvijas Republikas advokatūras darbības organizācija un pamatprincipi. Rīga, Latvijas Zvērinātu advokātu padome, 10 lpp. (2006)
16. Trechsel S. Human Rights in Criminal Proceedings. New York: Oxford University Press, 245 p. (2005)
17. European Court of Human Rights judgment in case Salduz v.Turkey, 36391/02, 26.04.2007 (2007)
18. Giannoulopoulos D. Strasbourg Jurisprudence, Law Reform and Comparative Law: A Tale of the Right to Custodial Legal Assistance in Five Countries. Human Rights

Law Review **16**(1), 103–129 first published online January 27, 2016, doi:10.1093/hrlr/ngv039 (2016)

19. European Court of Human Rights judgment in case Melnikov v.Russia, 23610/03, 14.01.2010 (2010)
20. European Court of Human Rights judgment in case Can v. Austria, 9300/81, 30.09.1985 (1985)
21. Eiropas Parlamenta un Padomes Direktīva (ES) 2016/343 (2016. gada 9. marts) par to, lai nostiprinātu konkrētus nevainīguma prezumpcijas aspektus un tiesības piedalīties klātienē lietas izskatīšanā tiesā kriminālprocesā, http://eur-lex.europa.eu/legal-content/LV/TXT/?uri=CELEX%3A32016L0343 (2016)
22. Priekšlikums Eiropas Parlamenta un Padomes direktīva par to, lai nostiprinātu nevainīguma prezumpcijas un tiesību piedalīties klātienē lietas izskatīšanā tiesā konkrētus aspektus kriminālprocesā, http://eur-lex.europa.eu/legal-content/LV/TXT/?uri=CELEX%3A52013PC0821

https://doi.org/10.1051/shsconf/20184001011

Legal consequences and problems of the servitudes of right of way established by administrative acts in Latvia

J. Dinsberga and K. Bite

Rīga Stradiņš University, Riga, Latvia

Abstract. The right to own a property is guaranteed by the Article 105 of the Constitution of the Republic of Latvia that in the same time stipulates that the said rights may be restricted only in accordance with law. This legitimate property right restriction is determined in Article 928 of the Civil Law: the ownership may be restricted both by private intent and by law. One of the restrictions of ownership rights is the servitude of right of way which is established by law, a contract, a court judgement, or a will. During the land reform, the State Forest Service, land commissions, and municipalities had the legal basis to establish the servitude of right of way also by an administrative act, taking a corresponding decision. However, many decisions were insufficient and legally incorrect. As a result, today the servitudes of right of way established by the administrative acts during the period of the land reform do not fulfil their purpose and cause a number of legal and practical problems both to owners of immovable property and to state institutions. In order to explore these problems, referring to the Article 7 of the Paragraph 105 of the protocol resolution No. 56 of the Cabinet of Ministers from 29 October 2013, the Ministry of Environmental Protection and Regional Development in cooperation with the Ministry of Transport, the Ministry of Justice, and the Ministry of Agriculture issued the Conceptual Report "On the Problems Occurring in Relation with the Servitudes of Right of Way Established during the Land Reform and Their Possible Solutions" (hereinafter – the Conceptual Report). In general there are 2 possible solutions, but for the second solution there are five courses of action, one of which – in cases when there is a road on private property that is needed for public use it is to be determined as an encumbrance of the immovable property "Public road". The aim of the research is by analysis of the problems of the servitudes of right of way, established by the administrative acts during the period of the land reform, to evaluate whether the solutions offered in the Conceptual Report are practically applicable and will give the desired result, as well as to offer the author's vision for solution of the existing problems. The author has used the scientific research approaches – descriptive, analytical, inductive and deductive, logical-constructive, formal-dogmatic, graphical and interpretation of legal norms – grammatical, systemic, and teleological.
Key words: servitude of right of way; land reform; public road; property rights.

1 Introduction

The right to own property is one of the fundamental rights of a human guaranteed by the Protocol No. 1, Article 1 of the Convention for the Protection of Human Rights and Fundamental Freedoms [1, 1], Article 105 of the Constitution of the Republic of Latvia that in the same time stipulates that the property rights are not absolute rights and may be restricted in accordance with law [2, 105]. This type of property right restriction is stipulated in the Civil Law. Article 927 of the Civil Law states "Ownership is the full right of control over property [...]" [3, 1130]. Full right of an owner is expressed as implementation of the right of possession, use and action. However, on basis of Article 928 of the Civil Law "[...] ownership may, pursuant to private volition, as well as pursuant to law, be restricted in various ways [...]" [3, 928]. One of the property right restrictions, causing problems to many owners of immovable property in Latvia, is the servitude of right of way. The essence of the servitude of right of way is revealed by Article 1130 of the Civil Law stating that "servitude is such right in respect of the property of another as restricts ownership rights regarding it, with respect to utilisation, for the benefit of a certain person or a certain parcel of land" [3, 1130]. The existence of servitude of right of way requires immovable property of a person, utilisation of which is determined for the benefit of another person or another immovable property that has the right to use the servitude of right of way. Thus, when establishing servitude of right of way the immovable property or a particular person for the benefit of which it is established must be precisely determined.

According to Civil Law servitudes shall be established by (1) law; (2) court order; (3) agreement or will. However, during the land reform, commenced in Latvia in 1990, within framework of which property rights of former owners or their heirs were resumed, or the land was granted for use or in property against payment, a range of laws and regulations were drawn up determining that a servitude can be established by state or municipal institutions issuing an administrative act. For example, the Law on Land Privatisation in Rural Areas, the Law on Land use and Land Survey, Regulation on Land in Rural Areas, Regulation on Land in Cities, etc. Above mentioned laws and regulations imply that during the land reform the State Land Service, Land Committees and municipalities had legal basis to establish servitude of right of way also by an administrative act passing a respective decision. A. Grūtups indicates: "The land reform takes place already since 1991, and should we only base on the Civil Law and the municipalities would not have the right to establish servitudes, the prospective owners as from 1991 would only have had two options: to conclude mutual agreements in each case, or, if they are unable to come to an agreement, to apply to a court" [4, 139]. We can unequivocally agree with the opinion of A. Grūtups, however, we have to emphasize that many decisions of the state and municipal institutions passed during the land reform were imperfect and legally incorrect. For example, the decision stated the servitude property (property encumbered with servitude of right of way) but did not state the dominant property (property the benefit of which the servitude is established for); the servitudes of right of way were not marked in title plans or the decisions were not supplemented with graphical attachments with servitude of right of way; the decisions establishing servitudes did not have particular references as to the width, length, etc. of the servitude of right of way. In the result of what the servitudes of right of way established during the land reform by an administrative act today do not perform their functions and cause a row of problems of legal and practical character both to the owners of the immovable property and the state institutions. In order to survey these problems on basis of Article 7 of the Paragraph 105 of the protocol resolution No. 56 of the Cabinet of Ministers from 29 October 2013, the Ministry of Environmental Protection and Regional Development in cooperation with the Ministry of Transport, the Ministry of Justice, and the Ministry of Agriculture issued the Conceptual Report. Mainly these are cases when there is a road on an immovable property that during the land reform has been established as servitude of right of way and that is necessary to provide access to national or municipal roads for one or even several hundreds of immovable properties. However, according to the current legal regulations and observing the principle of protection of a person's property rights there is no legal justification to use this road.

Aim of the research: The aim of the research is by analysis of the problems of the servitudes of right of way, established by the administrative acts during the period of the land reform, to evaluate whether the solutions offered in the Conceptual Report are practically applicable and will give the desired result, as well as to offer the author's vision for solution of the existing problems.

We have to note that in general there are 2 possible solutions in the Conceptual Report, but for the second solution there are five courses of action for solution of problems of the servitudes of right of way established by the administrative acts. The authors have planned to devote a separate article to each of the solutions and courses of action providing deeper analysis. The present article gives detailed analysis of the 2nd solution: Complex courses of action for solution of access problems; course of action: *Public road as an encumbrance of immovable property*, where the authors of the Conceptual Report offer establishment of a legislative property right restriction – *public road*, that would ensure use of private roads and national or municipal roads located on private land.

1.1 Research materials

The research is mainly based on analysis of the Conceptual Report and the land reform laws, which grant the state institutions the right to establish servitudes of right of way by administrative acts. The authors have studied the legal framework of the current valid Civil Law, as well as the legal framework of servitudes in historical view, analysing establishment of servitudes of right of way in Part III of the Collection of Local Laws that was valid in Latvia (except Latgale) until 1937. As secondary sources the authors have used the Study on Servitudes of Right of Way Established During the Land Reform ordered by the Ministry of Environmental Protection and Regional Development and prepared in 2014 by SIA "A&O konsultāciju pakalpojumi", opinions expressed in the legal literature of M. Auders and experts of the civil rights, for example A. Grūtups, J. Rozenfelds, G. Višņakova.

1.2 Methods

The following scientific research methods have been used during the research:

Descriptive – study of the legal framework of establishment of servitudes of right of way by administrative acts, gathering of information, provision of explanation and identification of problems on basis of the study. *Analytical* – study of laws and regulations, case law, opinions of authors to clarify the essence of servitude and the problems of servitudes of right of way established by administrative acts within framework of the studied issue. *Inductive and deductive* – for expression of separate judgements during the course of study. *Logical-constructive* – for expression of editions of particular amendments to legal provisions. *Formal-dogmatic* – for critical evaluation of the valid legal provisions on types of restrictions to immovable property and offering of amendments for their further determination. *Graphical* – arrangement of information in tables to ease the perception.

The following methods of interpretation of legal provisions have been used during the research:

Grammatical – study of the meaning and nature of concepts "servitude of right of way", "establishment of servitude", "making of a note in the Land Register". *Systemic* – study of the peculiarities of establishment of servitudes of right of way in the laws and regulations passed during the land reform in connection with the legal provisions of the Civil Law for establishment of servitudes of right of way. *Teleological* – study of the legislator's will when adopting laws and regulations by which the state institutions are granted the right to establish servitudes of right of way by administrative acts; to understand the social aim of it.

2 Results

Following the legal and political processes in Latvia in general, development of public and private rights, having analysed former studies, historical and currently valid laws and regulations, assessed opinions of case law and opinions of experts of civil rights regarding improvement, strengthening and modernisation of the legal framework the authors came to a conclusion that one of the most current problems in the area of property rights is connected with the servitudes of right of way established during the land reform. We have to admit that the problems with the servitudes of right of way established by administrative acts was topical more than ten years ago and the author J. Dinsberga studied a part of the problems related to restrictions of property rights arising from servitudes of right of way established during the land reform and provided possible solutions already in 2003 [6, 2003], as well as in 2012 when participating in the scientific conference organised by the Riga Stradiņš University "Topical Issues of Legal Problems *Pacta sunt servanda*" with the paper "Problems Related to Establishment of Servitude of Right of Way" [7, 2012].

Unfortunately we have to note that the laws and regulations regulating servitudes have almost not been amended since adoption. Moreover, the existing legal framework of servitudes given in the Civil Law is very similar to the legal framework provided in Part III of the Collection of Local Laws. The definition of servitude provided in Article 1089 [8], the list of types of servitudes of right of way in Article 1118 [8], and the list of basis for establishment of servitudes in Article 1250 [8] are practically identical to the current formulations provided in the Civil Law. Regardless the fact that the state and municipal institutions already at the beginning of implementation of the land reform had stated problems with legal registration of servitudes established by administrative acts in the Land Register and practical application thereof, these issues were disregarded for a long time. Therefore the range of unsolved issues related to servitudes increased year by year. The problem issues of abovementioned area became priority on a national level only on 29 October 2013 when the Cabinet of Ministers instructed the Ministry of Environmental Protection and Regional Development in cooperation with the Ministry of Transport, the Ministry of Justice, and the Ministry of Agriculture to draw up the above-mentioned Conceptual Report.

Of course, to speak about urgency of an issue it is not enough with problems and conflicts among several legal entities due to insufficient legal basis. In this case it is necessary to state the fact that problems of similar nature affect interests of a substantial part of legal entities and/or state or municipalities. It is also necessary to evaluate the legal consequences that might be caused if the problem is not solved in long-term. M. Auders [9] has studied servitudes of right of way established during the land reform by administrative acts, and his theses are included in the Conceptual Report [10]. The authors are of positive opinion of the development of the Conceptual Report and problems provided therein, but it has one substantial deficiency – information has been gathered only about one administrative unit of Latvia: Sigulda district, Sigulda Rural Municipality. The report states that: "The largest part of problematic cases today is formed by servitudes of right of way, where the dominant property has not been determined, as well as servitudes of right of way established for the benefit of particular immovable properties; however, without complete listing of the dominant properties" [10]. The table below provides information indicating which state institution and in how many cases in the Sigulda Rural Municipality has made a decision to establish servitude of right of way, but the decision fails to provide the dominant property for the benefit of which the servitude of right of way is established.

Table 1. Servitudes established by the Rural Municipality Councils, Land Committees and the State Land Service [10].

Rural Municipality Council		**Land Committee**		**State Land Service**	
Dominant immovable property has been provided	Dominant immovable property has not been provided	Dominant immovable property has been provided	Dominant immovable property has not been provided	Dominant immovable property has been provided	Dominant immovable property has not been provided
3	**15**	**39**	**29**	**52**	**19**

The information provided in the table indicates that all decision makers in Sigulda Rural Municipality have made mistake and failed to provide the dominant immovable property according to the Civil Law. Since there are 119 municipalities in Latvia – 110 districts and 9 cities [11, Attachments No. 1 and 2], it would be necessary to survey other administrative units of Latvia and to determine what is the situation and problems related to establishment of servitudes there (the point of view of the authors regarding this issue is provided in more detail in the conclusions of the present article). At the moment, while only the Sigulda Rural Municipality has been studied, it only gives an insight into the essence of the problem, but does not reflect the actual situation in the whole Latvia. Should the decision on solving of the problems related to servitudes of right of way established by administrative acts be based on these insufficient facts, the history might repeat and decisions that are unreasoned and inadequate for the local situation might be passed.

In addition to the mentioned problem when the decision of an institution does not determine the dominant immovable property the Conceptual report lists the following problems:

- "servitudes of right of way established by a decision of a municipality are registered in the Land Register not in the form of a record, but in the form of a note (*authors' note – servitude as an encumbrance (rights in Rem) must be recorded in the Land Register section of the servient immovable property as an encumbrance for the benefit of the dominant immovable property. Rights to an immovable property are registered in the form of a record, but securities and restrictions are registered in the form of a note. A note can be replaced by a record, if the owner or a respective interested party submits a request for registration in the Land register)*. Considering the fact that the dominant property has not been determined, it is impossible to identify the interested person, based on the request for registration of which the note might be replaced by a record. Therefore the rights of users of servitude are not protected [...];

- in order to get to the property several neighbouring *properties* must be encumbered, however, that has not been done and the road is only partly usable (*authors' note – neighbouring properties are immovable properties located next to each other and crossed by servitude of right of way. In case at least one of the immovable properties is not encumbered, nobody is entitled to use the road on that property and an interruption of the division of servitude of right of way is formed*);

- servitude is established, but the road is used by unlimited number of users or in the event of reorganisation of borders the road connects several roads of higher importance, thus it has become a public road;

- servitude is established, but the road in the National Real Estate Cadastre Information System (hereinafter – NRECIS) is registered as a construction and a municipal property, whereby there is a situation when the land is owned by one person, but the road located on the land is under jurisdiction of another person;

- servitude of right of way or a road marked "public road" is provided only in the title plan, but not in the administrative act. In fact this servitude of right of way is invalid; however, it might be registered in the NRECIS and probably in the Land Register in the form of a note". [12]

In order to find solutions to the particular problems, firs of all we have to state the cause of the problems and the reason of passing incomplete and legally incorrect decisions during the land reform. The Conceptual Report and the "Study on Servitudes of Right of Way Established by Administrative Acts during the Land Reform" do not provide in-depth research of this aspect. Therefore the answer can be found at the beginning of the land reform, when on 13 June 1990 the decision of the Supreme Council of the Republic of Latvia "On Agrarian Reform in the Republic of Latvia" was passed [13, 1990], determining that the government shall implement the land reform and the Council of Ministers was charged with preparation of the necessary laws and regulations. Having selected and evaluated the laws and regulations and amendments thereof passed during the land reform, we can state that the number is impressive and many laws and regulations were passed in the same time and over a rather short period of time. That means these laws and regulations were adopted in a hurry, affecting their quality that is indicated by the numerous amendments to the laws and regulations on the land reform.

Several laws and regulations that contained provisions regarding servitudes were passed before coming into effect of the Rights in Rem part of the Civil Law (1 September 1992) and they did not provide as detailed legal order for the servitudes as that provided in the Civil Law. For example, Article 32 of the "Law on Land Use and Land Survey" stated that "It is prohibited for natural persons and legal entities to use land tenures of other natural persons and legal entities (for walking, driving, movement of animals and other purposes) without agreement with the respective land user, except in cases when granting the land the servitudes related to use of the land have been established. Types of servitudes are determined by the Regulation on Land in Rural Areas and the Regulation on Land in Cities." [14] The Regulation on Land in Rural Areas [15] was passed on 25 November 1991, and several articles of the regulation included provisions regarding servitudes (Articles 1, 11.5; 11.6; 53–57). The legal framework of servitudes in the said regulation was incomplete and might have affected quality of the made decisions. However, also after coming into effect of the Rights in Rem part of the Civil Law a part of officials, when deciding on renewal of property rights or granting of land for use, or in property against payment and applying special laws and regulations did not take into account the special provisions of the Civil Law regarding servitudes. Thus by ignoring the definition of servitudes provided in Article 1130 of the Civil Law and other articles, the decisions did not stipulate immovable properties or persons, for the benefit of which servitudes of right of way were established, the location, width, length of servitude of right of way were not provided, the established servitudes were named public servitudes or public roads disregarding the fact that there is no such classification of servitudes in Latvia.

Having assessed the historical situation and the possible factors affecting the decision making, the authors have come to conclusion that the shortcomings of the made decisions before coming into effect of the Rights in Rem part of the Civil Law can be justified by the insufficient legal framework. But the shortcomings allowed after that are to be assessed critically and lack of competence and inability to pass legally correct decisions of the officials systematically applying provisions of the Civil Law and the valid laws and regulations regarding the institute of servitude has to be admitted. Expert of the civil rights G. Višņakova is of the opinion that when establishing servitude one should strictly base on the Civil Law [16, 140]. Article 1130 of the Civil Law is the general normative document that reveals the essence of servitude, but the special laws and regulations only stipulate in which cases and how servitudes are established.

Today, when drawing up and selecting a problem solution, we have to avoid the former mistakes made by the lawmakers and enforcers, wherewith we offer analysis of the draft protocol resolution of the Cabinet of Ministers and the solution offered in the Conceptual Report providing establishment of an encumbrance of the property rights – public road – to solve the problems caused by servitudes established by administrative acts.

At the moment the Ministry of Environmental Protection and Regional Development has drawn up a protocol decision of the Cabinet of Ministers supporting the solutions provided in the Conceptual Report, and Clause 4 of the protocol resolution states "The Ministry of Justice shall draw up and the Minister of Justice until 31 June 2017 in accordance with the procedure provided in the laws and regulations shall submit to the Cabinet of Ministers for review the draft Regulation "Amendments to the Cabinet of Ministers Regulation No. 305 "Provisions for determination of cadastral Value" of 18 April 2006" and the draft Regulation "Amendments to the Cabinet of Ministers Regulation No. 61 "On Establishment and Maintenance of Encumbered Territory Information System and Classification of Encumbrances of Immovable Property Objects" of 4 February 2014", including a new encumbrance "public road", as well as amendments to the Law on Information System of Encumbered Territories stating that the Information system of encumbered territories shall include information on territories that are encumbered with public roads. [17]. It means that part of the problems will be addressed by introduction of a new legal norm "public road".

Although the servitudes of right of way established during the land reform when passing decisions regarding renewal of property rights or granting of land for use, or in property against payment were marked "public roads" this concept in the laws and regulations did not exist. Also today there is no legal definition of public road in the laws and regulations of Latvia. Nevertheless this concept is mentioned in mandatory regulations of several municipalities. For example, in the section "Applicable Terms" of the Regulations for Territory Use and Construction of the Town of Valdemārpils with its Rural Territory the public road is explained as a road that is in possession of natural persons or legal entities and that is used to access several properties [18]. An identical description is provided in the Territory Planning of Roja District (2011–2023), Regulations for Territory Use and Construction of Roja District 2011 [19, Article 60]. In Clause 5.2.1 of the said regulations the public road is divided together with national motor roads, municipal roads, streets and roads as a separate category of roads determining respective protective zone on basis of intensity of use of the public road. Clause 304 of the Regulations for Territory Use and Construction of Jēkabpils of 4 March 2010 states "New local streets (as well as squares, pedestrian streets, public access roads, engineering communication corridors and separate bike lanes) and their red lines are determined in detail planning separating them from territories of other type of use" [20]. Article 1 Paragraph 11 and Article 44 Part eight of the Law on Expropriation of the Public Person Property use the term "public street (road)".

This allows concluding that the concept "public road" does not have an unequivocal description in the laws and regulations of Latvia. It is more used in everyday speech to designate roads that are used by more than one person for walking, driving or access to several properties.

> But Sect. 1, Article 3 Paragraph 11 of the Directive 2014/47/EU of the European Parliament and of the Council of 3 April 2014 on the technical roadside inspection of the roadworthiness of commercial vehicles circulating in the Union and repealing Directive 2000/30/EC Text with EEA relevance explains: "public road" means a road that is of general public utility, such as a local, regional or national road, highway, expressway or motorway [22]. The said explanation provides that the term "public road" includes roads of various categories used by public.

In Latvia, according to Article 3 of the Law on Motor Roads [23], motor roads are divided by their importance: national motor roads (main, regional and local); municipal roads; commercial roads; access roads. Access roads and commercial roads usually are in private property and the owners in accordance with Article 5 Part four of the Law on Motor Roads have the right to decide on a prohibition to use the road, restrictions or termination thereof having coordinated that with the local municipality. But the national and municipal roads are for free access to the public. Therefore the authors do not understand, why a separate restriction "public road" should be established, when it will have the same function as the existing national and municipal roads – it will be a road that can be freely used by any person. In addition, the Conceptual Report states that this restriction shall include [...] 3) the rights and obligations of the municipality to register in the municipal street and road registry a road owned by another person as an engineering construction under possession and maintenance of the municipality; 4) the obligation of the municipality to carry out the necessary maintenance of the road if the road is registered in the municipal street and road registry and the NRECIS [24].

In this situation a new term is introduced the necessity for which is doubtful, but which will cause more problems. Establishing a new type of public access road or encumbrance "public road", it is possible that there will be problems with understanding and interpretation of the concepts and terms. Thus, when speaking about separate category of property encumbrance – "public road", people will understand it as the already existing designation "public road" for national or municipal roads. Article 2 Part two of the Law on Motor Roads states that "a motor road is a complex engineering construction outside city borders that is used for road traffic with fixed speed, load and size determined in the laws and regulations" [25]. But Part two states that "Complex of a motor road includes: bedding, road, artificial constructions (for example, bridges, overpasses, tunnels, trestle bridges, culverts, drainage systems, support walls), road engineering constructions (for example, bus stops and pavilions, speed lanes, parking lots, parking places, recreation areas, snow protection walls, greenery, bike lanes and pathways, road communication and lighting lines) and technical means for traffic organisation (for example, road signs, traffic lights, signal poles, protective barriers, vertical and horizontal marks) [25].

Having assessed the proposal provided in the Conceptual Report that the municipalities should be granted the rights and obligations to register in the municipal street and road registry a road owned by another person as an engineering construction under possession and maintenance of the municipality, we have to mention that this solution would lead to breach of the principle provided in Article 968 of the Civil Law "A building erected on land and firmly attached to it shall be recognised as part thereof" [26, 968]. We have to note that this principle is not only applicable to buildings, but also to engineering constructions. J. Rozenfelds writes "physical integrity of a property means the legal restrictions for horizontal division of property or property fragmentation that ensures physical integrity of a property under ground and in the air space above the property, or the so called *ad caelum*. [...] During drawing up of the Civil Law in the 1930ies integrity of a property was very important thus emphasising that the Civil Law follows the general codification trends [...]. The requirement for physical integrity of a property is related to the regulation that any buildings and other constructions that are inseparably connected with the land are a part of it (*superficies solo cedit*). This connection is expressed in the way that *ad caleum*, which determines the sovereign power of the owner over the air and underground space could not exist, if at the same time there would be any separate property rights to any objects inseparably connected to that land." [27]

The Article 2 of the Law on Motor Roads cited above states that a road is an engineering construction closely connected with the bedding. Therefore it is recognised as a part of land. Of course, there always can be exceptions to the laws and regulations, as proposed by the authors of the Conception. Support to this solution will cause formation of shared property relations, where the land is owned by one owner, but the buildings by another. However, in period when the legal system of Latvia undergoes modernisation related to attempts to give up shared property [for example, see J. Dinsberga, I. Tiesniece (2016)], accepting of this solution would be irrational and unacceptable.

In the conclusion of the present article the authors would like to go back to the issue mentioned when evaluating the Conceptual Report with the data provided in Table 1. Since the Conceptual Report and the proposals therein are based on analysis of situation of only one municipality, there is a risk that problems of other municipalities have not been surveyed. Also there is no information of the situations that are in majority or minority in Latvia in general. The Conceptual Report states that the Ministry of Environmental Protection and Regional Development carried out survey of only 11 [28] municipalities regarding the criteria for establishment of public roads. During the survey the municipalities provided their opinion on solving of this problem, but it was not based on analysis of the actual situation of the particular municipality, which in its turn would be based on precise data. When adopting a decision of national importance and selecting a particular problem solution we cannot base on assumptions and incomplete situation analysis. The authors are of the opinion that each municipality of Latvia should carry out evaluation of the actual situation in their administrative territory, analyse the obtained data (on basis of single guidelines for data analysis) and draw up solutions for the problem situations. Each municipality after collection and analysis of data would be able to give concrete and not provisional assessment of the possible solutions taking into account the possibilities and human resources of the municipality. But the Ministry of Environmental Protection and Regional Development would carry out gathering of all data submitted by the municipalities, as well as aggregation and analysis of the solutions. It would allow understanding of the situation in Latvia in general and identifying the priority issues.

It is possible that finding of one solution suitable for all situations will be difficult. The results of the current study already show that each situation is individual and requires individual solutions. Therefore we have to agree with the opinion of the authors of the Conceptual Report that: "Considering the different points of view of the municipalities regarding the criteria for establishment of public roads, the competence of the municipalities to plan their administrative territory and to choose the most suitable criteria for establishment of public roads should be retained" [28]. The possible criteria, of course, should be drawn up and accepted by all municipalities, but each municipality would be able to choose the most suitable one.

Having conducted the research, the author states that:

1) The purpose of the servitude of right of way is to benefit the dominant immovable property. However, now there are many servitudes of right of way which do not fulfil their purpose in Latvia. Moreover, there is no regulation how to eliminate such imperfections.
2) Stability and efficiency of the legal base in the field of ownership rights in Latvia is endangered because of both the problems caused by the enactments regulating the land reform and the influence of globalisation. Because the legislators are unable to adjust and adapt quickly the legal system to fast development of economic, social, technological and political processes and to elimination of the past mistakes.
3) Despite the fact that the problems related to the servitudes of right of way, established during the land reform, could be detected and even eliminated at the beginning of the land reform, the process of solving the problems on the state level began only at the end of the reform.
4) The Conceptual Report provides several solutions for the detected problem, however, the negative risks are mentioned as well, which, in author's opinion, endangers further implementation of the Conception.
5) Survey of the problems with servitudes of right of way, established by administrative acts, included in the Conceptual Report and selection of the possible solution is based on generalised data that do not reflect the situation in Latvia.
6) Term "public road" does not have an unequivocal description in the laws and regulations of Latvia and introduction of another term for the servitude of right of way established by administrative act cannot be allowed. It is doubtful that the direction provided in the Conceptual Report to introduce a new encumbrance of immovable property "public road" would give the desired effect.

3 Conclusions

1) While developing policy planning documentation and regulatory enactments for elimination of problems caused by the servitudes of right of way established during the land reform, the legislators should: (a) analyse whether the developed regulation will allow achieving the legislative aim mentioned in the legal provision; (b) refer to general legal principles. The author considers that the legislator's decision would be against justice and good faith if the servitude of right of way was determined as common use way only because it is faster and cheaper than alienation of immovable property for the needs of society; c) take into consideration the interests of the property owner, avoiding delegation to immovable property owners the responsibility for mistakes made by the state institutions in the past; d) when choosing the direction of the solution we have to take into account the principle of integrity of land and buildings (including engineering constructions) stipulated in Article 968 of the Civil Law.

2) The Directive 2014/47/EU of the European Parliament and of the Council on the technical roadside inspection of the roadworthiness of commercial vehicles circulating in the Union and repealing Directive 2000/30/EC Text with EEA relevance states that "public road" includes roads of various categories used by public. Establishing a new type of public access road or encumbrance "public road", it is possible that there will be problems with understanding and interpretation of these concepts and terms. Moreover, it would lead to breach of the principle of integrity of land and buildings provided in Article 968 of the Civil Law.

3) The current study of the situation in one municipality of Latvia in relation to servitudes established during the land reform only gives an insight into the problem, but does not reflect the actual situation in Latvia. Should the decision on solving of the problems related to servitudes of right of way established by administrative acts be based on these insufficient facts, the history might repeat and decisions that are unreasoned and inadequate for the local situation might be passed. Therefore the authors provide the following proposals:

- Ministry of Environmental Protection and Regional Development shall charge all municipalities of Latvia to carry out study of the current situation, gathering and analysis of data, and to submit the obtained results to the ministry within the provided term;
- Ministry of Environmental Protection and Regional Development shall carry out aggregation and analysis of the all data and solutions submitted by the municipalities; draw up several courses of action and unified criteria, from which the municipalities will chose the respective course of action for solution of the particular problem, draw up respective drafts of laws and regulations based on the selected courses of action;
- grant all municipalities the right to choose the most appropriate course of action for solution of their problem situations.

References

1. *The Convention for the Protection of Human Rights and Fundamental Freedoms*. (1950) Protocol No. 1: international document. Latvijas Vēstnesis No. 143/144 (858/859), 13.06.1997
2. *The Constitution of the Republic of Latvia*. (1922) Article 105: Law of the Republic of Latvia. Latvijas Vēstnesis No. 43, 01.07.1993
3. *The Civil Law* (1937.): Law of the Republic of Latvia. Valdības Vēstnesis No. 44, 24.02.1937
4. Grūtups, A. *Tiesu prakse un komentāri*. Riga, Mans Īpašums (1994)
5. Conceptual Report *On the Problems Occurring in Relation with the Servitudes of Right of Way Established during the Land Reform and Their Possible Solutions*. Available at: `http://www.varam.gov.lv/lat/lidzd/attistibas_planosanas_dokumentu_projekti/` [viewed on 13.03.2017]
6. Dinsberga, J. *Servitūti un ar to nodibināšanu saistītās problēmas*. LPA Raksti **10** (2003)
7. Dinsberga, J. *Ar ceļa servitūta nodibināšanu saistītā problemātika*. Available at: `http://www.rsu.lv/studentiem/382-juridiska-fakultate/6296-macibu-konference-par-tiesisko-problemu-aktualiem-jautajumiem` [viewed on 02.01.2017]
8. *Collection of Local Laws*. Available at: `file:///D:/Document/2006_elektroniski_spriedumi_gramatas_likumi/Vietejo%20civilli kumu%20kopojums.III.dala.1928.g..pdf` [viewed on 13.03.2017]
9. Auders, M. *Zemes reformas laikā ar administratīvo aktu nodibinātie ceļu servitūti*. Available at: `http://www.varam.gov.lv/in_site/tools/download.php?file=files/text/Seminari/tap//5_cela_servituti_pascaron_valdibam_281114.pdf` [viewed on 28.11.2016]

10. Conceptual Report *On the Problems Occurring in Relation with the Servitudes of Right of Way Established during the Land Reform and Their Possible Solutions*. Available at: `http://www.varam.gov.lv/lat/lidzd/attistibas_planosanas_dokumentu_projekti/` [viewed on 13.03.2017]
11. *Law on Administrative Territories and Populated Areas* (2008) Attachments No. 1 and No. 2: Law of the Republic of Latvia. Latvijas Vēstnesis No. 202, 30.12.2008
12. Conceptual Report *On the Problems Occurring in Relation with the Servitudes of Right of Way Established during the Land Reform and Their Possible Solutions*. Available at: `http://www.varam.gov.lv/lat/lidzd/attistibas_planosanas_dokumentu_projekti/` [viewed on 13.03.2017]
13. Decision of the Supreme Council of the Republic of Latvia *On Agrarian reform of the Republic of Latvia* (1990): Legislative document of the Republic of Latvia. Ziņotājs, 29, 19.07.1990
14. *Law on Land Use and Land Survey* (1991): Law of the Republic of Latvia. Ziņotājs No. 31/32, 15.08.1991. (Expired: 30.12.2009)
15. *Regulations of Rural Area Land* (1991): Resolution of the Council of Ministers of the Republic of Latvia. Ziņotājs No. 10, 05.03.1992
16. Grūtups, A. *Tiesu prakse un komentāri*. Riga, Mans Īpašums (1994)
17. Cabinet of Ministers Meeting Protocol, Conceptual Report *On the Problems Occurring in Relation with the Servitudes of Right of Way Established during the Land Reform and Their Possible Solutions* (2013): Latvijas Vēstnesis No. 216, 05.11.2013
18. *Regulations for Territory Use and Construction of the Town of Valdemārpils with its Rural Territory*. Available at: `www.valdemarpils.lv/file.php?id=17` [viewed on 26.11.2016]
19. *Territory Planning of Roja District (2011 – 2023):* Regulations for Territory Use and Construction of Roja (2011). Available at: `file:///D:/Downloads/Rojas%20AN.pdf` [viewed on 26.11.2016]
20. *Regulations for Territory Use and Construction of Jēkabpils* (2010). Available at: www.jekabpils.lv/sites/default/files/universalais/2013/01/1023.../apbuve2010.doc [viewed on 26.11.2016]
21. *Law on Expropriation of the Public Person Property* (2002): Law of the Republic of Latvia. Latvijas Vēstnesis No. 168 (2743), 19.11.2002
22. *Directive 2014/47/EU of the European Parliament and of the Council of 3 April 2014 on the technical roadside inspection of the roadworthiness of commercial vehicles circulating in the Union and repealing Directive 2000/30/EC (Text with EEA relevance)*: OJ L 127, 29.4.2014, p. 134–218
23. *Road Traffic Law* (1992): Law of the Republic of Latvia. Ziņotājs No. 13/14, 02.04.1992
24. Conceptual Report *On the Problems Occurring in Relation with the Servitudes of Right of Way Established during the Land Reform and Their Possible Solutions*. Available at: `http://www.varam.gov.lv/lat/lidzd/attistibas_planosanas_dokumentu_projekti/` [viewed on 13.03.2017]
25. *Road Traffic Law* (1992): Law of the Republic of Latvia. Ziņotājs No. 13/14, 02.04.1992
26. *The Civil Law* (1937.) Article 968: Law of the Republic of Latvia. Valdības Vēstnesis No. 44, 24.02.1937

27. Rozenfelds, J. *Īpašuma fiziskās vienotības (ad caelum) entropija Latvijas tiesībās*. Jurista Vārds. **51** (750) (2012)
28. Conceptual Report *On the Problems Occurring in Relation with the Servitudes of Right of Way Established during the Land Reform and Their Possible Solutions*. Available at: http://www.varam.gov.lv/lat/lidzd/attistibas_planosanas_dokumentu_projekti/ [viewed on 13.03.2017]
29. Dinsberga, J., Tiesniece. I. *Problems Related to the Abolition of Divided Real Estate Ownership (2016)*. Available at: https://www.degruyter.com/view/j/jec.2016.13.issue-2/jec-2016-0021/jec-2016-0021.xml [viewed on 12.03.2017]

SHS Web of Conferences **40**, 01012 (2018) https://doi.org/10.1051/shsconf/20184001012
Int. Conf. SOCIETY. HEALTH. WELFARE. 2016

State officials and illicit asset-grabbing: The Roman approach

A. Apsitis and J. Joksts

Rīga Stradiņš University, Riga, Latvia

Abstract. The article reflects author's findings regarding the regulation found in Roman legal sources, which is directed against corruptive activities of persons in public state positions, in particular in relation to unlawful seizure of assets belonging to citizens. Legal mechanisms are examined in relation to cases of force (vis-Latin) and fear (metus-Latin) application. The Code of Justinian (Codex Iustinianus) and The Digest (Digesta) contained regulation in relation to interpretation and application of The Julian Law on Extortion (Lex Iulia repetundarum, 59 B.C.) in cases of all types of extortion and bribery with the involvement of public office administering persons, including judges and arbitrators, are examined.

Key words: Roman law; Ancient Rome; asset-grabbing

1 Introduction

Possible corruption of persons in public state positions is a serious threat to sustainable development and efficient functioning of the state. This particularly holds true for corrupting activities of the officials and/or persons belonging to the judicial system in relation to unlawful seizure of assets belonging to the state citizens.

Particular hazards for sustainable development of the state are caused by unlawful seizure of material benefit creator–entrepreneur assets. Functioning of entrepreneurial environment in accordance with the principles of lawfulness is regarded as an essential prerequisite for sustainable development. State power, inter alia, must be able to ensure that valuables owned by an entrepreneur or those under any law-ful control are adequately protected against any threats and unlawful claims, as well as guarantee inviolability of an entrepreneur's life and health protecting him against any unlawful acts [1].

The key ingredient of success factors for sustainable entrepreneurship is motivation of entrepreneur that could be increased by various political measures aiming to rise initiative of people to run existing or start new business ventures in secure and sustainable way [2].

Namely, the basis of sustainable development of any country is formed by, inter alia, a stable, efficient and fair system of state administration and judiciary which counterbalances the interests of the whole society and its individual elements.

The origins of legal framework for such kind of system, just like of many other contemporary legal thought phenomena, go back to Roman law – as it is known, within the ancient Roman civilization a rather complex and very successful legal basis was set up, so the large and for that time highly developed empire could successfully function. Also, it is necessary to point out that legal principles created by Romans, in particular in areas of private or civil rights have proven to be so successful that they still are the basis of private law of the West and the whole globalized world [3].

Therefore, understanding of the mentioned legal concept is impossible without a thorough study of the primary sources of Roman law. The aim – to facilitate in-depth knowledge of contemporary legal norm makers, enforcers and implementers about the legal framework for stable, efficient and fair system of state administration and judiciary by studying information found in the primary sources of Roman law.

2 Materials and methods

Within the research there were performed the studies and analysis of the primary sources of Roman Private Law (The *Digest* (*Digesta*) (Krueger; Mommsen 1928) [4] and *The Code of Justinian (Codex Iustinianus)* (Krueger 1906) [5]), mostly by applying the inductive, deductive and comparative methods.

3 Results and discussion

The sources of Roman law recognise and define the rights of the Roman magistrates (elected officials) to duly use powers or force in compliance with the law and legal powers of their position (D 4.2.3.1).

However, if a magistrate of the Roman people or governor of a province (*populi Romani magistratus vel provinciae praeses* – Latin) committed a wrongful act, for instance, extorted money by threatening with death or whipping (*mortis aut verberum terrore pecuniam alicui extorserit* – Latin), the mechanisms provided for in the laws for the cases of force and fear were applicable against them (D 4.2.3.1).

Equally, if someone who, due to the pressure on the part of the administrative officers, i.e. through the intervention of the governor's attendants (the apparatus) (*apparitione praesidis interveniente* – Latin), without the judge's knowledge (*sine notione iudicis* – Latin), was compelled to pay (*coactus est dare* – Latin) a non-existing debt which he was coerced to undertake by force (*per vim* – Latin), he could claim the judge's (*iudex* – Latin) decision on restitution[a] of what was illegally extorted (*inciviliter extorta restitui* – Latin)

[a] What is done through fear can never be ratified or confirmed (*ratum habebit* – Latin) on the part of a state official (D 4.2.21.1, D 4.2.1). The victim, i.e. someone who was compelled to act against his will, could restore his earlier status through restitution (D 4.2.3 pr.). Respectively, full restoration (*restitutio* – Latin) of the initial situation based on the authority of the court that, depending on a specific case (see D 4.2.7.1, D 4.2.8 pr., D 4.2.9.2, D 4.2.9.3, D 4.2.9.4, D 4.2.9.7, D 4.2.9.8, D 4.2.10, D 4.2.11, D 4.2.14.6, D 4.2.14.8, D 4.2.22, D 4.2.23.1, D 4.2.23.3), could be manifested, for instance, as a return of a thing when said thing was given to someone under duress together with an obligation to reimburse the victim for deceit (*dolo* – Latin) (D 4.2.9.7, D 4.2.9.5). If the defendant did not apply for voluntary restoration (restitution) of the initial situation and intervention of court was necessary, the victim was authorised to claim a fourfold repayment of the losses incurred by him, i.e. four times more than what would be voluntarily recovered (restituted) (D 4.2.14.1, D 4.2.14.3, D~4.2.9.7, D 4.2.9.6). (After a year, the victim only had the right to claim repayment of the simple value (*simplum actionem* – Latin), not always, but only on cause being shown – the process was connected with investigation of the case (*sed causa cognita* – Latin) (D 4.2.14.1)). Fourfold payment included the restitutable value itself; therefore, the restitutable amount to which a penalty of threefold the amount was added was actually supposed to be paid (D 4.2.14.10, D 4.2.14.9). To calculate the fourfold payment, the restitutable value itself was taken into consideration together with all fruits (*cum fructibus et omni causa* – Latin) (D 4.2.14.7). Determining the amount of payment, only the loss actually incurred by the victim was valuated (D 4.2.21.2). It is specified that the fourfold payment

(D 4.2.23.3). Debts could only be enforced in compliance with the respective procedural order (D 4.2.13); therefore, non-procedural intervention on the part of state management attendants was regarded as a wrongful act: 'But if he satisfies a debt by [getting] a simple order (*Quod si debitis satisfecit simplici iussione* – Latin) rather than [through] a court investigation (*et non cognitione habita* – Latin), even though recovery under extraordinary proceedings has not been performed (*quamvis non extra ordinem exactionem fieri* – Latin), but [an order] has been civilly executed (*sed civiliter oportuit* – Latin); however, settling debts [in such a manner]… (*tamen quae solutioni debitarum…* – Latin) [their] recovery is unlawful (*revocare incivile est* – Latin)' (D 4.2.23.3).

Invalidity of transactions made as a result of thrusting someone into prison with the aim to extort something from him ('…to extort something (*…ut aliquid ei extorqueret* – Latin)') were pointed out (D 4.2.22), as well as the officials' duty to ensure ('…the governor of the province will restore the matter to its rightful position (*…res suae aequitati per praesidem provinciae restituitur* – Latin)') that the sold property of the person threatened with court investigation will be restituted (D 4.2.23.1).

Simultaneously, obviously with the purpose to reinforce the authority of the state power, it was actually acknowledged that unlike people not holding any office, the officials and administrative officers had greater possibilities to protect themselves from force and fear should an attempt to employ them be made: 'It is impossible that a person who alleged that he held rank (*qui claram dignitatem se habere praetendebat* – Latin)[b] was compelled in the city groundlessly to pay a non-existing debt (*Non est verisimile compulsum in urbe inique indebitum solvisse eum* – Latin). [Since] he could have referred to the public law[c] (*cum potuerit ius publicum invocare* – Latin) and turn to someone having authority (power) (*et adire aliquem potestate praeditum* – Latin) who could prohibit subjecting him to force (*qui utique vim eum pati prohibuisset* – Latin). But presumption of this kind can be rebutted [only with] the most manifest proof of violence (*sed huiusmodi praesumptioni debet apertissimas probationes violentiae opponere* – Latin)' (D 4.2.23 pr.).

It should be noted that the Roman law contained provisions which are rather directly against alleged corrupt activities from persons in public state positions. In this context *Lex Iulia repetundarum* (59 B.C.) – last and severest law of Roman republic regarding extortion (*repetundae* – Latin) should be mentioned. The law was adopted by the proposal – legislative initiative of Julius Caesar (*C. Iulius Caesar*, 100 B.C. – 44 B.C.) in meeting the obligations of consul. It was still in force in accordance with Roman law codification realized by the Emperor Justinian (*Flavius Petrus Sabbatius Iustinianus Augustus*, c. 482 AD – 565 AD) in the period from 529 A.D. to 534 A.D. (see D 48.11; C 9.27) and covered all types of bribery cases involving public office administering persons, including judges and arbitrators [6].

Responsibility was requested from the person who, guided by corrupt considerations, had done more or less (*magis aut minus* – Latin), than his official duties requires. (D 48.11.4)

---

should be regarded only as interest due to the litigant (*interest quadruplari solum* – Latin) (D 4.2.14.14) [1].

[b] Respectively, alleged with reference to his high office or administrative officer's position; to read more about administrative officer hierarchy – ***ordo dignitatum*** (Latin), please see C 12.1 tit.; C 12.8 tit.; C 1.52 tit.

[c] Respectively, to refer to ***Ius publicum*** (Latin), i.e. legal norms in relation to existence, organisation and functioning of a state – the norms "which cannot be changed by the contracts signed between private individuals"; please see D 2.14.38; D 50.17.45.1, see also: A. Berger,. *Encyclopedic Dictionary of Roman Law*. 532 (The American Philosophical Society, Philadelphia, 1953, 1991) [6].

In addition, charges of public crimes (*crimen publicum* – Latin) could have put not only to the recipient of the bribe, but also the giver. (*non modo adversus accipientem, sed etiam adversus dantem* – Latin). (C 9.27.6.1)

The Julian Law on Extortion (*Lex Iulia repetundarum* – Latin) was applied to the money received by someone who took the magistrates office (*magistratu* – Latin), was endowed with power (*potestate* – Latin), carried out administrative duties (*curatione* – Latin), was appointed legate (*legatione* – Latin) or executed any other official work or public profession. The law was also applicable to the associates (*cohorte* – Latin) of the mentioned above. (D 48.11.1 pr.)

For example, sources speak of a provincial administrator (the head of a military district) – Duke (*Dux* – Latin) who was caught in corruption, "who acted badly" (*ducem qui male egit* – Latin) – from "our residents of the province" (*provincialibus nostris* – Latin), "extorted or took away" (*rapuit aut sustulit* – Latin) (C 9.27.1), the superintendent of imperial buildings (*comite domorum* – Latin), who could have extorted money from his inferior subordinate officials – e.g. from procurator (*procurator* – Latin, official of the administration), from chief of subaltern officers in imperial female apartments/imperial garment factory (*praepositus gynaecei* – Latin, see C 11.8), from tabularius (*tabularius* – Latin, a subordinate official in the fiscal administration, clerk, tax accountant, archivist, see C 10.71), from collector of taxes (*susceptor* – Latin, see C 10.72; C 11.70), from the colon (*colonus* – Latin) or any other (C 9.27.5 pr.), city magistrates (*urbani magistratus* – Latin), who must refrain from any tarnishment of reputation (*ob omni sorde se abstineant* – Latin) and who, during one year (*in anno* – Latin), can't receive more gifts and bounty than in value of one hundred gold coins (*aureorum centum* – Latin) (D 48.11.6.2) etc.

According to the Julian Law on Extortion, nobody with corresponding powers was allowed to accept money for the purpose of recruiting or releasing soldiers from the service, and nobody was allowed to take money for giving opinion in Senate or the public council. (D 48.11.6.2)

Responsibility for the crime of extortion was requested (*crimine repetundarum postulantur* – Latin) from the person who, hoping to receive money, withdrew from public duty (office) execution. (D 48.11.9)

It was possible to hold persons in occupying positions responsible for corrupt practices committed by their subordinates – such as already mentioned Duke (*Dux* – Latin) was responsible for anything which the members of his household (*domesticus* – Latin), his soldiers (*manipularius* – Latin), and servants (*minister* – Latin) illegally received.(C 9.27.1)

Furthermore – subordinates also were materially responsible for the extortion of their superiors, unless they had reported such fact in accordance with the prescribed deadlines. (C 9.27.5.1)

According to the Julian Law on Extortion, claim rights (*actio* – Latin) against the heirs (*heredes* – Latin) of officer/official caught in corruption were given, but only within a year after (*intra annum dumtaxat* – Latin) the death of the accused (*a morte eius qui arguebatur* – Latin). (D 48.11.2)

The law foresaw exceptions to those, from whom receiving money was allowed: cousins (*sobrinis* – Latin), the nearest blood relatives (*propioreve gradu cognatis* – Latin) and wives (*uxore* – Latin). (D 48.11.1.1)

Also, sources indicated that the Julian Law on Extortion included exhaustive list in accordance with the persons responsible – those who were excluded from the law, were allowed to receive money without restriction, but those listed therein were not permitted to catch (*capere* – Latin) anything from anyone. (D 48.11.7.1)

From the officers/officials, at least in theory, impeccable reputation and excellent moral and ethical qualities were requested – "... let those men (*viros* – Latin) of the provincial management get power, who honour the brilliant, not ambitions or search for goods ([*viros*]...*qui honoris insignia non ambitione vel pretio* – Latin), but fair / honest life ... let them serve as evidence of elevation ... (*sed probatae vitae et ... testimonio promoveri* – Latin)" (C 9.27.6 pr.).

Similarly, parties to the legal proceedings had to behave with dignity and honor – "All representatives of the parties (*cognitores* – Latin) and the judges (*iudices* – Latin) from the [litigants] money (*pecuniis* – Latin) and property (*patrimoniis* – Latin) hold further [their] hands (*manus abstineant* – Latin) nor consider others quarrels as [a source of] their prey (*neque alienum iurgium putent suam praedam* – Latin)." (C 9.27.3)

Thus, according to the Julian Law on Extortion, it was due to hunt down the one who, being endowed with some sort of power (*qui, cum aliquam potestatem haberet* – Latin), received money for the judgment or decision making or not making. (D 48.11.3)

The law stated, that no one should have received anything for making of judge's judgment or arbitrator's decision, nor for changing of sentence, nor for not making judgment, nor for not changing of sentence, nor for not giving the order, nor for enclosure of human (*hominem* – Latin) in public prison, nor for binding with obligations, nor for ordering to put into chains, nor for ordering to free from chains, nor for conviction or justification of human (*hominem condemnandum absolvendumve* – Latin), nor for assessment of imposed amount, nor for sentencing or not sentencing death or money penalty. (D 48.11.7 pr.). Nobody with the corresponding powers was allowed to accept money for someone's prosecution or not prosecution (*accusandum vel non accusandum* – Latin). (D 48.11.6.2)

According to the Julian Law on Extortion, judge's assistants (*comites* – Latin) also could be persecuted. (D 48.11.5.)

Convicted, according to the Julian Law on Extortion, was forbidden (*prohibetur* – Latin) to be a judge (*iudex* – Latin), case initiator, someone's representative in court or testify publicly (*testimonium publice dicere* – Latin). (D 48.11.6.1)

Thus, witnesses involved in the proceedings were also responsible – "[In accordance] with the same law accountable (*Eadem lege tenentur* – Latin) are those, who receive money (*pecuniam acceperint* – Latin) for referring or not referring the testimony (*denuntiandum vel non denuntiandum testimonium* – Latin)." (D 48.11.6 pr.)

Similar to the situation with the officers/officials, in accordance with the Julian Law on Extortion, responsibility was required from the successors of the judges caught in corruption – "We let the judges know, that penalties imposed for their caused damage can be collected not only from themselves, but also their heirs (*Sciant iudices super admissis propriis aut a se aut ab heredibus suis poenam esse repetendam* – Latin)." (C 9.27.2)

The law foresaw reporting obligation of officials and citizens on noticed lawlessness in judges actions – if any occupying honorary position (*honoratorum* – Latin), decurion (*decurionum* – Latin), possessor of property (*possessorum* – Latin), and finally even colon (*colonorum* – Latin) or anyone occupying ordinary position (*ordinis* – Latin) was intimidated or shaken (*concussus* – Latin) from the judge, or if he became aware that the judgment of the court is sold, or of he knew that the penalty was dismissed/ revised for a fee (*poenam vel pretio remissam* – Latin), in immoral greed hoping for a reward, or if he later could confirm judge dishonesty at some time in past cases, he, either during administration or after administration (*vel administrante eo vel post administrationem* – Latin), had to notify about the crime submitting evidence. If the accusation was confirmed, the reporter was honoured and glorified (*et victoriam reportaturus et gloriam* – Latin). (C 9.27.4)

In addition to the rules, which were directed against the corrupt activities from the persons who occupied public state positions, the Julian Law on Extortion contained a rather strict requirements regarding public procurement and orders for the needs of state or municipality – it was determined that no enforceable public work (*opus publicum* – Latin) is not acceptable as finished, grain crops or food (*frumentum* – Latin) for public distribution is not considered delivered or procured, buildings are not considered renewed, before mentioned public projects are not completed, delivered, accepted according to the law. (D 48.11.7.2)

Internal aversion of position occupants to corruption as a phenomenon was attempted to cultivate with sacramental activities – mandatory assistance oath: "…so that those who are honoured with the post…giving the oath [could] declare, that for entrusted administration they have not given, or will not give later, themselves or through intermediaries (*sive per se sive per interpositam* – Latin), in violation of the laws or personal oath, or [by conclusion of] donation's, sale's title, or any other basis contract, and therefore, with the exception of salary and absolutely nothing [else] being in administration, also after the administration (*post depositum officium* – Latin) …will not be accepted." (C 9.27.6 pr.)

As it is known, fear from the anger of Supremacy can serve as a fairly strong incentive element – "…nobody in fear of God [anger] will not flout the oath, thus forgetting about his own soul salvation, preferring any profiting… (*...neminem divini timoris contemnendo iureiurando arbitramur immemorem, ut saluti propriae ullum commodum anteponat...* – Latin)". (C 9.27.6.1) [d]

It must be noted, that Roman legislators were perfectly aware of the importance and need of secular incentive measure, thus laying down a very effective sanctions mechanism against persons who were possibly involved in corrupt activity – "…however, to fear about the salvation of the soul danger [of punishment] needs to be added so, if anyone dared to behave recklessly against given sacrament/oath (*sacramenta neglegere* – Latin), four time penalty (*quadrupli poena* – Latin) to the one who is convicted must be achieved by all means." (C 9.27.6.1). Thus the offender was due quadruple punishment of the received or given bribe value.

It was directly stated on the general preventive aim of the penalty – with a certain type of intimidation measures to ensure compliance with the law and to discourage public office holder / person executing duties from committing criminal offence – "May the punishment (*poena* – Latin) for one be an example for a lot, we determine that the Duke, who acted badly (*ducem qui male egit* – Latin), must be arrested and sent to his subordinate province, where what his household members (*domesticus* – Latin), soldiers (*manipularius* – Latin) and servants (*minister* – Latin) have received, and he himself, has extorted or taken away (*rapuit aut sustulit* – Latin), is forced to pay in quadruple" (C 9.27.1)

In addition, according to the law on extortion (*ex lege repetundarum* – Latin), regarding those responsible acted in accordance with extraordinary proceedings (*extra ordinem* –

[d] It is necessary to mention, that during times of Justinian, Christianity took over the Roman empire, according to it's religiously ethical principles, as it's known, manifestations of bribery and corruption is condemned, especially according to decision of the court: "*nec accipias munera quae excaecant etiam prudentes et subvertunt verba iustorum*" – Latin [7] ("And thou shalt take no gift: for the gift blindeth the wise, and perverteth the words of the righteous." [8]/ "You shall take no bribe, for a bribe blinds those who have sight and perverts the words of the righteous." [9]) (Exodus 23:8); "*nec in alteram partem declinent non accipies personam nec munera quia munera excaecant oculos sapientium et mutant verba iustorum*" – Latin [7] ("Thou shalt not wrest judgment; thou shalt not respect persons, neither take a gift: for a gift doth blind the eyes of the wise, and pervert the words of the righteous." [8]/ "You shall not pervert justice. You shall not show partiality. You shall not take a bribe, for a bribe blinds the eyes of the wise and perverts the words of the righteous." [9]) (Deuteronomy 16:19); "*maledictus qui accipit munera ut percutiat animam sanguinis innocentis et dicet omnis populus amen*" – Latin [7] ("Cursed be he that taketh reward to slay an innocent person. And all the people shall say, Amen." [8]/ "Cursed is he who takes a bribe to kill an innocent person. All the people shall say, Amen." [9]) (Deuteronomy 27:25).

Latin) and often punished with either exile (*vel exilio puniuntur* – Latin), or even more severe (*vel etiam durius* – Latin), according to the assessment of the committed offense.

If the guilty had received money for killing people (*hominem necandum pecuniam acceperint* – Latin), or, even if they had not received money, however, in anger had killed either innocent or one who was not to be punished (*vel innocentem vel quem punire non debuerant* – Latin), they were "… Punished with death (*Capite plecti debent* – Latin) or at least sent to the island (*vel certe in insulam deportari* – Latin) as most of them have been punished (*ut plerique puniti sunt* – Latin)". (D 48.11.7.3)

As previously mentioned, if someone was convicted according to the Julian Law on Extortion, he was forbidden (*prohibetur* – Latin) to be a judge (*iudex* – Latin), the initiator of a case, representative in the court or to publicly testify (*testimonium publice dicere* – Latin). (D 48.11.6.1)

The Julian Law on Extortion foresaw specific consistency in the field of private law circulation – the law stipulated that the sale or lease contract (*venditiones locationes* – Latin), which, due to the pressure of those occupying positions, has been concluded for a higher or lower price than that which would have been fair, becomes invalid (*irritas facit* – Latin) and acquisition on the basis of prescription (*usucapionem* – Latin) is prevented, before things have not arrived back at it, from which they have departed, or to his heir. (D 48.11.8.1) Similarly: "What, contrary to the law on extortion, has been donated to proconsul or praetor, (*Quod contra legem repetundarum proconsuli vel praetori donatum est* – Latin) is not obtainable on the basis of prescription (*non poterit usu capi* – Latin)." (D 48.11.8.pr.)

Anti-corruption obligations provided in the law tied the office holders also after they had passed their office powers – ceased to hold office: "…So they… [could] declare… that … being in administration and after passing the office (*post depositum officium* – Latin) they will not accept absolutely nothing [other] than wages." (C 9.27.6 pr.)

Similarly worked reporting obligations of the officials on noticed lawlessness in actions of judges: "… [We demand that he] must report about the crime either during his administration or after the administration time (*vel administrante eo vel post administrationem* – Latin)." (C 9.27.4)

Also, the responsibility in relation to the Julian Law on Extortion rule violation could be required from officers/officials who had already left the office: "…When he, who accepted the money, has left [his] administrative duties (*administratione decesserit* – Latin)…"(C 9.27.5 pr.)

The Julian Law on Extortion contained certain prescription periods. Claims (*actio* – Latin) against the heirs (*heredes* – Latin) of the accused were exercisable only within a year (*intra annum dumtaxat* – Latin) after the death of the accused (*a morte eius qui arguebatur* – Latin). (D 48.11.2) Claim rights against the officer/official expired during one year after leaving the post (C 9.27.5 pr.) – if his former subordinates – colleagues had not reported the lawlessness of the former chief in a timely manner, they themselves became materially responsible: "If, however, from the moment when [he] has left the administrative powers, [claim lifting] allotted time has [already] passed, no voice in defence [can't] appear (*nulla vox advocationis emergat* – Latin), but those procurators (*procuratores* – Latin), chiefs (*praepositos* – Latin), colons (*colonos* – Latin), tabulariuses (*tabularius* – Latin), tax collectors (*susceptores* – Latin) are [themselves] responsible for the payment, which they want to challenge." (C 9.27.5.1)

4 Conclusions

In the sources of Roman law a significant role is played by the legal regulation intended to combat malice on the part of state officials, administrative officers and judges.

The sources recognise and define the rights of the Roman magistrates (elected officials) to duly use powers or force in compliance with the law and legal powers of their position (D 4.2.3.1). However, if a magistrate, administrative officer or any other official committed a wrongful act, for instance, extorted money (D 4.2.3.1), compelled somebody to pay a non-existing debt (D 4.2.23.3) or extort something else from him (D 4.2.22), the mechanisms provided for in the laws for the cases of force and fear were applicable against them (D 4.2.3.1).

The victim, i.e. someone who was compelled to act against his will, could restore his earlier status through restitution (D 4.2.3 pr.; D 4.2.23.1 etc.).

Simultaneously, it was actually acknowledged that unlike people not holding any office, the officials and administrative officers had greater possibilities to protect themselves from force and fear should an attempt to employ them be made (D 4.2.23 pr.).

The law of Ancient Rome contained provisions which are rather directly against alleged corrupt activities from persons in public state positions. In this context law relating to extortion (*repetundae* – Latin) should be mentioned – The Julian Law on Extortion (*Lex Iulia repetundarum*, 59 B.C.). The law was adopted by the legislative initiative of Julius Caesar in meeting the obligations of consul. It was still in force in accordance with Roman law codification realized by the Emperor Justinian (see D 48.11; C 9.27) and covered all types of bribery cases involving public office administering persons, including judges and arbitrators.

Responsibility was requested from those who guided by corrupt considerations had done more or less (*magis aut minus* – Latin) than their official duties required (D 48.11.4.), it was possible to prosecute not only the recipient of the bribe, but also the giver of the bribe (C 9.27.6.1). Persons in occupying positions also could be held responsible for their subordinate corruptive activities (C 9.27.1), subordinates were responsible materially for their superiors once extorted, unless such fact was reported (C 9.27.5.1). Claim rights were given against caught corrupt officer's/official's heirs. (D 48.11.2) The law foresaw exceptions to those from whom it was allowed to receive money (D 48.11.1.1), also sources indicated, that the Julian Law on Extortion according to it includes exhaustive list of the persons responsible (D 48.11.7.1). Excellent reputation and excellent moral and ethical qualities were required from officers/officials. (C 9.27.6 pr.)

Similarly, the trial participants – judges, representatives of the parties, etc. had to behave with dignity and honour (C 9.27.3). Thus, under the Julian Law on Extortion, those, who received illegal payments for performed or not performed activities of judicial proceedings, were persecuted (48.11.3; D 48.11.7 pr., D 48.11.6.2). Judge's assistants could be persecuted for corruptive activities (D 48.11.5.), witnesses (D 48.11.6 pr.), involved in the proceedings were responsible and responsibility was demanded also from the heirs of the judges caught in corruption. (C 9.27.2) The law foresaw official and citizen reporting obligation on noticed lawlessness in judge's activities. (C 9.27.4)

The Julian Law on Extortion contained rather strict requirements regarding public procurement and orders for the needs of state or municipality. (D 48.11.7.2)

Internal aversion of position occupants to corruption as a phenomenon was attempted to cultivate with sacramental activities – mandatory assistance oath, thus emphasizing the motivational impact of fear from God's wrath. (C 9.27.6 pr.; C 9.27.6.1)

At the same time Roman legislators were perfectly aware of the importance and need of secular incentive measure, thus laying down a very effective sanctions mechanism against persons who were possibly involved in corrupt activity – the offender was due quadruple punishment of the received or given bribe value (C 9.27.6.1; C 9.27.1), often they were punished with either exile, or even more severe, according to the assessment of the committed offense – especially if offenders had received money for killing people, then death penalty was possible (D 48.11.7.3). The general preventive aim of the penalty was not forgotten – to ensure law enforcement and discourage public office / executing the obligations of persons from a criminal offense (C 9.27.1). According to the Julian Law on Extortion convict was forbidden to be a judge, representative in the court or to publicly testify. (D 48.11.6.1)

The Julian Law on Extortion foresaw specific consistency in the field of private law circulation – the law stipulated that the sale or lease contract, which has been concluded due to the pressure of those occupying positions, becomes invalid (D 48.11.8.1) and acquisition on the basis of prescription is prevented (D 48.11.8.1; D 48.11.8.pr.).

Anti-corruption obligations provided in the law tied the office holders also after they had passed their office powers – ceased to hold office. (C 9.27.6 pr.; C 9.27.4; C 9.27.5 pr.)

The Julian Law on Extortion contained certain prescription periods (they expired within 1 year). (D 48.11.2; C 9.27.5 pr.; C 9.27.5.1).

References

1. A. Apsītis, J. Joksts, J. Antanavičienė, J. of Sec. and Sust. Iss., *Threats to sustainable development: asset grabbing phenomenon and the legal concept of Force and Fear in Roman Law.* **6**(2), 289, 290, 293, 294 (2016) http://dx.doi.org/10.9770/jssi.2016.6.2 (8)
2. J. Raudeliūnienė, M. Tvaronavičienė, I. Dzemyda, (2014) J. of Sec. and Sust. Iss., *Towards economic security and sustainability: key success factors of sustainable entrepreneurship in conditions of global economy.* **3**(4), 77 http://dx.doi.org/10.9770/jssi.2014.3.4 (7)
3. A. Apsītis, Socrates RSU JF el. jur. zin. r. žur., *Romiešu sabiedrības (societa) līgums un romiešu tiesiskie principi mūsdienu Latvijas tiesībās* [*Roman Partnership Agreement (Societa) and Roman Legal Principles in the Law of Modern Latvia*] **3**(3), 92 (2015) http://www.rsu.lv/images/stories/dokumenti/publikacijas/socrates_3_2015.pdf
4. P. Krueger, T. Mommsen, eds., *Corpus Iuris Civilis. Digesta.* [*Body of Civil Law. Digest.*] 60, 80–82, 857, 921 (Weidmannos, Berlin, 1928) http://fama2.us.es/fde/ocr/2006/corpusIurisCivilisT1InstitutionesDigesta.pdf, pp. 143, 163–165, 940, 1004
5. P. Krueger, ed., *Corpus Iuris Civilis. Codex Iustinianus.* [*Body of Civil Law. The Code of Justinian.*] 89, 384–385, 453–454, 456 (Weidmannos, Berlin, 1906) http://fama2.us.es/fde/ocr/2007/corpusIurisCivilisCodexIustinianus.pdf, pp. 119, 414–415, 483–484, 486
6. A. Berger, *Encyclopedic Dictionary of Roman Law.* 532, 555 (The American Philosophical Society, Philadelphia, 1953, 1991)
7. *Biblia Sacra Vulgata.* http://vulgate.org/ot/exodus_23.htm, http://vulgate.org/ot/deuteronomy_16.htm, http://vulgate.org/ot/deuteronomy_27.htm, http://vulgate.org/
8. *The King James Bible.* http://www.gutenberg.org/files/10/10-h/10-h.htm#The_Second_Book_of_Moses_Called_Exodus, http://www.gutenberg.org/files/10/10-h/10-h.htm#The_Fifth_Book_of_Moses_Called_Deuteronomy, http://www.gutenberg.org/files/10/10-h/10-h.htm#The_Fifth_Book_of_Moses_Called_Deuteronomy

(The Project Gutenberg EBook, 2011)
http://www.gutenberg.org/files/10/10-h/10-h.htm

9. *World English Bible with Deuterocanon*.
http://ebible.org/web/EXO23.htm,
http://ebible.org/web/DEU16.htm,
http://ebible.org/web/DEU27.htm, http://ebible.org/web/

Interest group impact on E-health implementation in Latvia

I. Bikava and I. Kreituse

Rīga Stradiņš University, Riga, Latvia

Abstract. Democratic state governance should provide the society with mechanisms to impact policy development and implementation. Society members are presented by interest groups that can use internal and external ways to have impact on the government and reform implementation processes.
This article represents research that was made on E-health implementation in Latvia – identifying main obstacles and reasons of delayed implementation, and analyses active interest groups and their impact on the implementation process.
The research results showed that the main reason of delayed implementation was poor governance from the responsible authority and lack of interest group involvement in the reform development and implementation process.
Key words: E-health; interest groups; healthcare reforms; change implementation.

1 Introduction

Since 1991 different reforms in healthcare system in Latvia were planned and carried out with the aim to change previously existing system and to create a new system – modern, effective and accessible to citizens.

Democratic state governance requires society participation in the decision making and reform implementation process. Citizen activity and engagement in political processes are promoted to improve development of a civil society. Since 2005 society involvement is favoured not only by changes in the jurisdiction, but also by providing different sources of funding available to NGO [1, 2].

According to Interest Group theory – organised groups or individuals ensure mediator role between the society and state and provide the opportunity to participate in policy making and implementation [3]. According to Lobbying theory there are two ways how interest groups can influence processes: internally – in cooperation with the authority to share their knowledge and best practice to gain better results for the whole society. Or externally – through demonstrations, protests or raising alarm and pressure in the society, block or delay implementation of reforms [4, 5].

Interest groups acting in the healthcare sector are as follows: the dominating are different organizations of doctors and medical specialists; the second interest group endowed with most of resources – pharmacy and medical equipment manufacturers; the third is healthcare institutions – hospitals, ambulatory care centres, General Practitioners (GP); the fourth – patient associations and the last – state regulatory organizations [6, 7].

The aim of the work was to conduct research on E-health implementation process in Latvia from 2003 till November, 2016 – to identify reasons why the reform was not implemented in the originally planned time and what impact interest groups left on the implementation process. To identify the impact of interest groups, analysis of articles from different interest groups published by news agency LETA was made.

2 Overview of E-health implementation in Latvia

The development of E-health as a tool that can improve healthcare system, providing effectiveness and efficiency, better accountability and necessary data for decision making processes, started in 2003 when the Ministry of Health established a group of researchers to develop guidelines of E-health implementation in Latvia [8]. On June 14, 2004 the developed guidelines were presented to the society for public consultation [9].

The main implementation risks were the result of the infrastructure at that period – only 20–30% of GP had a computer in their practices, they had limited access to the Internet, and there were weak skills of working with information technologies [10, 11]. Obstacles that can evoke public resistance to the system implementation were mentioned: accountability achieved by informatics systems can cause doctors' dissatisfaction because everyone could see how many patients have been cured and how much money received. And secondly, patients can be worried about the confidentiality of their personal information [12].

Publicly expressed risks and doubts about E-health have not stopped the process and guidelines were accepted by the Cabinet of Ministers on August 17, 2005. To provide further development of the system during the next four years monetary funding was granted in amount of 400 000 LVL. In accordance with the guidelines E-health project was a long term project, planned to be developed from 2006–2016 and with the total amount of funding approximately 25 million LVL. Concept consisted of a common EHR system that included electronic patient data, information about clinical data treatment – orders and results of visits to specialists, laboratory and radiology results as well as designated treatment. The other modules were: E-prescription, E-administration, E-booking, Tele-medicine, E-services and different types of data exchange with authorities [13, 14].

The next step in the implementation process was to develop a detailed plan of implementation activities with the deadline June 1, 2006 [13]. In accordance with the state budget for year 2006, 600 000 LVL were assigned for E-health activities [15].

After the guidelines were accepted new obstacles were identified in the society – low income of the medical staff could lead to the situation that efficiency benefits declared in E-health concept would not be achieved, because for institutions it would be cheaper to hire some more employees working with paper-based process than to involve modern technologies and upgrade employees' information technology skills. Doubts that introduction of technical support can lead to negative impact in the treatment process, i.e., reducing direct contact between the patient and the specialist [16].

The implementation plan was not developed by the original deadline and was extended to March 1, 2007 [17]. The plan for the period 2008–2010 was developed and accepted by the Cabinet of Ministers on October 24, 2007. To provide planned activities, the required amount of funding was 29.6 million LVL. The state budgetary funding was granted in the year 2007 in the amount of 681 800 LVL. The other funding sources were expected to be obtained from European Union Funds and private investors [18].

At the end of 2007 the Ministry of Health carried out a project to improve skills and knowledge of computer and the Internet use for more than 100 medical specialists [19].

In accordance with the guidelines there were tasks that had to be achieved by the year 2008: 60% of healthcare institutions use EHR according to common data standard (but there was even no such standard approved till March 11, 2014) and implemented central EHR system (in reality there was even no responsible authority till March 11, 2014 and there was lack of equipment and software till August, 2009) [13, 20, 21].

The main causes for the implementation delay in the period 2004–2008 as stated in the State Audit Office review are delayed development of implementation activities, large scale project with unclear scope, no funding provided for the project development, no regulatory framework. The main reasons for that were weak governance by the Ministry of Health, no compliance with the stated control plan, no risk mitigation plan, inefficient use of resources [22].

In 2008 infrastructure had changed and according to the European Commission research: internet infrastructure conforms to an average European level, 88% of GP practices use personal computer and 85% of GP have access to the Internet [23].

Year 2009 was a breakout year in E-health implementation because the project funding from the European Union was obtained, the purchase tender for equipment and technology necessary for data processing and storage for developing E-health platform was announced [21, 24]. The hindering factor in 2009 was the change of the institution responsible for E-health implementation. Due to institutional reform a new institution was established – Health Economy Centre and responsibility of E-health development was transferred to the newly established institution [25].

In 2010 all previously accepted pilot projects were developed and available to residents. Patients' access to data of healthcare services paid by the state uncovered problems of accountability in GP practices – there were cases when services were declared by GP while patients reported that they did not use or get the service [26]. Transparency and accountability provided by this service showed the problem that was declared as a risk of E-health implementation in early 2004, i.e., that doctors could be against the system because everyone could see what services have been promoted and paid for.

In 2010 Health Economy Centre announced that a significant obstacle in the further system development was the lack of normative regulation on data and process. The work on normative regulation was started in the Ministry of Health, but due to institutional changes further development of regulations was put on hold, so Health Economy Centre had to start work on regulation from the very beginning [27].

In 2010 two tenders were announced: for Electronic Patient Data system development which was successful, but the tender result on E-prescription was contradicted and successful result of tenders was achieved only in November, 2011. Failures in tenders led to hinder the whole development process and originally planned tests of designed system were changed several times [28–31].

In autumn 2011, the responsible authority was changed again – two previously existing institutions Health Account Centre and Health Economic Centre were merged in the National Health Service [32].

The regulatory framework draft on processing Health data was developed and in July, 2012 was opened for public discussions [33]. New technical risks arose – there were different IT systems used in healthcare institutions and to promote the system integration investments had to be made in commercial systems. This led to actions against E-health implementation because of the expenses to manage systems already used [34–36].

The regulatory framework on processing Health data was confirmed only on March 11, 2014 [20]. Only from this date there was a clear vision on responsibilities, data exchange standard, requirements for secure data process and storage.

In 2015 the debates on E-health became more active – doubts rose in the society about successful implementation. The risk of losing ERAF funding arose due to the approaching project deadline and the lack of system implementation [37]. The State Audit Office report awoke dissatisfaction in the government and society about the project – announcing that during nine years 14.5 million EUR were spent, but none of the services was available [22, 38, 39] Third round of E-health project development process started by ensuring ERAF funding in amount of 36 million EUR [40].

Six month before compulsory use of E-health system, Latvia Hospital Association asked the Ministry of Health to change the date of compulsory use and give six-month period, because of technical difficulties of hospital IT systems to provide data exchange with E-health [41]. At the same time GP raised pressure in media about doubts on system quality, patient data security and changes in their daily practice. In response to that compulsory date of E-health use was changed to December 1, 2016 [42, 43].

Finally in 2016 E-health system was developed and there were no risks of losing ERAF funding [44]. In the summer E-health portal was available to residents [45]. According to National Health Centre information by October, 2016 there were more than 3000 specialists that had attended special programmes and had been educated in work with E-health and the training process will be continued [46]. On October 26, 2016 the Minister of Health announced that E-service "Data on incapacity to work" is working properly, but E-prescription need to be improved and could be used voluntarily [47]. One of the biggest pharmacy networks "BENU" announced that their systems were ready to use E-prescriptions [48]. On October 28, 2016 the E-prescription project was successfully implemented: E-prescription was drawn up in one of the leading hospitals – PSKUS and the medicine was received in BENU pharmacy based on E-prescription [49].

E-health was designed and implemented before the compulsory use date – December 1, 2016. But due to earlier protests of GP associations and a call to sabotage contract signing on E-health use, there were doubts about the possibility of patients to be able to register their incapacity to work through their GP, because by November 1, 2016 only 260 agreements were signed [50–52]. As a result, the Ministry of Health initiated the change of the date of E-health system compulsory use [53].

According to data of the Ministry of Health by November 29, 2016, 45 records on incapability to work and 21 E-prescription have been discharged. The Cabinet of Ministers accepted the initiative to shift compulsory use of E-health system to July 1, 2017. But on the mentioned date, not only those two services: E-prescription and Data on incapability to work have to be used, but also a plenty of other services – referrals to get ambulatory or stationary service, responsibility to register in E-health system results and reviews of consultation and treatment process as well as other information of treatment process and documentation [54, 55].

3 Analysis of interest group activities during E-health implementation

Implementation of E-health system has an impact on patients, medical staff, healthcare institutions, pharmacy and health insurance companies as well as institutions that govern healthcare policy. Interest groups in Latvia which represent those affected groups are:

1. Patients – patient organisations and the Patients' Ombudsman Office.
2. Medical staff – professional organizations of doctors which are widely known – Association of Doctors, two Associations of General Practitioners, The Union of Health and Social Care Employees of Latvia.
3. Healthcare institutions – Latvia Hospital Association.
4. Pharmacy – Pharmacy Owners' Association and Pharmacists' Society of Latvia.
5. Health insurance companies are represented by Latvian Insurers Association.

According to publicly accessible information and the State Audit Office review, during the policy formation process interest groups mentioned above were not engaged in the process. Only interests of one of the groups was represented in this process – employees of the Ministry of Health and institutions which are subordinated to the Ministry of Health, i.e., healthcare policy developers, executers, managers and administrators [22].

During the long implementation process which started in 2004, the first opinions publicly expressed by interest groups appeared only in 2010. At that time there were debates in the government about the progress of E-health implementation and doubts were expressed about risks that the system would not be implemented [57].

For more than five years during which the work in E-health implementation was in progress, there was no discussion in the society and interest groups were inactive, that can be explained by the lack of information on the reform as well as by the fact that interest groups were not engaged in the processes.

The interest group that shows strong support to E-health implementation is pharmacy represented by Pharmacy Owners Association. They understand the benefits of the system – the system will ensure a possibility to control accurate and proper use of medicine and using E-prescription will decrease bureaucratic burden on pharmacists' daily work. The system implementation will also decrease the number of cases when there are mistakes in prescriptions and pharmacists have to contact the specialist to find out the reason for the mistake [58].

Pharmacies were ready for E-health implementation from January 1, 2016. Delay of compulsory use of the system led to disappointment because their systems were adapted to use E-health with the new normative regulation. As E-health compulsory use date was changed, but normative regulation was not changed, pharmacies had to change their system again [59]. The other interest groups representing pharmacy – pharmaceutical industry and pharmacists had not been active in E-health discussions.

The other group that supports E-health implementation is private insurance companies. The problem that insurance companies have and which implementation of E-health system could resolve is that different systems that healthcare service providers use are not connected, so a patient can go and have analysis and treatment in one healthcare institution today and the same service or treatment in another institution next day. As the result, this can lead to unnecessary treatment episodes that are paid for twice – by the state and the insurer, thus being ineffective [60].

The interest group that had a neutral attitude to E-health implementation was healthcare service providers – hospitals and health care centres. Their only doubts were – how the systems already used in their institutions could maintain requirements to provide data exchange with centralised E-health system [61]. Healthcare institutions had a neutral attitude, but in comparison to pharmacy, they were not ready to implement E-health on the planned date – January 1, 2016.

Interest groups that showed the discountenance to E-health implementation were representatives of the medical staff. The first criticism came from the Chairman of the Board of the Association of Doctors who implied that development of E-health is for pharmacy and software developer interests, but not in patients' interests [62].

The other institution protecting the rights of the medical staff – The Union of Health and Social Care Employees of Latvia in August, 2012 asked The Minister of Health to stem the implementation of E-health system since medical data of patients that will be collected in the centralised system poses a risk to the protection of those data [63]. When GP called to sabotage the system, the union expressed their support, indicating that implementation of unfamiliar system in a compulsory mode will threaten working environment, and that the use of E-health has not only benefits, but has also side effects like burnout syndrome to medical staff, less time dedicated to patient and hindering of all working process [64].

The most resistant to E-health implementation are interest groups representing GP. They mention different obstacles according to which the system should not be implemented or should be used voluntarily, not compulsory:

- For practising GP who has not grown up in E-environment, data input in electronic ambulatory patient card during the consultation process is cumbersome and time-consuming. So they will still use paper based process and somebody will input those data electronically in the system afterwards which leads to double work and is not effective [34].
- Practices of GP are not technically ready to use information technologies as there are still practices without a computer, especially in rural areas – there are some doctors who use library computers to input data needed for the National Health Service. There is also uncertainty of computer using skills, especially in the group of elderly GP – about 10% of GP are of the retirement age and the pressure to use E-health can lead to the situation that doctors decide to leave their practice and retire [43, 65].
- There are doubts about patient data security. GP are confused about data confidentiality that has to be kept between the patient and the doctor, but the ability to get data from E-health will be promoted for other institutions as well, for example, Health Inspectorate [42, 50].
- GP are not familiar with developed E-health system, so they are concerned about how the system will work and how it will impact their daily practice and relations with their patients [51].

Associations of GP not only expressed their doubts publicly, but also called their members to sabotage the implementation of the system by not signing the contracts of the system use [42, 51].

After the interest group expressed its opinion, doubts about threats to patient data security, the hindering of the whole working process in practices and threat that about 10% of GP could leave their practices in the existing conditions, where there is already shortage and unmet needs for GP in rural areas – those activities raised an uproar in the media and pressure in the society. Their activities led to change the date of compulsory use of E-health system twice – first time, when the Minister of Health Guntis Belēvičs initiated date change from January 1, 2016 to December 1, 2016, and second time – when the Minister of Health Anda Čakša initiated date change to July 1, 2017.

4 Conclusions

Results of research on E-health implementation process in Latvia showed that during the period 2005–2015 the main reasons for inability to implement the system was weak project governance and that only from 2015 there had been an impact of interest groups on the implementation data delay.

The main conclusion which the State Audit Office announced in their review was that the Ministry of Health had not worked efficiently enough on E-health implementation process supervision and control [22]. Also the current Minister of Health publicly announced that in the E-health implementation process there were a lot of omissions that are still observed today and there have been a lot of inappropriate decisions made [56].

Representatives of impacted interest groups were not involved in the policy formation process – neither in developing guidelines nor in the implementation plan or acceptance test process.

Interest groups whose members will be affected by the implementation of E-health system have been active only in the last phase of the implementation. Patients and pharmacy industry did not take part in the process at all.

Interest groups that showed strong support to E-health implementation were pharmacy owners and private insurance companies. That can be explained by two reasons: first – in most pharmacies and insurance companies some kinds of information technology systems are used, so the pharmacists and insurance companies are familiar with the use of personal computer in their daily practice. And the second – pharmacy owners and insurance companies know about the benefits they will gain from E-health and those benefits will directly improve and facilitate their work.

The neutral attitude of healthcare providers can be explained by the same reasons as in previous cases – they are familiar with information technology use in their institutions and know about benefits in accountability and transparency the systems can provide. For the healthcare institutions there are no clear benefits they will gain by implementing E-health. This could be the reason why they didn't hurry to invest resources in the system change to improve compatibility with E-health.

Interest groups that discounted and even sabotaged the implementation process were organizations which represent the interests of doctors. According to the previously mentioned counteraction purposes, the main reasons are – no information about data processing mechanisms and no information about benefits the doctors will gain from the implementation of the system. Another reason is the lack of education and training, using computer and information technologies, unfamiliarity with the new system that has to be used.

The counteraction could have been less if interest groups would have been involved in the developing process from the early beginning. As showed the experience of Estonia E-health implementation – from the point when GPs were involved in the development process and there were information and training campaigns carried out, the resistance to the new system noticeably decreased [66].

The research showed that interest groups involved in healthcare in Latvia are active – they use lobbying strategies and have significant impact on healthcare development and change implementation.

Although the presence of interest groups in political activities conforms to a democratic governance, because of the weak governance of E-health system by the Ministry of Health and lack of involvement of the main interest group in the policy forming, development and implementation planning process, the result was that interest groups did not act internally – they did not cooperate with the government institutions to create policy that meets the needs of the society, but their activities were only external – they acted just like disincentive to the policy implementation.

Result of the absence of a constructive cooperation between authority and interest groups acting in healthcare is that the implemented changes in the healthcare system do not conform to a long term development policy, but are chaotically made according to a lobbied demand without assessment how the change will impact the whole system and the society.

References

1. Special Assignments Minister for Social Integration Affairs of Latvia. *Policy Guidelines for Strengthening Civil Society 2004–2015*
2. Society Integration Foundation. *Grant scheme "Strengthening civil society and promotion of the integration of society" source of Funding "EEA Financial Mechanism 2004-2009"*

3. B.G. Peters. *Institutional Theory in Political Science. The "New Institutionalism.*, Contiinum Londond and New York (2000)
4. G. Jordan and W.A. Maloney. *Democracy and Interest Groups. Enhancing Participation? Palgrave Macmillan* (2007)
5. H. Kluver. *Lobbying in the European Union. Interest Groups, Lobbying Coalitions, and Policy Change.* Oxford University Press (2013)
6. Y. Yishai. *Participatory Governance in Public Health: Choice, but no Voice.* Oxford Handbook of Governance, Oxford University Press (2012)
7. R.M. Kaplan and Y.M. Badad. *Balancing influence between actors in healthcare decision making.* BMC Health Services Research. DOI: 10.1186/1472-6963-11-85 (2011)
8. Veselības ministrija. *Rīkojums Nr.302 "Par darba grupas izveidi e-veselības attīstības stratēģijas izstrādāšanai" (03.11.2003)*
9. A. Ādamsone. *Sabiedrību aicina izteikt viedokli par e-veselības pamatnostādņu projektu,* LETA (14.06.2004)
10. A. Ādamsone. *Ģimenes ārsti: Tikai neliela daļai ārstu prakšu nodrošinātas ar datoriem,* LETA (02.02.2005)
11. A. Ādamsone. *Moderno IT ieviešanu veselības jomā kavē vairāki faktori,* LETA (07.04.2005)
12. L. KĻaviņš. *Microsoft Latvia pārstāvis prognozē grūtības veselības aprūpes sistēmas datorizācijā,* LETA (10.03.2005)
13. Ministru kabinets. *Rīkojums Nr.560 "Par pamatnostādnēm "e-veselība"Latvijā"* (17.08.2005)
14. Veselības ministrija. *Pamatnostādnes "e-veselība Latvijā" (informatīvā daļa) (17.08.2005)*
15. A. Ādamsone. *"E-veselības" plānu īstenošanai nākamgad atvēlēti 600 000 latu,* LETA (31.12.2005)
16. A. Ādamsone. Informācijas tehnoloģiju plašāku izmantošanu veselības aprūpē var kavēt vairāki šķēršļi, LETA (21.08.2005)
17. Ministru kabinets. *Rīkojums Nr.430 "Grozījums Ministru kabineta 2005.gada 17.augusta rīkojumā Nr.560 "Par pamatnostādnēm "e-Veselības Latvijā"* (09.06.2006)
18. Ministru kabinets. *Rīkojums nr. 660 "Par pamatnostādņu "e-Veselība Latvijā"īstenošanas plānu 2008.-2010.gadam (24.10.2007)*
19. S. Sproģe. *Veselības aprūpes speciālistus reģionos apmācīs darbam ar datoru.,* LETA (06.12.2007)
20. Ministru Kabinets. *Noteikumi Nr.134 "Noteikumi par vienoto veselības nozares elektronisko informācijas sistēmu"* (11.03.2014)
21. I. Mežiniece. *E-veselības projekta ieviešanas iekārtām tērēs Ls 58 122.,* LETA (16.08.2009)
22. Latvijas Republikas Valsts Kontrole. *Revīzijas ziņojums: Vai projekts "E-veselība Latvijā" ir solis pareizajā virzienā. Lietderības revīzija "Informācijas sistēmas veselības aprūpē",* Rīga (2015)
23. L. Grundule. *Pētījums: Latvijā elektroniskie pakalpojumi veselības aprūpes jomā tiek izmantoti ierobežotā apjomā.* LETA (25.04.2008)
24. I. Mežiniece. *VM plāno izstrādāt pilotprojektus vairākiem e-veselības pakalpojumiem,* LETA (14.01.2009)
25. Ministru kabinets. *Noteikumi Nr.1110 Veselības ekonomikas centra nolikums* (29.09.2009)
26. I. Mežiniece. *Gavars ar atbildīgajām institūcijām pārrunās ar e-veselību saistītus jautājumus,* LETA (15.09.2010)

27. K. Pots. *Ieviest "e-veselību"traucē normatīvā regulējuma neesamība"*, LETA (18.06.2010)
28. K. Pots. *Veselības ekonomikas centrs pirks konsultācijas par 107 000 latiem; "e-receptes" izstrādes konkurss pārsūdzēts.* LETA (16.11.2010)
29. K. Pots. *VEC "e-veselības" izstrādei slēdz līgumus par 2.96 miljoniem latu"*, LETA (01.03.2010)
30. LETA. *E-receptes informācijas sistēmu izstrādās "In-Volv Latvia" un "ABC software"*, LETA (22.12.2011)
31. K. Miļūna. *E-veseslības sistēmas testēšanu ar reāliem pacientu datiem sāks decembrī,* LETA (17.10.2012)
32. Ministru Kabinets. Noteikumi Nr. 850 "Nacionālā veselības dienesta nolikums" (01.11.2011)
33. D. Kļanska. *Nodrošinās centralizētu pacientu veselības informācijas uzglabāšanau,* LETA (17.07.2012)
34. E. Rutule. *Ministrija: Medicīnas nozarē izmanto dažādas IT sistēmas, bet to salāgošanaiar e-veselības portālu būs pārejas periods,* LETA (05.12.2012)
35. A. Ādamsone. *Apdrošinātāji cer uz e-veselības ieviešanu,* LETA (10.10.2013)
36. A. Sauka. *ARS vadītājs: Ieviešot e-veselību, būtu jāparedz valsts līdzekļi privāto sistēmu salāgošanai ar valsts sistēmu,* LETA (04.04.2015)
37. K. Pētersons. *NVD vadītāja: Nepastāv risks, ka Latvijā neizdosies ieviest e-veselības sistēmu,* LETA (11.05.2015)
38. E. Valtmane. *Valsts kontrole informēs par e-veselības sistēmas revīzijā konstatēto,* LETA (07.09.2015)
39. E. Valtmane. *VK:Valstij var nākties apmaksāt neizdevušos e-veselības projektu,* LETA (08.09.2015)
40. K. Pētersons. *Nesaņemot papildus līdzekļus e-veselības ieviešanai, VM projekta ieceres realizes mazākā apjomā,* LETA (08.09.2015)
41. K. Pētersons. *Pirmajā pusgadā ārstniecības iestādes e-veselības sistēmu tomēr varēs lietot brīvrātīgi,* LETA (16.09.2015)
42. LETA. *Ģimenes ārsti valdību lugs aicināt pārstrādāt e-veselības projektu; gatavojas neslēgt līgumus par tās lietošanu,* LETA (28.11.2015)
43. K. Pētersons, E. Valtmane. *Pēc plašām diskusijām valdība vienojas e-veselības sistēmas ieviešanu atlikt uz gadu.,* LETA (01.12.2015)
44. K. Pētersons. *NVD tomēr izdevies izstrādāt e-veselības projektu, un drīzumā sistēmas lietošana tiks piedāvāta ārstniecības iestādēm,* LETA (04.01.2016)
45. K. Pētersons. *Darbu sācis e-veselības portals,* LETA (02.07.2016)
46. K. Pētersons. *NVD organizes seminārus veselības nozares speciālistiem par e-veselības sistēmas ieviešanas aktualitātēm,* LETA (12.10.2016)
47. LETA. *Čakša: Elektroniskā slimības lap air gatava darbam; ar e-recpti sarežģītāk,* LETA (26.10.2016)
48. A. Sauka. *"BENU aptiekas"sāk e-recepšu apkalpošanu,* LETA (27.10.2016)
49. Preses relīze. *Aptiekā izsniegtas zāles pret pirmo e-recept.,* LETA (28.10.2016)
50. K. Pētersons. *Bažījoties parpacientu datu aizsardzību un savām iespējām strādāt ar e-veselību, ģimenes ārsti sistēmai vēlas pievienoties brīvprātīgi,* LETA (04.03.2016)
51. A. Antanoviča. *Ģimenes ārstu asociācija vienojas neparakstīt līgumus par e-veselības sistēmas izmantošanu,* LETA (21.10.2016)
52. K. Pētersons. *NVD: Parakstīšanas procesā ir 2100 līgumi par e-veselības lietošanu, kas ir vairāk kā puse no plānotā.,* LETA (01.11.2016)
53. K. Pētersons. *Čakša rosinās pagarināt laiku, kurā ārsti brīvprātīgi varēs lietot gan e-recepti, gan darbnespējas lapu,* LETA (08.11.2016)

54. K. Pētersons. *NVD noslēdzi 733 līgumus par e-veselības lietošanu, tai skaitā 128 – ar ģimenes ārstiem,* LETA (29.11.2016)
55. Ministru Kabinets. *Noteikumi Nr.748 "Grozījumi Ministru kabineta 2014.gada 11.marta noteikumos Nr. 134 "Noteikumi par vienoto veselības nozares elektronisko informācijas sistēmu",* LETA (29.11.2016)
56. K. Pētersons. *VM sākusi pārbaudi saistībā ar e-veselības informācijas sistēmu ieviešanu,* LETA (08.11.2016)
57. E. Kozule. *Saeimas komisijā vērtēs e-medicīnas ieviešanas progresu un lietderību,* LETA (17.02.2010)
58. A. Blumberga. *Latvijas farmaceiti pozitīvi vērtē e-recepšu ieviešanu,* LETA (02.04.2010)
59. A. Sauka. *Asociācija: E-veselības neieviešanas dēļ aptiekām būs jāatliek IT sistēmu modernizācija; pieaugs rindas,* LETA (01.12.2015)
60. A. Ādamsone. *Mudina nodrošināt labāku caurskatāmību un racionālāku resursu izmantošanu veselības aprūpē,* LETA (21.07.2012)
61. A. Sauka. *ARS vadītājs: Ieviešot "e-veselību", būtu jāparedz valsts līdzekļi privāto sistēmu salāgošanai ar valsts sistēmu,* LETA (04.04.2015)
62. K. Pots. *Apinis: "E-veselības" projektu īsteno biznesa, nevis pacientu interesēs,* LETA (02.06.2010)
63. K. Miļūna. *Mediķu arodbiedrība aicina apturēt VM izstrādāto e-veselības projekta virzību,* LETA (07.08.2012)
64. K. Pētersons. *Mediķu arodbiedrība atbalsta ģimenes ārstu lēmumu neparakstīt līgumu par e-veselības sistēmas izmantošanu,* LETA (24.10.2016)
65. K. Pētersons. *Ģimenes ārsti aicina e-veselības ieviešanai noteikt sešu mēnešu izmēģinājuma periodu,* LETA (08.11.2016)
66. A. Kinca. *Reportāža no Igaunijas: Kopš e-veselības ieviešanas papīra kaudzes ārstu kabinetos aizstāj dators,* LSM (19.11.2016)

Evaluation of healthcare policy in Latvia

I. Bikava and A. Skride

Rīga Stradiņš University, Riga, Latvia

Abstract. Performance of healthcare system of Latvia has been criticised for several years, due to the problems of accessibility, weak efficiency and insufficient funding. Politicians, foreign and local experts all agree upon the need of reforms to improve the performance of the healthcare system. The aims are defined in long-term policy documents, but there are no defined tasks to be made and aims to be achieved in mid-term and short-term policy.
The aim of the research was to make evaluation of performance of Latvia healthcare system as well as evaluation according to "Health 2020" framework recommendations.
The results showed that changes in the healthcare system in Latvia are made on "ad-hoc" basis without a strategic long-term plan, and despite the fact that guidelines on the public health refer to "Health 2020", the changes that are made and reforms that are planned are not in compliance with "Health 2020". Work on social determinants (equity, universal coverage, accessibility) hasn't been effective, the aim "health in all policies" hasn't been achieved and the primary healthcare due to its unaccessibility doesn't provide expected efficiency in the whole healthcare system.
Key words: Healthcare policy; healthcare reforms; healthcare governance.

1 Introduction

Researchers, politicians, local and foreign experts all agree upon the need of healthcare reforms in Latvia to improve inhabitants' access to healthcare and provide sustainability of healthcare resources.

In the case of Latvia important factors are – urban and rural regional differences, high degree of income disparities and high proportion of the people who are at the risk of poverty.

Since 1991, when Latvia regained its independence, different reforms in Latvia healthcare system have been made. First reforms began in 1993 with Healthcare financing reform, effective healthcare provider network development and public health policy development. In 1996 the reform to implement primary and secondary healthcare providers began. To improve health of the society, primary healthcare based on General Practitioners (GP) was promoted to provide available and cost effective healthcare services to patients as close to the patients' homes as possible. The reforms in secondary healthcare were made to optimize network of service providers and to improve the quality, cost efficiency and availability to patients. Financing model change was made in 1997 by implementing health insurance principles in the healthcare. The public health policy development started in 2001 by developing the "Public Health Strategy". The second change on financing model took place in 2004 by refusing of the insurance principle and with return to healthcare financing from the state budget [1, 2].

During 1997–2015 the number of hospitals has decreased from 156 to 67, the number of healthcare institutions providing outpatient services has increased from 1127 to 4192, the number of general medical practices (GP, pediatricians, internists) has increased from 361 to 1275, the number of other outpatient healthcare institutions has increased from 75 to 729 [3].

The reforms have impact not only on the number of institutions, but also on medical personnel. During 1997–2015 the total number of physicians per 10 000 inhabitants has increased from 34.7 to 38.7, but the number of nurses has significantly decreased from 46.2 to 38.7 with the lowest number of nurses in 2009 – 37.6 [4, 5]. The decrease in numbers can be explained by two main reasons – low income that leads to the situation that nurses are working in another, a better paid field, and secondly – open borders of the European Union that lead to the situation that a significant amount of the medical personnel chooses to live and work in another European country. In accordance with the European Commission report: "The number of practising doctors per capita is just below the EU average. However, the number of practising nurses per capita is among the lowest in EU. The age of doctors and nurses is rapidly increasing threatening the sustainability of providing healthcare services." [6]

There have been reforms implemented and a lot of changes made in institutional network, normative regulations and treatment guidelines with the general aim to provide service availability to patients and to make the system more effective and efficient. Each Minister of Health had made changes in regulations and short-term planning documents, had nominated priorities for the period and done some work to achieve their nominated goals. However, in accordance with the opinion of the society – the situation is not satisfactory.

One of the indicators reflecting the level of service availability is the range of unmet needs that are measured in Latvia since 2005. During 2005–2014 the range of unmet needs has decreased from 29.4% to 19.2%, however the proportion of unmet needs is still significantly high. In the first quintile group (a group of people where household income is at its lowest), the rate of unmet needs has been significantly high during the whole period – 34.5% to 28.4%, with the highest rate of 35.3% in 2006. The main reason for unmet needs during the whole period was "could not afford to (too expensive)". The range in the whole population was 55.5% in 2005, it reached its the lowest point of 34.9% in 2007 and increased again after economic crises reaching the highest point of 65.5% in 2011 [7].

The Ministry of Health in 2016 was still working on the nominated priorities and implementation of different reforms that should provide better access, efficiency and effectiveness in healthcare.

The reforms that are planned to be made in healthcare system should be coherent to the society needs and in accordance with healthcare long-term policy.

Long-term policy document in Latvia is "Sustainable Development Strategy of Latvia until 2030". The main priorities in healthcare according to the document are investments in human capital – quality and availability of health and social services. Investments in human capital have to be made to ensure the participation of all potential human resources, particularly the groups of inhabitants subjected to the risk of poverty and social exclusion in the labour market [8].

Mid-term policy document for healthcare in Latvia is "Public Health Guidelines 2014–2020". As written in the document, guidelines are a mid-term policy planning document, which is based on the World Health Organization (WHO) Regional Strategy for Europe "Health 2020". The document "Public Health Guidelines" identifies main problems in Latvia healthcare system: insufficient financial support from the state and high level of direct payments that has significant impact on the access to healthcare. Therefore healthcare system solves neglected health problems as patients do not always have the possibility to find a doctor in a due time. The supreme goal of the policy is to increase the number of healthy life years of the inhabitants and to prevent premature death by preserving, improving and restoring health [9].

The aim of the research was to evaluate health policy in Latvia according to the society needs and changes in the healthcare system performance during 2008–2015 and to analyze developed reforms in accordance with the health policy framework "Health 2020" on which health policy guidelines of Latvia for 2014–2020 are based.

WHO policy framework "Health 2020" focuses on improvement of the citizens' health and reduction of health inequalities through better leadership and governance for health. "Health 2020" suggests shifting mindset from combating illness to promotion of health and well-being. "Health 2020" policy framework is based on values: universality, solidarity and equal access and includes – fairness, sustainability, quality, transparency, accountability, gender equality, dignity and the right to participate in decision making [10].

The aim of the research was to evaluate the healthcare system performance in Latvia and analyze the outcome of reforms made in the recent years as well as to evaluate the reforms in accordance to the recommendations given in the health policy framework "Health 2020".

2 Healthcare system performance in Latvia 2008–2015

Information about the healthcare system performance is essential to evaluate healthcare policy because it provides ability to measure the outcome and compare it in a selected time period or in comparison to other countries. An overall system performance measurement within a time period can be an indicator of the policy outcome – to determine how implemented reforms influence the performance.

Evaluation of the healthcare system performance in Latvia is made based on Irene Papinicolas and Peter C. Smith described performance assessment framework used to compare healthcare systems in different countries. In accordance with the noted in introduction, their work summarizes the current "state of the art" of health system comparison and is based on the performance comparison indicators used by the European Commission, Organisation for Economic Co-operation and Development, the Commonwealth Fund and the WHO. The framework includes analysis of five key elements of the health system architecture – service provision, financing, resource generation, leadership/governance and risk factors. To evaluate the healthcare policy outcome and performance of healthcare system, five segments have to be evaluated – population health, health service outcome, equity, fairness in financing and responsiveness [11].

Identified problems in evaluation process of each segment are analyzed in accordance with the recommendations in "Health 2020".

2.1 Health of the population

Population's health is the domain, which can be evaluated basing on the commonly available data – life expectancy, mortality and morbidity.

Life expectancy at birth in Latvia, as can be seen in Table 1, during the period 2008–2015 has increased in total from 72.0 to 74.7 years. Life expectancy for males has increased by 3.2 years and for females – 1.9 years.

Table 1. Life expectancy at birth (years).

Year	2008	2009	2010	2011	2012	2013	2014	2015
Total	72.0	72.7	73.1	73.7	74.0	74.2	74.3	74.7
Males	66.5	67.5	67.9	68.6	68.9	69.3	69.1	69.7
Females	77.4	77.6	77.9	78.5	78.7	78.7	79.3	79.3

Data source: Central Statistical Bureau of Latvia [13].

Rate of infant death to 1000 live birth, as can be seen in Table 2, has decreased from 6.6 to 4.1. Perinatal mortality which is more sensitive to the quality of medical care has decreased from 9.0 to 6.7 and neonatal mortality rate has also decreased significantly from 4.6 to 2.4.

Table 2. Perinatal, neonatal and infant death rates.

Year	2008	2009	2010	2011	2012	2013	2014	2015
Infant deaths (from 0 days to 1 year)	6.6	7.6	5.6	6.6	6.3	4.4	3.8	4.1
Perinatal mortality (stillbirths and 0–6 days old)	9.0	9.3	7.9	9.1	8.5	7.3	7.2	6.7
Neonatal mortility (0–27 days old)	4.6	4.9	3.4	4.3	4.2	2.5	2.8	2.4

Data source: Central Statistical Bureau of Latvia [14].

In accordance with these indicators, overall environment and medical care provide possibilities to live longer, but do not represent the quality of the healthcare system. Mortality by cause provides clearer image on the public health.

The leading cause of death, as can be seen in Table 3, is diseases of the circulatory system and the death rate in this group of diseases has increased from 750.7 in 2009 to 806.10 in 2014. The leading cause of circulatory system diseases is Ischemic heart diseases (IHD) and the death rate from IHD has increased from 401.2 to 405.6.

Table 3. Leading causes of death by cause group (100 000 inhabitants).

Death rates by cause/year	2009	2010	2011	2012	2013	2014
I00-I99 Diseases of the circulatory system	750.7	776.1	760.9	801.9	812.8	806.1
I20-I25 Ischemic heart diseases	401.2	409.6	406.2	418.6	412.0	405.6
C00-D48 Neoplasms	281.5	292.1	289.6	299.7	300.4	304.9
V01-Y89 External causes of morbidity and mortality	100.8	100.5	89.3	92.9	89.6	93.8
F01-F99 Mental and behavioural disorders	11.2	12.2	15.5	14.8	16.9	15.3

Data source: Centre for Disease Prevention and Control of Latvia [15].

The second leading cause of death is neoplasms where the rate has increased from 281.5 to 300.4. Next leading cause group are external causes that have decreased from 100.8 to

93.8. Leading causes of mortality are common to the whole situation in region, where the leading cause is circulatory diseases.

The data indicates that the situation has not improved, because the leading group of diseases is IHD, which is an indicator of premature mortality in industrialized countries and it has increased. IHD are attributed to prevalence of risk factors such as smoking, high cholesterol and hypertension. Persisting high level of mortality from IHD usually indicates systematic problems that cover the entire course of the disease – from primary prevention and health promotion to treatment [11].

There are also some specific groups of diseases that can be analyzed to evaluate performance of healthcare system – diabetes treatment, mental and behaviour disorders and neoplasms on which there are screening programmes run by the state.

Death rate from diabetes, as can be seen in Table 4, has increased from 23.7 in 2009 to 25.3 in 2014. The rate of mortality from mental and behavioural diseases has also increased from 11.2 to 15.3 and in 2013 has reached the rate of 16.9. The increase indicates that availability of the primary care is not sufficient.

Table 4. Death rates by cause group (100 000 inhabitants).

Death rates by cause/year	2009	2010	2011	2012	2013	2014
C61 Malignant neoplasm of prostate	35.6	37.8	39.5	42.3	39.9	44.74
E10-E14 Diabetes mellitus	23.7	25.6	23.6	23.1	22.8	25.3
C50 Malignant neoplasm of breast	20.7	20.3	21.7	19.9	21.7	22.2
C54 Malignant neoplasm of corpus uteri	11.1	10.5	11	13.2	10.8	9.7

Data source: Centre for Disease Prevention and Control of Latvia [15].

There are three screening programmes for neoplasms provided by the state in Latvia – for females: breast cancer and cervical cancer and for males – prostate cancer. Despite the programmes, mortality rate from these diseases is high: mortality from breast cancer, as can be seen in Table 4, has increased from 20.7 to 22.2, for prostate cancer from 35.6 to 44.74 and only from cervical cancer mortality has decreased from 11.1 to 9.7. In accordance with the indicators the screening campaigns and the state run programmes have not performed successfully.

Unfortunately, indicators that are widely used to evaluate the health of the population – years of life lost, disability adjusted life years (DALYs), avoidable morbidity, amenable mortality and preventable mortality – are not available for the selected period in Latvia. However, in accordance with the data available on the population mortality – healthcare system is not performing successfully.

To improve the public health the recommendation given in "Health 2020" is to create social and physical environment that fosters healthy behaviour: tobacco control programmes (raising tax, encourage smoke-free environment, banning advertisements, etc) changes in alcohol policies (raising tax, restricting access, enforcing bans on advertising, restricting drink-driving), prevent road crashes (one-way streets, mandatory speed limits, radar speed enforcement programmes) [9].

According to normative regulations on tobacco and alcohol, as well as on traffic safety proposed changes recommended by WHO had already been implemented in the legislature in Latvia. Those amendments had positive impact on the death rate decrease from external causes from 100.8 to 93.8.

Another group of recommendations given by WHO are actions to promote healthy eating – reformulating processed food, decreasing salt, trans-fatty acids and saturated fat and promoting healthier food – more fruit and vegetables for children. As well as promoting physical activity through social media campaign and also initiating changes in the transport system and urban design [9].

There have been campaigns to promote healthier eating and physical activities, especially for school-age children, also changes in legislature have been made, providing restrictions on beverages that could be sold in schools. Also changes in urban designs can be observed – especially the popularity of bicycle roads in the city. However, death rate from IHD has increased as well as obesity and overweight rate in children and youth which indicates that the situation is not satisfactory and more changes have to be made to impact a healthy lifestyle and behaviour factors in the inhabitants [12].

2.2 Healthcare service outcomes

Healthcare service outcomes has to be analyzed using information about the performance of different areas of healthcare services – standardized hospital mortality rates, case fatality rates for acute myocardial infarction (AMI) and stroke, the postponement of death, readmission rates, patient safety indicators, administrative convenience [11]. Unfortunately, there is no data publicly available on these indicators in Latvia.

Another healthcare evaluation aspect is measuring the contribution of primary healthcare to healthcare system performance, because primary healthcare system is an essential element for an efficient and effective health system and cornerstone on health systems in the 21st century [9, 11].

To evaluate proper resource usage, medical rescue service calls have been compared – the number of acute cases to the number of inappropriate cases when a patient could get the needed medical help in the primary healthcare.

Table 5. Emergency medical rescue service usage.

EMR episode/year	2008	2009	2010	2011	2012	2013	2014	2015
Total EMR episodes	427 881	371 209	432 360	461 258	460 229	451 972	436 245	443 215
Episodes without results	31 531	25 479	31 563	37 114	27 881	28 855	30 713	30 834
Episodes without results	7.3 %	6.9%	7.3%	8.0%	6.1%	6.4%	7.0%	7.0%
Critical cases	301 960	260 765	279 752	281 804	293 286	286 676	259 864	243 356
Critical cases	70.6%	70.2%	64.7%	61.1%	63.7%	63.4%	59.6%	54.9%
Non critical cases	94 410	84 986	121 072	142 370	139 091	136 470	145 700	169 062
Non critical cases	22.1%	22.9%	28.0%	30.9%	30.2%	30.2%	33.4%	38.1%
CC – injuries, accidents	15%	14%	14%	14%	15%	15%	16%	15%
CC – sudden morbidity, acute cases	46%	47%	43%	40%	42%	42%	37%	34%
CC – childbirth and pregnancy pathologies	1.2%	1.3%	1.1%	0.9%	0.9%	1.7%	1.9%	1.7%
CC – transportation	8%	8%	6%	6%	6%	5%	5%	5%

NCC – morbidity	20%	21%	27%	30%	29%	29%	32%	37%
NCC – transportation	3%	2%	1%	1%	1%	1%	1%	2%

Data source: Centre for Disease Prevention and Control of Latvia [16].

In accordance with the data from the use of emergency medical help episodes, as can be seen in Table 5, the number of acute cases has decreased from 71% to 55% of the total count of episodes, the number of cases associated with injuries and accidents is stable – about 15%, but the number of sudden morbidity and acute cases has decreased. However, the number of non-critical cases, when emergency service is used, has significantly increased from 22% to 38% as well as non-acute morbidity cases have risen from 20% to 37%.

The data shows that resources are not used efficiently and the performance of primary care is not sufficient. The main problem, as announced by the Head of the RAKUS (the largest hospital in Latvia), is that daily the Emergency department receives in average 314 patients out of which 200 are not hospitalised. In the Emergency department those patients get blood tests, cardiograms, computer tomography and other diagnostic services, receive consultations and treatment recommendations [17].

Allocated efficiently, expressed by share of GDP, is the government spending on the healthcare sector relative to other sectors can be another indicator of the resource use [11]. In Latvia the public resource allocation is unsatisfactory, because the state expenditure on health is about 3% of GDP or about 10% of the state budgetary expenditure, that is relatively low compared to other EU countries.

As mentioned in "Health 2020", the concept "Health in all policies" is designed to make governance for health and well-being a priority for more than just the health sector. It has to work in both directions – ensuring that all sectors understand and act on their responsibility for health while recognizing how health affects other sectors [9].

As can be observed in the case of Latvia, there is no doubt that healthcare system is insufficiently financed and that healthcare sector has problems. But there are no activities taken – healthcare is no priority in the government. Despite all expert advice, the expenditure on healthcare from the state is only about 10% instead of at least 14% and the healthcare as priority is declared only for the year 2018. But for now, Latvia is a country with the second lowest expenditure on healthcare among developed countries per OECD data [18].

As outlined in "Health 2020", in the conditions of restricted state funding, effective resource usage is one of the main subjects that Ministry of Health and authorities have to work on. Targeting public expenditures according to the social need protects the poor and vulnerable people. The recommended directions to work in are – health technology assessment that can provide cost effective use of medicine and seeking efficiency gains by rationalizing the service delivery structures [9].

As previously mentioned, there are problems with availability for services in primary healthcare for the poor and vulnerable people, but there are no reforms planned in the near future that could provide better accessibility to primary healthcare. The current Minister of Health is working on the priorities called "green passageway for oncology patients" and "the health of mother and child", but not to improve accessibility to GP as the cornerstone of effective and efficient healthcare system.

Health technology assessment, as recommended by WHO, to provide cost effective healthcare, is not used since 2009 when the Centre of Health Economy of Latvia has been shut down and merged into the National Health Service.

Work on the development of effective service providers has started, following the report of The World Bank, but there are doubts about the outcome of the reform, because of the political setting that no hospital should be closed.

2.3 Equity

Evaluation on the third domain "equity" is made based on the distribution of health, progressivity of financing system and distribution responsiveness of health services: equity in health status, including socio-economic, demographic factors and genetics [9].

In the case of Latvia equity is declared as a normative value, but there is no equal access to healthcare for all inhabitants. In accordance with socio-economic situation, there is a high proportion of people that are at poverty and social exclusion risk as mentioned before.

The proportion of this group of people, as can be seen in Table 6, is high in the whole society with variance from 30% to 40% of the total population. The highest proportion of people at poverty and social exclusion risk is in the age group 55 and older, i.e., the group that has the highest need of healthcare.

Analysis of unmet needs in the age group 55 and older indicated that in the households with the lowest income the main reason for unmet needs is "cannot afford to (too expensive)" [7].

Table 6. People at risk of poverty or social exclusion.

Year	2008	2009	2010	2011	2012	2013	2014	2015
Total	34.2%	37.9%	38.2%	40.1%	36.2%	35.1%	32.7%	30.9%
Age: under 16	31.5%	37.8%	42.0%	43.7%	39.1%	37.0%	34.5%	30.3%
Age: 16–24	27.7%	33.8%	39.7%	44.2%	39.1%	38.7%	32.7%	33.5%
Age: 25–54	26.3%	31.2%	36.2%	39.7%	35.0%	32.9%	28.6%	25.1%
Age: 55+	50.4%	49.4%	38.3%	37.2%	35.4%	36.0%	37.1%	37.7%

Data source: Eurostat [19].

2.4 Fairness of financing

The fourth domain that has to be evaluated is "fairness of financing". In case of Latvia, the share of out-of-pocket payments is very high and varies from 32.1% to 36.1% of the total expenditure on health.[20] That indicates that the financing system is more regressive than progressive.

Because of insufficient funding of healthcare from the state there are several groups of people for whom healthcare is not accessible due to financing burdens, especially for elderly people and patients with rare diseases.

As mentioned in the section "Healthcare System Outcome" there are no significant changes planned to improve accessibility to healthcare for the poor and vulnerable people and there is no initiative in the change of the state budgetary assignment or to provide progressivity in taxation system.

2.5 Responsiveness

The fifth domain "responsiveness" that evaluates patients' satisfaction, choice availability, respect and dignity for patients – cannot be evaluated in Latvia because there is no data publicly available. In some hospitals there is a quality management system implemented, but there is no data available on the whole healthcare system evaluation.

3 Conclusions

During the research evaluation of five main segments of healthcare system has been made. In accordance with the research results the main problems of healthcare system performance are in the domain of healthcare system outcome. The main problems that have been recognized are:

1. Based on a high level of people at poverty and social exclusion risk and high rate of unmet needs with the leading cause "cannot afford it" it can be concluded that work on social determinants had not been done and the main values – universal coverage, equity and accessibility have not been reached.
2. The state has not increased funding to the healthcare, healthcare and health promotion have not become the priority of the government, despite all the recommendations from local and foreign experts. The concept "health in all policies" has not been implemented.
3. Primary healthcare in accordance with the data of emergency department usage and publicly expressed problem about quota on healthcare service availability – indicates that healthcare services are not provided effectively and it leads to poorer society health, inability to get medical examination and treatment timely.
4. Although the legislature to provide environment that fosters healthy behaviour is accepted, the social campaigns haven't been effective and there are still actions that have to be taken to improve public health, decrease morbidity and mortality of diseases related to unhealthy life-style: healthier food, enhance physical activities, reduce usage of tobacco and alcohol.
5. Improvement of healthcare governance is the main direction where the changes have to be made – to work on the guidelines and policies that provide more appropriate use of resources according to the society needs, improve efficiency of resource usage, providing technical assessment and rationalizing the network of service providers.
6. According to the evaluation and analysis of healthcare system performance and health policy in Latvia, priorities declared by the Ministry of Health are not coherent to the society needs, based on statistical indicators. Although neoplasm is the second leading death cause in Latvia, activities provided by the Ministry of Health in oncology field couldn't provide benefits for the whole society. The main problems according to the results of the research are primary care availability (especially for the poor and vulnerable people), improvements in primary health availability could lead to efficiency and better system outcome.
7. The second direction that Ministry of Health has to work on is to provide more effective funding usage – providing technology assessment and appropriate network of service providers.
8. Overall conclusion is that changes in the healthcare system in Latvia are made on "ad-hoc" basis due to priorities seen by the current minister and without a strategic long-term plan, there are no defined tasks and metrics that have to be achieved to provide available and sustainable healthcare system. The reforms are made without impact assessment on the society and public health, they are not in accordance with the main values expressed in WHO policy "Health 2020" – universal coverage, equity in access, effective and efficient governance.

References

1. Ministru Kabinets. Rīkojums Nr.547 "Par veselības aprūpes reformas vidusposma un tālākās attīstības ndorošināšanas pamatnostādnēm (30.09.2002)
2. Ministru Kabinets. Noteikumi Nr.1026 "Veselības aprūpes organizēšanas un finansēšanas kārtība" (21.12.2004)
3. Central Statistical Bureau of Latvia. VAG01. Basic indicators of health care services at the end of the year, Period: 1997–2015
4. Central Statistical Bureau of Latvia. VAG15. Physicians by speciality at the end of the year. Physicians total, Period: 1997–2016
5. Central Statistical Bureau of Latvia. VAG16. Medical personnel with secondary medical education at the end of the year. Nurses, Period: 1997–2015
6. European Commission. Commission staff working document. Country Report Latvia 2016. Brussels, 26.2.2016
7. Central Statistical Bureau of Latvia. VPG07. Need to undergo medical examination or treatment at health care provider (except dentist) and reasons for unmet need for medical care for persona aged 16 or over (%), Period: 2005–2014
8. Saeima of the Republic of Latvia. Sustainable development Strategy of Latvia until 2030, June, 2010
9. World Health Organization Regional office for Europe. Health 2020. A European policy framework and strategy for the 21st century (2013)
10. The Cabinet of the Ministers of Latvia. Public Health Guidelines 2014–2020, 30.09.2014
11. European Observatory on Health Systems and Policies, edited by I. Papanicolas and P.C. Smith. Health System Performance Comparison. An Agenda for policy, information and research, 1–29 (2013)
12. Centre for Disease Prevention and Control of Latvia. Latvijas Skolēnu veselības paradumu pētījums 2013./2014.mācību gada aptaujas rezultāti un tendences. Rīga, 2015
13. Central Statistical Bureau of Latvia. IMG07. Average life expectancy at birth (years) Period: 2008–2015
14. Central Statistical Bureau of Latvia. IMG051. Perinatal, neonatal and infant death rates. Period: 2008–2015
15. Centre for Disease Prevention and Control of Latvia. Statistikas dati par 2015.gadu. Mirstība. Iedzīvotāju mirstība uz 100 000 iedzīvotājiem.Dati atjaunoti 31.03.2016
16. Centre for Disease Prevention and Control of Latvia. Pārskats par neatliekamās medicīnas palīdzības darbību
17. A. Slokenberga. Universitātes slimnīcai nav jāstrādā lieljaudas poliklīnikas režīmā. RAKUS/ Aktualitātes, 29.09.2016
18. Organisation for Economic Co-operation and Development. OECD. Stat Government schemes and compulsory contributory health care financing schemes. Data extracted on 22.01.2017
19. Eurostat. People at risk of poverty or social exclusion by age and sex. Data last updated 15.12.2016
20. The World Bank. World Development Indicators. Out-of-Pocket health expenditures (% of total expenditure on health) Latvia. 2008–2014

Latvian health care competitiveness in relation to its infrastructure and available resources

A. Kokarevica, A. Villerusa, D. Behmane, U. Berkis, and V. Cauce

Rīga Stradiņš University, Riga, Latvia

Abstract. Resources are one of the essential indicators for the functioning of the health care system. Better health care provision is an essential prerequisite for the export of services. Traditionally a competitive health care system is linked to a number of factors (price, quality, reliability, products and services) largely determined by the new technologies, innovations and implementation the new methods. The authors of this article analyzed and collected data from the European Commission Eurostat and OECD data. Current situation in health care in Latvia is characterized by populations' restricted access to health care services, high out-of-pocket payments and poor health outcomes of the population. More than 10% of Latvian population can't afford medical care. The ratio of public funding for healthcare in Latvia is among the lowest in EU countries. Latvia spends 5.3% (USD PPP 1217) of GDP on health, lower than the OCED country average of 8.9% (USD PPP 3453). Latvia is facing a dramatic gap between the availability of hospital beds and long term care beds and the lowest prevalence of general medical practitioners among all Baltic States 321.6 per 100 000. These mentioned factors may hinder the development of health care in Latvia and reduce the ability to participate in international health service market.

1 Introduction

Good health benefits in all sectors of economy and the whole of society are making it a valuable resource. Good health is essential for economic and social development and a vital concern to the lives of every single person, all families and communities. As stated in the World Health organizations (WHO Constitution): "The enjoyment of the highest attainable standard of health is one of the fundamental rights of every human being". To achieve this goal it is necessary to target the health inequities and inequalities in the country. The Health 2020 strategy stresses that investing in health makes sense, especially by growing the GDP for health. Of the increase in health spending, 92% correlates with positive growth in gross domestic product (GDP). In recent years European health systems are being required to improve their performance and respond to new challenges.

Prioritizing investment in public health institutional arrangements and capacity building, and efforts to strengthen health protection, health promotion and disease prevention can have important cost-effective benefits. Reconfiguring services and responsibilities, redesigning incentives and payment structures, and being attentive to return on investment can result in improved value for money [1]. The WHO in their Framework for Action towards coordinated/integrated health services states, that to improve the health of the Europeans and to achieve the Health 2020 goals there must be shifting to people centered health systems and services. From the perspective of the system, patient-centred services delivery must strive to provide comprehensive, coordinated, quality, and equitable care and it is in the pursuit of these performance gains that the system must act. Importantly, the provision of services takes direction from the system-at-large and is derived from the conditions it creates. Thus, the success of transformations towards patient-centred health services rests on an understanding of both the root causes (determinants) of poor performance of the function itself and the contributions of the health system underpinning the provision of health services [2].

In most countries, universal health coverage provides financial protection against the cost of illness and promotes access to care for the whole population. Health spending now accounts for about 9% of GDP on average in European countries, and exceeds 10% in several countries. Higher health spending is not a problem if the benefits exceed the costs, but there is sample evidence of inequities and inefficiencies in health systems which need to be addressed. There is also a need to achieve a proper balance between spending on disease prevention and treatment [3]. Despite these improvements, important questions about how successful countries are in achieving good results on different dimensions of health system performance remain.

Contemporary health systems are faced with an important paradox. On the one hand, consistent increases in health care investments over past decades and major developments in biomedical research have resulted in an extraordinary expansion of knowledge, technologies, techniques, skills and resources that make it possible to tackle many major health problems more effectively than ever before. On the other hand, many recent attempts to reform the health care sector have had only limited success in the quest to develop a more effective, efficient, safe and equitable delivery system that achieves the fundamental goal of improving population health [4].

One of the main challenges and risks that are influenced by insufficient health infrastructure is a gap between countries in health status and health inequalities. Latvian health system has one of the lowest levels of funding in the EU for medical expenses. As regards most health system performance criteria, such as health status, financial risk protection, and patient satisfaction, Latvia still lags behind not only western EU countries, but also other countries that joined the EU in 2004. The Latvian health care system is based on general tax-financed statutory health care provision, with a purchaser–provider split and a mix of public and private providers. The Ministry of Health is responsible for national health policy and the overall organization and functioning of the health system. The independent National Health Service (NHS) institution implements state health policies and ensures the availability of health care services throughout the country. Different ownership structures characterize health care provision in Latvia. Smaller hospitals and some bigger regional hospitals are usually owned by some of the 119 municipalities, whilst larger tertiary hospitals (university hospitals) and single specialty (monoprofile) hospitals (e.g. psychiatric hospitals) are owned by the state. Most primary care physicians have the legal status of an independent professional, and almost all dental practices and pharmacies are private [7].

In recent years, the economic crisis has constrained the financial resources available and thus increased Member States' difficulties in ensuring their health systems' sustainability. In turn, this could jeopardize Member States' ability to provide universal access to good quality healthcare. Ensuring universal access to cost-effective healthcare, including health promotion and prevention, in a financially sustainable manner is essential both for social cohesion, and for long-term economic growth, as a healthier population will improve labour market participation and labour productivity.

There are three widely accepted goals in the organization of health care in Europe: care should be of high quality, it should be comprehensive and it should be made available without financial barriers to access. These are not the only important goals, but they define a set of core policy purposes. Public policy works well when it combines the three core purposes in acceptable ways; however, achieving such combination becomes ever more challenging as the tensions between them increase under pressure of background trends such as an ageing population, technological innovation and a global rise in healthcare costs [8].

The aim of this report is to analyze the Latvian health care performance in relation to its infrastructure and available resources, in comparison with Baltic and some Scandinavian countries. The main objective is to discover the main differences and issues in order to improve health care delivery for local population and facilitate competitiveness in region. This will be part of a bigger picture involving the market structure for healthcare; Latvia has pretty liberal legislation for healthcare professions, the powers of the certified medical professional are very substantial. This indicates that the system can nurture significant competitive order in the Euckenian sense, and develop market pressure, optimizing the costs. To initiate rational studies of market power in healthcare, uncovering of the market mechanism is challenging in case where on demand side very strong rationing persists. In this case it is not easy to estimate the equivalent market power to unrestricted competition, as will be the case for the external markets. An intermediary stage will be the system based on procurements, which is slowly developing in Latvia.

2 Materials and methods

In this article a secondary data from EU level data sources, reports and literature analysis is used. The authors of this report analyzed and collected data/information using the following keywords: health care performance, resources, health infrastructure, health inequalities, health care expenses, medical personnel and technology. The data used for the analysis in this report is extracted for the European Commission Eurostat data base and Organization for Economic Co-operation and Development (OECD) data.

3 Results and discussion

The public sector plays a major role in the financing of health services: in two thirds of EU Member States, more than 70% of health expenditure is funded by the public sector. In 2014 in the EU-28, total expenditure of general governments on "health" amounted to 7.2% of GDP. "Hospital services" accounted for 3.6% of GDP, "outpatient services" for 2.2% of GDP and "medical products, appliances and equipment" for 1.0% of GDP [9]. The ratio of public funding for healthcare in Latvia is among the lowest in EU countries. Latvia spends 5.3% (USD PPP 1217) of GDP on health, lower than the OCED country average of 8.9% (USD PPP 3453). Accessibility (including affordability) of healthcare services is a topical policy issue in Latvia. The most important causes of the limited accessibility of healthcare services are: lack of funding, lack of human resources, high out-of-pocket payments, regional disparities in service provision and low solvency of patients. The share of health expenditure as a percentage of total general government expenditure declined from 10.6% in 2008 to 9.2% in 2010 and the Ministry of Health budget dropped by 12.6% in 2009 and by 1.5% in 2010, public health expenditure constituting around 3% from GDP. The growth of recent years is insufficient. In 2017, funding for the health sector only by approximately 30 million will exceed the pre-crisis level in 2008. Fiscal adjustments led to a broad spectrum of reforms affecting hospital and secondary ambulatory services as well as public health programs. As shown below (see Fig. 1), than Latvia has the lowest GDP expenditure on health among all Baltic states and one of the smallest among other EU-28 states. The biggest spending from the GDP goes to medical and outpatient services (see Fig. 1).

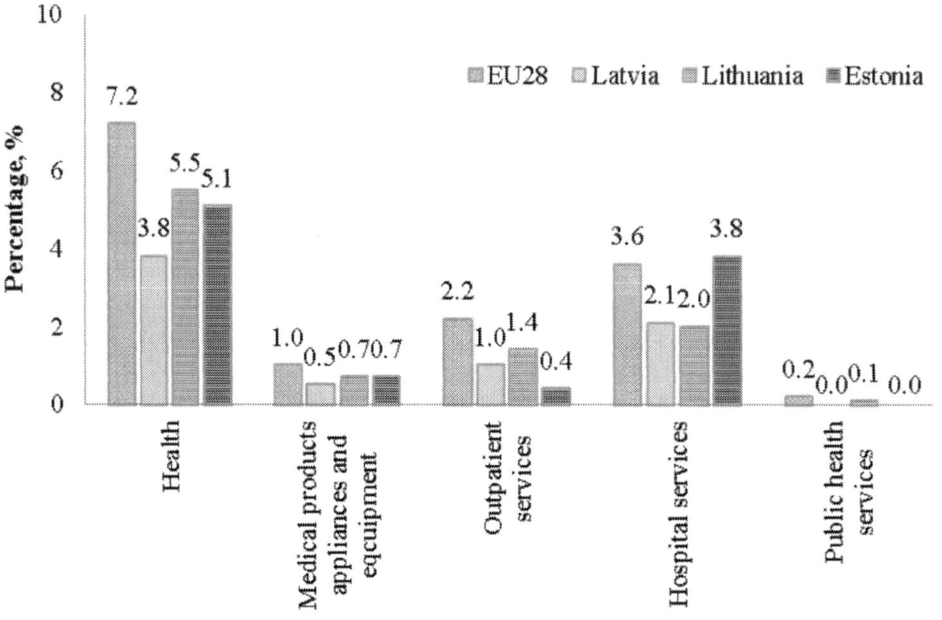

Fig. 1. GDP per country and service (%) (Data of 2014). Source: Eurostat, 2014.

Growth in health spending in Latvia has been modest compared to overall economic growth. Real GDP growth in Latvia averaged about 9% in 2002–07, and then concentrated sharply by 3.6%, 14.3% and 3.6% between 2008 and 2010, finally picking up again at around 3% annual growth until recent slowdown. Prior the global financial crisis, between 2005–2009 health spending in Latvia increased by 3% per year, slower the OCED average of 3.4% annual growth. Between 2009 and 2013, Latvian health spending growth was no different to the OCED average at 0.6% per year [6]. As shown in the Fig. 2. Latvia has the lowest per capita spending on health between the Baltic and selected Scandinavian countries.

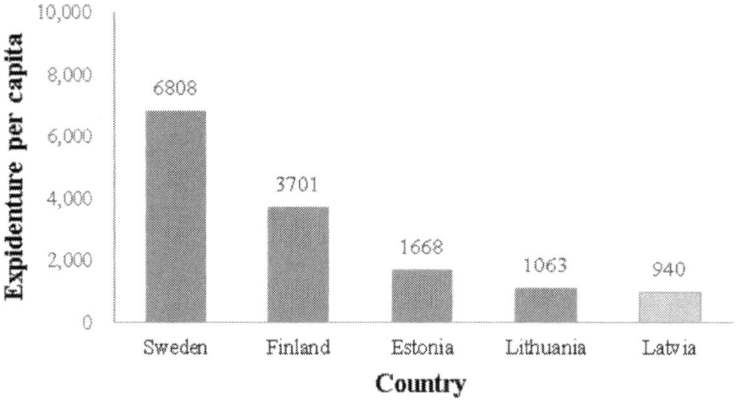

Fig. 2. Health expenditure per capita per country (PPP 2011 $). Source: OCED, 2014.

Latvia is facing a dramatic gap between the availability of hospital beds and long term care beds. Per 100 000 inhabitants there are only 60.6 long term beds available in the country (see Fig. 3). Facing the ageing of the population and the increasing demand, Latvia is facing serious difficulties to meet this demand, and it is predicted that in the future there will be an even greater risk to the population not to receive this service.

In 2014, more than three quarters (76.0%) of all beds in EU-28 hospitals were for curative care, 14.0% for rehabilitative care and the remaining 10.0% for long-term care and other purposes.

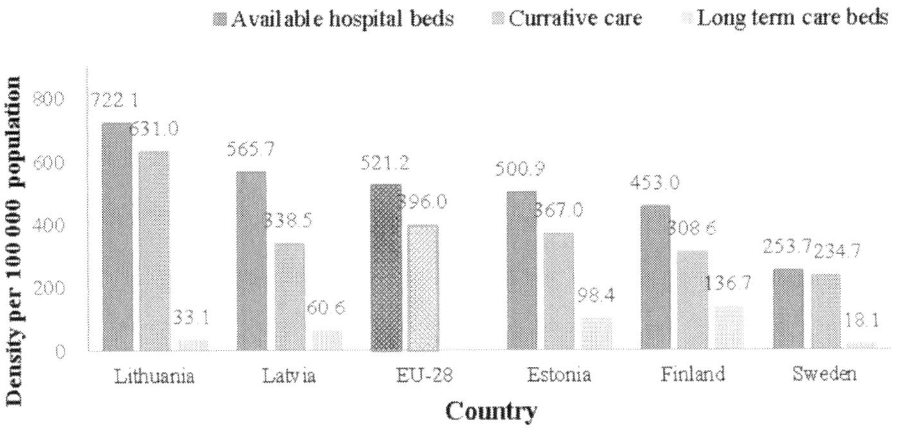

Fig. 3. Hospital beds by type of care in 2014 (per 100 000 inhabitants). Source: Eurostat, 2014.

The importance of human resources for the sustainable development of health sector has been neglected during the reform path since the start of economic crisis in Latvia. The "Basic Statement on Development of Human Resources for Health Care" was adopted by the Cabinet of Ministers in 2006. The policy envisaged the adoption of staffing requirements for physicians and nurses per bed and patient by specialty in hospitals. In addition, the document suggested a plan for the increase in health care personnel remuneration levels to promote recruitment of new doctors and retain existing staff. However, as a result of financial constraints in the context of the economic crisis, salaries of health professionals have, in general, been reduced rather than increased in recent years. The adopted Workforce strategy (adopted in 2006) was suspended in 2010 and the strategic vision regarding the supply, structure and remuneration of human resources has been lost.

In 2013 in Latvia there were 6172 practicing physicians, or 3.19 per 1000 inhabitants, which is on the EU average level. Of these 1313 or 21.3% were family doctors. In overall, the total number of medical doctors during 10 years period has decreased by 4%, but number of practicing medical doctors by 3%. Shortages of some professional groups are anticipated. There are 3.1 practicing doctors per 1000 population, close to the OCED average of 3.2, but many doctors are retiring in coming decades. The average age of Latvian GPs is 54. More than two-thirds of practicing GPs are aged 50 and over, and more than a quarter of them are aged 60 and over [6].

Latvia has two higher medical education faculties, which is a relatively large number for a country of this size. The number of medical practitioners who have left the healthcare sector is substantial. This silent reserve has been tried to activate e.g. via European Social fund activities for nurses. Latvia possesses purely non-governmental system of medical certification, and this is an important aspect in the forecasts. Particular evidence exists about medical practitioners leaving the sector in case they cannot secure anymore public quotas for rationed services purchases, which is of high significance for sparsely populated rural regions where out-of-pocket purchases of services is rare. Furthermore, competition for residency places is very high among university graduates

In comparison in the EU – 28 the "health and social work" sector is the sector which saw the largest rise in employment in recent years (and notably between the second quarter of 2008 and the second quarter of 2015) with over 2.6 million new jobs. With regard to absolute figures across the EU, the "human health and social sector" accounted for 23 400 500 employees in the second quarter of 2015. The majority of them – 13 293 700 employees – were employed in the "human health" subsector. Workers in the health and social work sector have an education level far above the average of all sectors [5]. The three most common types of physicians across the EU Member States are generalist medical practitioners, the medical group of specialists and the surgical group of specialists [9]. Latvia has the lowest prevalence of general medical practitioners (physicians) among all Baltic countries 321.6 per 100 000 (see Fig. 4). The data could be explained and linked to the low wage rates among the medicines in Latvia – in year 2013 only 13.74% in the health sector employed earned a wage more than 1000 EUR. Comparison of the remuneration of health professionals between countries substantiates that the state policy in Latvia has been very shortsighted and there is an enormous gap between the publicly regulated doctor salaries in Latvia (annual salary 13 936 EUR) and the EU countries [6].

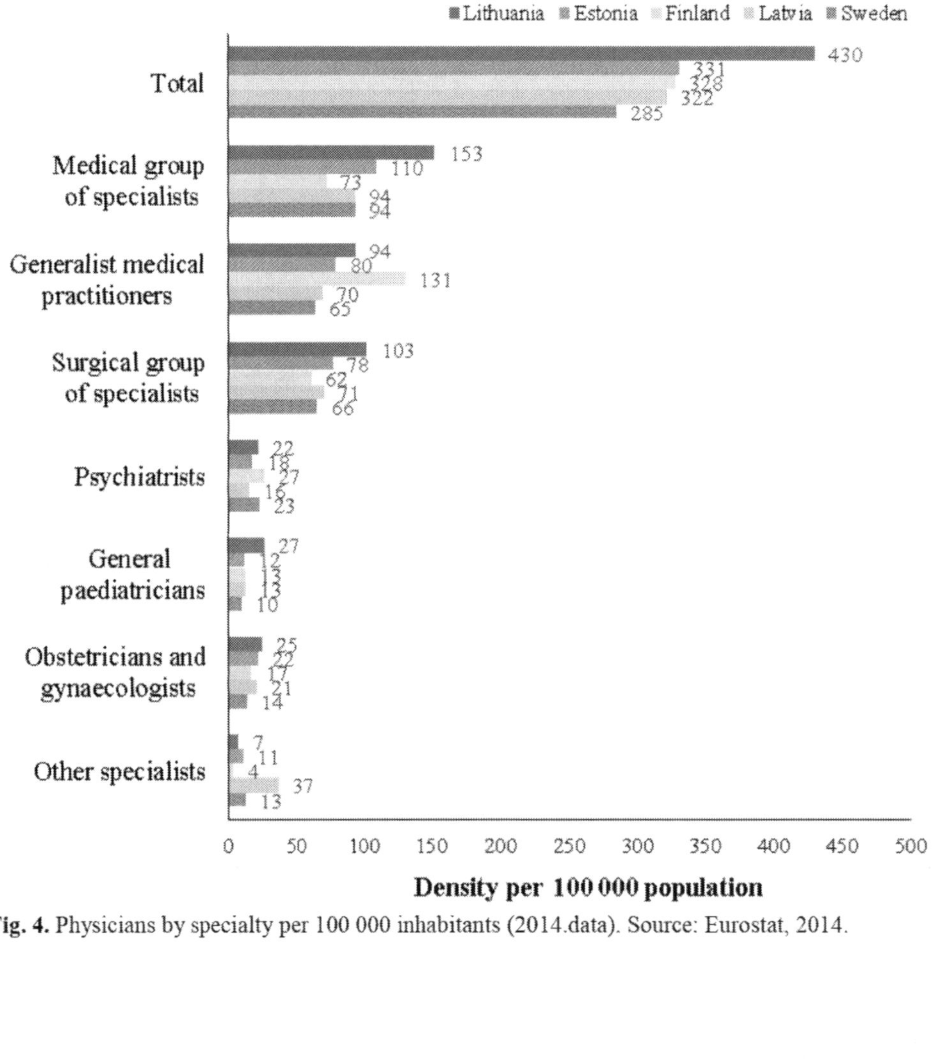

Fig. 4. Physicians by specialty per 100 000 inhabitants (2014.data). Source: Eurostat, 2014.

Recent analysis shows that the overall situation regarding health personnel in Latvia is an important bottleneck for system quality and efficiency. The number of medical graduates is 14 per 100 000 of population, which is slightly above the EU average and is not a hurdle for the supply of labour in the sector. Nevertheless that educational system provides good basis for professional medical education, it does not correlate with the number of practicing doctors evidently highlighting the problem of retention of medical graduates in the country (see Fig. 5).

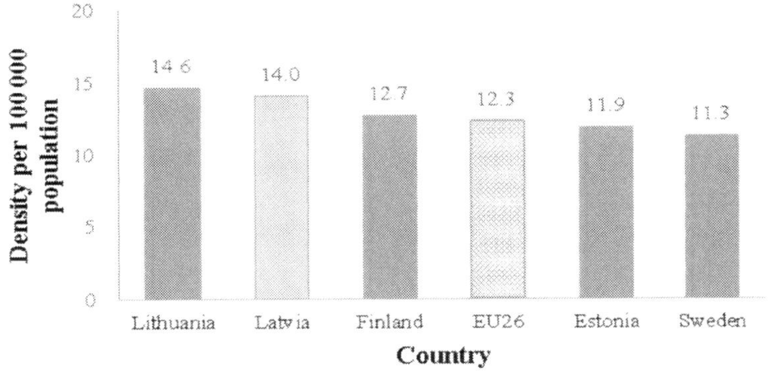

Fig. 5. Medical graduates per 100 000 inhabitants, 2013. Source: OECD (2016), Medical graduates.

The number of practicing nurses is even a greater concern. The age of nurses and other supporting professionals is also increasing, making it challenging to secure their future supply [6]. The supply of middle level workforce in comparison to doctors is very low. The number of nurses is on the bottom level between EU/EEA countries: 4.8 nurses per 1000 inhabitants in 2013, which is around 2 times less than the average EU countries (see Fig. 6).

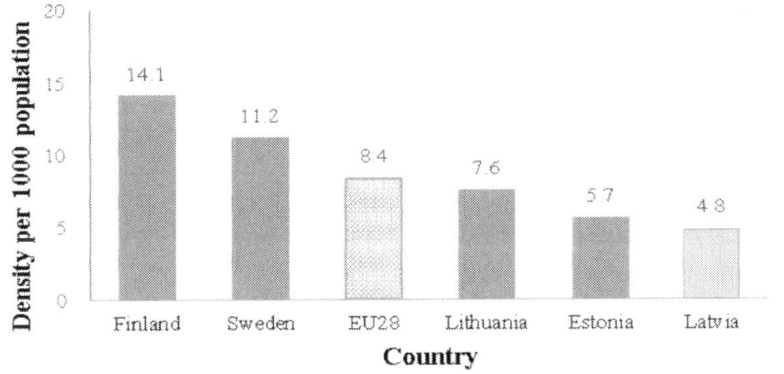

Fig. 6. Practicing nurses per 1000 population, EU countries, 2013. Source: OECD (2016), Practicing nurses.

The ratio of nurses to physicians in Year 2015 in Latvia was 1.5 in comparison with the EU 28 average – 2.5, in Finland – 4.7, Sweden – 2.7. Very close, but slightly less than the other Baltic countries: Estonia – 1.7 and Lithuania – 1.8.

Much better situation in Latvia is with the high-tech availability ratio per 100 000 inhabitants, which demonstrates the possibilities and capacity of a countries health care force. The availability of equipment for diagnosis has increased rapidly over recent decades in Latvia, which could be a positive prerequisite for service volumes increase, including foreign patient's attraction. Latvia has a high availability of CT scanners and MRI units

(3.6 and 1.3 per 100 000 inhabitants) (2014.data), what demonstrates an opportunity to compete not only on local (Baltic countries) level, but also internationally.

Table 1. Availability of medical technology – imaging equipment – (2014) per 100 000 inhabitants. Source: Eurostat, 2014.

Country	CT scanners	MRI units	Gamma cameras	Angiograp hy units	Mammo graphy	PET scanners
Latvia	3.6	1.3	0.3	0.6	2.3	0.1
Lithuani a	2.2	1.1	0.3	0.8	1.4	0.1
Estonia	2.0	1.1	0.2	0.9	1.0	0.2
Finland	8.1	2.3	0.8	2.0	3.1	0.2
EU 28	2.1	1.5	...	...	...	...

In most of all cases Latvia is the leading country in Baltic States in the availability of new technologies per 100 000 inhabitants. The frequently used and available technology is CT scanners and MRI units (see Table 1).

An indicator frequently used to show barriers in access to health care is patient self-reported unmet needs based on surveys. Reasons given for not receiving care include: excessive treatment costs, long waiting times, or having to travel too far to receive care [5]. Current situation in health care in Latvia is characterized by populations' restricted access to health care services, high out-of-pocket payments and poor health outcomes of the population. Overall in the EU-28, the most common reason for not having a medical examination or treatment was that it was too expensive; this reason alone accounted for one third of all the people who reported an unmet need for medical care, equivalent to 2.4% of the population. As shown in Table 2, than Latvia has the highest unmet need prevalence between all Baltic States and between all EU-28 countries (in 2014, 12.5% as opposed to 3.7% for the EU-28 average; EUROSTAT, 2014) (Year 2014.data) [9]. The cost of healthcare service is the main factor restricting the accessibility of healthcare services [10, 11]. The high rate of people who can't meet her need for health treatments are showing that Latvia is lacking both – finances and infrastructure for the health system. The most common reason for unmet need are the expenses for health care (12.5%) people need to cover from their own budget which in most cases (most of the people who reported that they can't afford the health care they need where over 70 years of age) is not possible. More than 10% (10.5%) of Latvian population can't afford medical care (see Table 2).

Table 2. Share of persons aged 16 and over reporting unmet needs for medical care, by detailed reason (2014 or latest available) (%). Source: Eurostat, 2014.

Countries	All reasons	Health related reasons	Of which		
			To expensive	To far to travel	Waiting list
EU 28	6.7	3.6	2.4	0.1	0.1
Latvia	19.2	12.5	10.5	0.4	1.6
Lithuania	5.4	3.7	0.7	0.3	2.7
Estonia	13.1	11.3	0.5	0.7	10.1
Finland	4.7	3.3	0.1	0.0	3.1
Sweden	9.2	1.5	0.5	0.1	1.0

The self-reported unmet medical needs should also be seen in the context of objective measures of the use and expenditure on healthcare, such as the level of public, private and "out of pocket spending" on health care, which also provide information related to the financial protection of the population against the risks of ill-health, as well as the actual use of health services. As shown in Fig. 7, than Latvia has the highest rate of reported private-household out of pocket expenditures in all Baltic States and 17% than in EU-28 member states.

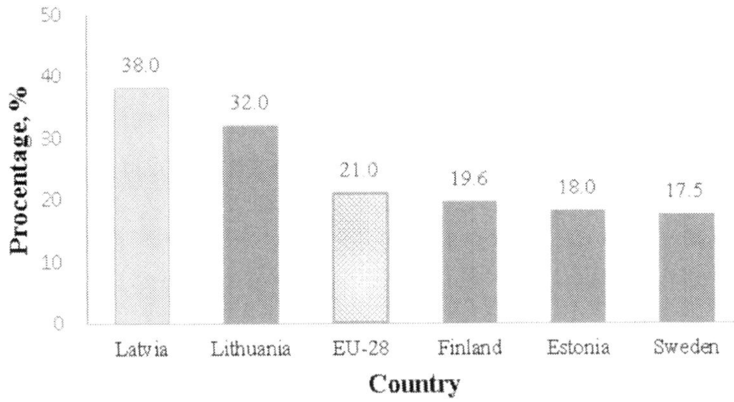

Fig. 7. Private household out-of-pocket expenditure as % of total current health expenditure, 2013 or latest available. Source: Eurostat, 2014.

To understand the competitive health care options and potentially necessary investments in health care and health care workforce, as well as to complete understand the consequences of under-investment in health care and prevention system of a country, it is need to be clear that in the last decade health expenditure of a state is of interest from two perspectives. Firstly, increasing costs for labor and constant inefficiencies in the organization of healthcare production push overall expenditures of healthcare. Secondly, the way production of healthcare is organized affects the productivity effect of investments in new technology and innovation. Preferably, a higher capital share in the production will increase productivity, partly by substituting labor and old technology, partly by making production more efficient. But this is not necessarily the case with investments in any sector, and it seems safe to say that healthcare have shown particularly small effects on productivity by investments. Healthcare service is largely a non-tradable sector. It is integrated with world markets in some aspects on the input side, but the service output is in most countries detached from world markets. Yet increasing elements of health-care services have grown more tradable. However technological progress, like teleradiology, and increased mobility of labour and capital have fostered greater interests for cross-border healthcare services, although still only at the margin [9].

4 Conclusions

- Assessing the available data and information, one can see that Latvia has great opportunities, capability and potential for further improvement among European Union member states in the field of health care. The high availability of high-tech and low personnel costs, favorable geographical location may provide the potential for attracting medical tourists from the West and the East as well.

- Nevertheless the development of the Latvian health care system performance and its competitiveness is negatively affected by the relatively low level of public investment from GDP in the sector. To develop the health care enough fit the needs of the country's population the government needs to invest at least 6% of GDP. High levels of cost sharing for almost all service users create some financial barriers to access and large proportion of unmet needs of patients.
- Improvement in workforce structure and planning is essential for reaching Health 2020 goals – as improvement of population's health, reducing inequalities and to ensure patient centred health system. These mentioned factors may hinder the development of health care in Latvia and reduce the ability to compete with other countries.
- Latvia demonstrates a complex interaction between demand and supply side on the healthcare market in case of strong rationing of public demand combined with demand insufficiency of privately funded services, exaggerated by regional disparities and still severe consequences from 2008 crisis.

The research has been supported by National Research Programme "EKOSOC-LV" project 5.2.1 "Research on Latvian enterprises competitiveness in external markets and its improvement", subproject "Healthcare industry external competitiveness".

References

1. World Health organization, *Health 2020 – European Policy framework and strategy for the 21st Century*, World Health organization, 5–25 (2013)
2. World Health organization, *Health Services Delivery Programme Division of Health Systems and Public Health,* World Health organization, 10–15 (2014)
3. C. Etienne, A. Asoma-Baah et al., *Health systems financing, the path to universal coverage,* The World Health report, World Health organization, 8 (2010)
4. R. Bernd, C.-A. Dubois, M. Mckee, *the Health care Workforce in Europe, Learning form experience,* European Observatory on Health Systems and Policies, 15–20 (2006)
5. European Commission, *Questions and answers on Solidarity in health: Reducing health inequities in EU,* http://europa.eu/rapid/press-release_MEMO-09-467_en.htm?locale=en (2009)
6. Organization for Economic Co-operation and Development, *OCED Reviews of health systems: Latvia,* OCED Publishers, 14–30 (2016)
7. U. Mitenbergs, M. Taube, J. Misins, E. Mikitis, A. Martinsons, A. Rurane, W. Quentin, *Health Systems in Transition, Latvia – Health system review*, European Observatory, **14**, No. 8, 22–40 (2012)
8. M. Marmot, R. Belt, P. Goldblatt, et al., *The Future of Healthcare in Europe – a social determinants perspective*, The Future of Healthcare in Europe, London's Global University, 7–15 (2012)
9. F. Erixon, M. nav der Marel, *What is driving the rise in Health care expenditures? An Inquiry into the Nature and Causes of the Cost Disease,* Ecipe working paper **5**, 2–5 (2011)
10. European Commission, *Health and health systems* http://europa.eu/epc/pdf/joint_healthcare_report_en.pdf, 1–10 (2016)
11. European Commission, *Changing the funding of the Latvian compulsory healthcare system: For better or for worse?* ESPN Flash Report 2016/24, 1–2 (2016)

Health literacy as a challenge for health education

I. Stars

Rīga Stradiņš University, Riga, Latvia

Abstract. Scientific literature demonstrates a well-established correlation between education and health literacy; education and health outcomes; health literacy and health outcomes. Health literacy has a mediating role between education and health. Health literacy is known as a complex concept, it has multiple definitions, underlying meanings and conceptual models. Diversity of health literacy can be a challenging motivator for health education. In-depth study of scientific literature was carried out to highlight the link between health literacy and health education and to emphasize health literacy challenges for health education. Five health literacy challenges for health education were identified: (1) health literacy as an outcome related to health education interventions; (2) multi-layered health literacy or different levels of health literacy; (3) the concept of health literacy and its comprehensiveness; (4) shift from a passive to more interactive and empowering health education; (5) cultural context of health literacy and health education. Enhanced health literacy definitions and conceptual models require more complex, comprehensive, socially oriented, culture sensitive, participatory health education.

1 Introduction

The association between education and health is well researched, well established and strong [1]. It has received much attention and interest in scientific discourse [1] and a large body of evidence establishes a positive correlation between education, health and health outcomes [2]. The link between education and health has different potential explanations and interpretations [3], therefore this relationship is defined as a complicated one and mechanisms or pathways underlying "education – health" association have not been conclusive and accomplished [4]. Consequently, understanding the ways by which education affects individual and population health, health-related behaviour and health outcomes can be a challenging and not an easy task.

General education positively influences people's knowledge and skills (reading, writing, listening, speaking, numeracy, etc.), also promotes development of multiple literacy (for example, science, media, information, civil, cultural, etc. literacy), among them – health literacy [5]. Health literacy is defined as the cognitive and social skill that determines the motivation and ability of individuals to gain access, to understand and use information in ways that promote and maintain good health [6]. It is described as a discrete form of literacy [7], which is a context and content of specific literacy [8] that stands in its own right and has proved itself as a self-sufficient issue in public health, health promotion and health education discourses.

Health literacy as a distinctive concept was first used in 1974 in publication on health education as a policy issue and indicated a school health education in relation to health literacy development. Schools were seen as responsible for delivering fundamental health information on different health topics, such as safety, nutrition, physical activity, hygiene, etc. [9]. This early use of the concept of health literacy demonstrates an interconnection between health education and health literacy [10]. Currently, health literacy is conceptualized as a possible pathway by which the link between education and health can be researched, comprehended and interpreted [11]. There is evidence that low health literacy is associated with poorer health outcomes and poorer use of health care services [12, 13], higher medical costs [14] and lower self-reported overall health [15]. As health literacy affects health, health-related behaviour and health outcomes and comes to be a strong predictor of health, there is a need to develop, improve and promote health literacy among population. Health education interventions are critical to improve health literacy and increase health-related knowledge and competencies. Health education and health literacy are closely interconnected. Health education is directed towards improving health literacy and is presented as one of the key outcomes of health education policies, activities and interventions [16]. At the same time, health literacy is an ambiguous concept that has become a remarkable issue of today's studies [17] in health and public health sciences. Endeavour to define, redefine, conceptualize and measure health literacy leads to the condition that *"the term 'health literacy' itself has come to mean different things to various audiences and has become a source of confusion and debate"* [17]. The concept and field of health literacy has evolved [18] and has expanded in its meaning [19], scope and depth [17]. Thus, multiple definitions, conceptualizations and different conceptual models of health literacy now are available [20]. It is further argued to revise the concept of health literacy and consider for a new – more comprehensive and integrative – definition of health literacy [21]. This theoretical and conceptual diversity of health literacy concept can be proposed as a challenging factor for public health and especially – for health education theory and practice.

2 Aim and methods

The aim of this theoretical research is to highlight a mutual relationship between health literacy and health education and identify health literacy challenges for health education. For this purpose, an in-depth scientific literature analysis was made.

3 Results

The literature on health literacy is gradually and constantly developing, and the quantity of scientific knowledge in theoretical aspects of health literacy is increasing. An in-depth scientific literature analysis allowed us to recognize the following health literacy challenges for health education:

3.1 First challenge. Health literacy as an outcome related to health education interventions

Nutbeam [16, 18] notes that from public health perspective health literacy can be proposed as an outcome to health education and health communication activities. Health education is focused towards developing, improving and promoting health literacy [16]. More specifically, health literacy can be conceptualized as a learning outcome in schools and five core components of health literacy should be taught, learned and reached – theoretical knowledge, practical knowledge, critical thinking, self-awareness, citizenship – regarding health and health-related behaviour [22].

When defining health literacy as an outcome of health education, some aspects of outcome-based education can be used in this context. An outcome – education directs every part of the educational process and educational system around the proposed and desirable outcomes – clearly state the expected knowledge, skills, attitudes and competencies. It makes the learning process more targeted and well organized because it is oriented toward educational experience and persuasion that each student should have achieved the goal [23]. Apart from this thoughtfully and clearly defined learning outcomes help to plan, organize and guide educational programs (content) and learning process (methods) [23].

Health literacy therefore offers a framework for defining learning outcomes of health education that expects to be delivered from the health education interventions and identifies intended outcomes of health education programmes. Health literacy explains the purpose of health education, also clarifies details of proposed health literacy domains. However, demands for more extensive and integrative definition of health literacy, challenge health education to be more diverse as well.

3.2 Second challenge. Multi-layered health literacy and different levels of health literacy

The field of health literacy has expanded and evolved noticeably over recent decades [16, 17]. The early conceptions of health literacy focused predominantly on the functional health literacy and on an individual's ability to understand and comply with health instructions and basic health information [16]. Nowadays more than one health literacy level is outlined, for example, Nutbeam distinguishes and describes three levels or dimensions relevant to health literacy [16]:

1) **basic functional health literacy** is conceptualized as sufficient reading, writing, speaking, numeracy skills to be able to function effectively in everyday health contexts and situations [16];

2) **communicative or interactive health literacy** – more advanced cognitive and literacy skills which, together with social skills, can be used to actively participate in everyday health activities, use different forms of communication and to apply new information to changing circumstances regarding health context [16];

3) **critical health literacy** – the ability to critically analyse and use health information, to make informed health decisions. It is a set of skills, which enable people to participate more actively in political and social decisions affecting their health [16].

Conceptualized this way, health literacy comes to be more comprehensive and multi-layered. Consequently, health education programmes and interventions should address all these dimensions of health literacy.

3.3 Third challenge. The concept of health literacy and its comprehensiveness

The concept of health is a multi-facetted and represents a complex system, which includes not only physical and biological well-being, but also social, mental, spiritual well-being. Health is much more than preventing ill – health (disease, disability, etc.) in individual level. It is an ambiguous and dynamic concept, which includes a mutual relationship between individuals and their social, political, cultural and economic environment [24] or the social determinants of health. Health education needs to recognize and consider the complexity of health and surrounding determinants of health [25].

More broad and comprehensive conceptualization of health concept determines that the definition of health literacy has been broadened and become more comprehensive as well. The field of health literacy has expanded in its scope and depth [26]. The uncertainty about the concept of health literacy can be a challenge for health education. Today health literacy goes beyond its narrow conceptualization and is much more than an individual's ability to perform basic reading, writing and numeracy tasks required for an adequate and effective functioning in the health care environment and settings [27]. New concept of health literacy means much *"more than being able to read pamphlets and successfully make appointments"* [16]. There is an assumption that health literacy should be considered not only as an individual issue, but also as a feature of intense and mutual reciprocity between personal health literacy, his or her health and wider social conditions and general environment [26]. Now health literacy includes more advanced and complicated skills needed to function, participate and communicate in health context, also to think critically, evaluate and use information, make informed decisions to achieve a greater control over health circumstances [16, 18]. World Health Organization insists that health literacy should become more complex, namely, an individual behaviour-oriented communication and collaboration should be combined with the environmental, political and social factors that affect individuals and public health [25]. To avoid *"'blindness' to social aetiology of health and illness"* [28], there is a need to supplement health education with more comprehensive understanding of social determinants of health. Health education, in this more comprehensive approach, should consider the complexity of health and social determinants of health – social, cultural, economic, political and environmental factors [26] influencing individual and community health context and health outcomes. The awareness of the social determinants of health can encourage and empower people and community for reasoned actions to influence and change these social, cultural, economic, political and environmental determinants of health [24, 25].

3.4 Fourth challenge. From a passive to more interactive and empowering health education

Dynamic and comprehensive approach to health literacy requires more interactive health education that not only gives information in a passive and directive way, but also empowers people by enhancing their ability to think of healthy behaviours and lifestyle habits, seek, critically evaluate and use health information, and also motivates people to take actions to improve their health [29] and allows them to look at health in a holistic way [24]. Such health education should address health in personal, as well – social arenas [30], as Nutbeam concludes: *"Health literacy is critical to personal empowerment"* [16].

Empowerment is essentially important in the context of critical health literacy, because it leads to personal and social change, it encourages people for more active and profound participation in group actions and social dialogues, it enhances control and beliefs in ability to change own lives [30]. It promotes awareness of the fact that world and life is created and lived by people [24], it contributes to a wider ability to achieve control over the determinants of health [31].

New health literacy requires more active individual participation in health education activities instead of passive receiving of health-related information diffused in a directive, top-down approach. Health education should employ methods that allow interaction, participation, critical reflections and analysis. While traditional health education programmes concentrate predominantly on providing information in a transmission approach and see learners to be simply as passive recipients, new health education has the advantage of enabling to engage with health information [24]. World Health Organization points out: *"Such health education leads to health literacy, leading to personal and social benefit, by enabling effective community action, and by contributing to the development of social capital"* [25].

3.5 Fifth challenge. Health literacy, health education and cultural context

Literacy, health and health literacy do not exist in a vacuum or neutral zone. It is because people develop their ideas and conceptions of literacy, health, illness and health literate person in a cultural framework that is influenced by the family and community issues, linguistic context, social and cultural factors [32]. Culture gives the meaning to health beliefs, its explanations and interpretations. Understanding of the cultural context is necessary for an in-depth understanding of health, health related-behaviour, health literacy and related meanings. Cultural and social environment and family experience affect people's beliefs about healthy lifestyle, also form the contexts in which people communicate and debate about health and health-related themes [32]. Health literacy is also affected by belief systems and communication styles of a particular community and its culture, as well as people's understanding and response styles to health-related information [32]. Health literacy as a social construct requires understanding of the local cultural and linguistic context because culture and language form the context for the acquisition and application of health literacy skills [32].

Thus, health promotion and health education interventions must consider and respond to the local cultural context [29]. Culturally and linguistically appropriate health education interventions are required. Health education must be reflective to community cultural issues regarding health and ill-health, must take into account that culture, language, and learning are closely interrelated and *"to be effective, health education must be conducted in both culturally and linguistically appropriate formats to address the increasingly diverse multicultural and multilingual population"* [32]. There is a need in culturally responsive health education policy and practice. Health educators must recognize this cultural and linguistic diversity of health literacy and health education: *"In an environment of high diversity, the interconnections between health literacy, culture and language must also be considered in policies, practice and training"* [33]. Culturally sensitive health education is respectful of the health beliefs, interpretations, practices, experience and needs of different people [32] and their worldview [33].

4 Conclusions

Concept of health literacy indicates to a reciprocal and meaningful association between education, literacy and health.

An impressive progress characterizes the field of health literacy. The concept of health literacy has evolved and now presents itself as a complicated and ambiguous theoretical issue that challenges traditional health education and its theoretical and methodological underpinnings.

Diversity and comprehensiveness of health literacy concept can be perceived as a challenging aspect for health education. Health literacy is not any more an individual trait or even more – deficiency. Health literacy is not simply health knowledge. Health literacy gains more wide and comprehensive definitions and conceptualizations. Multi-layered models of health literacy require more complex health education. Functional health literacy predominantly focuses on the communication of health-related information, interactive health literacy is oriented towards development of personal skills that are required in health environment, critical health literacy focuses particularly on personal and community empowerment [16]. Besides the functional aspects of health literacy there should be also information and awareness of the social determinants of health included into health education policies and interventions. New health education should encourage critical thinking, motivated and reasoned action and information based choices, active participation in promoting health and wellbeing.

Health literacy is an active and dynamic phenomenon, because societies and communities are changing constantly, they are not static. People from different cultural and linguistic backgrounds form different belief systems, communication styles, and understanding and response to health information.

Enhanced health literacy definitions and conceptual models require more complex, comprehensive, socially oriented, culture sensitive, participatory health education to plan more targeted health education interventions and to avoid failures in health education.

References

1. Zimmerman, E., Steven Woolf, H.S. Understanding the relationship between education and health (2014). Discussion Paper, Institute of Medicine, Washington, DC. Accessed at http://www.iom.edu/understandingtherelationship (on 16.12.2016)
2. Feinstein, L., Sabates, R., Anderson, M.T., Sorhaindo, A. and Hammond, C. What are the effects of education on health? Measuring the Effects of Education on Health and Civic
3. Engagement Proceedings of the Copenhagen Symposium (2006). Accessed at https://www.oecd.org/edu/innovation-education/37437718.pdf (on 02.12.2016)
4. Suhrcke, M., de Paz Nieves, C. The impact of health and health behaviours on educational outcomes in high income countries: a review of the evidence (2011). Copenhagen, WHO Regional Office for Europe
5. Cutler, M.D., Lleras-Muney, A. Education and Health: Evaluating Theories and Evidence. NBER Working Paper No. 12352 June, JEL No. I1, I2 (2006)
6. Education Matters for Health. Issue Brief 6: Education and Health, page 1–11(2009) Accessed at http://www.commissiononhealth.org/PDF/c270deb3-ba42-4fbd-baeb-2cd65956f00e/Issue%20Brief%206%20Sept%2009%20%20Education%20and%20Health.pdf (on 02.12.2016)
7. Nutbeam, D. Health promotion glossary. Health Promotion International **13**, 349–364 (1998)
8. Kickbusch, S.I. Health literacy: addressing the health and education divide. *Health Promotion International* **16**(3), 289–297 (2001)
9. Nutbeam, D. Defining, measuring and improving health literacy. *HEP* **42**(4), 450–455 (2015)
10. Simonds, S.K. Health education as social policy. *Health Education Monograph* **2**, 1–25 (1974)
11. Selden, C., Zorn, M., Ratzan, S.C., Parker, R.M. Health literacy. Bethesda (MD): National Library of Medicine, Feb (2000)
12. van der Heide, I., Wang, J., Droomers, M., Spreeuwenberg, P., Rademakers, J., Uiters, E. The relationship between health, education, and health literacy: results from the Dutch Adult Literacy and Life Skills Survey. *Journal* of *Health Communication* **18**(Suppl. 1), 172–184 (2013)
13. DeWalt, D.A., Berkman, N.D., Sheridan, S., Lohr, K.N., Pignone, M.P., *et al* Literacy and health outcomes: a systematic review of the literature. *Journal of General Internal Medicine* **19**, 1228–1239 (2004)

14. Berkman, N.D., Sheridan, S.L., Donahue, K.E., Halpern, D.J., Crotty, K. Low health literacy and health outcomes: an updated systematic review (2011). *Annals of Internal Medicine* **155**(2), 97–107
15. Howard, D.H., Gazmararian, J., Parker, R.M. The impact of low health literacy on the medical costs of Medicare managed care enrolees (2005). The American Journal of Medicine **118**, 371–377
16. Baker, D., Parker, R., Williams, M.V., Nurss J. The Relationship of Patient Reading Ability to Self-Reported Health and Use of Health Services. *American Journal* of *Public Health* **87**(6), 1027–1039 (1997)
17. Nutbeam, D. Health literacy as a public health goal: A challenge for contemporary health education and communication strategies in the 21st century. *Health Promotion International* **15**, 259–267 (2000)
18. Baker, D.W. The meaning and the measure of health literacy. *The Journal of General Internal Medicine* **21**(8), 878–883 (2006)
19. Nutbeam, D. The evolving concept of health literacy. *Social* Science & Medicine. **67**(12), 2072–2078 (2008)
20. Zarcadoolas, C., Pleasant, A., Greer, D.S. Understanding health literacy: an expanded model. *Health Promotion International* **20**(2), 195–203 (2005)
21. Sørensen, K., Van den Broucke, S., Fullam, J., Doyle, G., Pelikan, J., Slonska, Z., Brand, H. Health literacy and public health: a systematic review and integration of definitions and models. *BMC Public Health* **12**, 80 (2012)
22. Pleasant, A., R.E. Rudd, C. O'Leary, M.K. Paasche-Orlow, M.P. Allen, W. Alvarado-Little, L. Myers, K. Parson, and S. Rosen. Considerations for a new definition of health literacy. Discussion Paper, National Academy of Medicine, Washington, DC (2016)
23. Paakkari, L., Paakkari, O. Health literacy as a learning outcome in schools. *Health Education* **112**(2), 133–152 (2012)
24. Spady, W. Outcome-based Education: Critical Issues and Answers (1994)
25. Matthews, C. Critical pedagogy in health education. *Health Education Journal.* **73**(5), 60–609 (2014)
26. WHO, Track 2: Health literacy and health behaviour. 7th Global Conference on Health Promotion: track themes. Accessed at `http://www.who.int/healthpromotion/conferences/7gchp/track2/en/` (on 10.01.2017)
27. Ishikawa, H., Kiuchi, T. Health literacy and health communication. *Biopsychosoc Med* **4**, 18 (2010)
28. American Medical Association; Ad Hoc Committee on Health Literacy for the Council on Scientific Affairs. Health literacy: report of the Council on Scientific Affairs. *JAMA* **281**, 552–557 (1999)
29. Tones, K. Radicalism and the Ideology of health education. *Health Education Research* **8**(2), 147–150 (1993)
30. Jahan, A. Health Literacy. Promoting health literacy: a case study in the prevention of diarrhoeal disease from Bangladesh. *Health Promotion International* **15**(4), 285–289 (2000)
31. Wallerstein, N., Bernstein, E. Empowerment education: Freire's ideas adapted to health education (1988). *Health Education Quarterly*. Winter **15**(4), 379–394
32. Estacio, E.V. Health literacy and community empowerment: it is more than just reading, writing and counting (2013). *Journal of Health Psychology* **18**(8), 1056–1068
33. Institute of Medicine, Committee on Health Literacy. Health Literacy: A Prescription to End Confusion (2004)

34. Lowell, A., Schmitt, D., Ah Chin W., Connors, C. Provider Health Literacy, cultural and communication competence: towards an integrated approach in the Northern Territory. Department of Health Library Services (2014)

Young adult health literacy on health promotion as social determinant of health: Challenges for health education practice in Lithuania

J. Cesnaviciene, A. Kalinkeviciene, and S. Ustilaite

Lithuanian University of Educational Sciences, Vilnius, Lithuania

Abstract. A synthesis of theory, research, and practice shows that education is linked to better health through individuals' increased health knowledge and healthy behaviour. Health literacy is put forward as the central mediator between education and health. Scientific studies have shown a strong association between the levels of health literacy and health outcomes. The aim of the study is to identify the levels of health literacy on health promotion among Lithuania's young adults. The anonymous self-reported questionnaire survey was carried out in the period of 2014–2015. The research sample involved 842 young adults aged 18–29. Health literacy was measured using an HLS-EU-Q-47 questionnaire. The research studies suggest that less than half of Lithuanian young adults have sufficient or excellent health literacy on health promotion, which is essential in maintaining good health and the quality of life. These findings are important evidence that health education should become a priority area at all levels of educational process at every stage of the life: from childhood through adulthood.
Key words: health education, health literacy, young adults, Lithuania.

1 Introduction

Health is one of the most important multidimensional agents in the age of globalization, which determines the welfare of the society and the individual as well as the processes of state advancements. It is the high level of public health which is a prerequisite for the sustainable development of the country and the accumulation of human capital. It is acknowledged that health condition is largely influenced by the lifestyle of people themselves. Investing into the development of healthy lifestyle habits is more beneficial than paying for the treatment of an ill person. Therefore, efforts must be made that each individual would realise the influence of his lifestyle on his health and to encourage responsibility for his own health and that of others, to strengthen the motivation to develop healthy living habits. The foundations of behaviour and lifestyle, which remain for a lifetime, which in the future will determine health and quality of life of everybody, are built up in the childhood and adolescence. School, after the family, is the most important arena for health education. Coordinated formal and non-formal health education, which is based on continuity, builds students' knowledge, skills, and positive attitudes to health and healthy lifestyle.

1.1 Health education in the Lithuanian education system

Health education in Lithuanian schools has come a long way over the past quarter of the century. When, after the Restoration of the independence of the Republic of Lithuania, the education reform was commenced, an important goal was established – to bring up a healthy young generation, each member of which is responsible for his health and that of other people, is able to choose an appropriate healthy lifestyle, is aware of the need to take care of his health. During the first stage of the education reform (1988–1997) the greatest attention was paid to the restructuring of the content of the education. In 1992 the first Health Education Programme [1] was prepared, which later on (in 1994 and 2000) was improved and was repeatedly published. The content of the programme covered ten topics: know yourself; personal hygiene; me and others; food and nutrition; activity, rest and physical activeness; body care; environment and health; harmful habits; substance use and abuse; family and health; diseases and their prevention. These topics were consistently repeated during the whole period of studying. When implementing the Health Education Programme it was sought to ingrain a holistic perception of health, to provide the pupils' knowledge and to develop skills, assisting in promoting their health and that of others, to develop health protection and promotion related value orientations and healthy lifestyle attitudes. The implementation of the programme was based on the principles of consistency and regularity, the differentiation of health education content, taking into consideration a pupil's age and the experience he possesses, maximum integration into the content of all the subjects of general education and non-formal education.

During the later stages of the education reform, when renewing the content of general education the mentioned Health Education Programme [1] was not renewed, however, three additional programmes were prepared. One of them is Life Skills Education Programme [2], aimed at developing the personal and social skills in pre-school children and the pupils of 1–10th forms and preparing for the living outside school and adult life. Other two programmes (Programme for Prevention of Use of Alcohol, Tobacco and Other Psychoactive Substances [3] and Preparation for Family Life and Sexuality Education Programme [4]) were compulsorily implemented at all levels of education from nursery up the 12th form.

In 2008, when the renewal of the general education content was commenced, not only special, but also transversal competences (learning to learn, communication competence, cognitive competence, social competence, sense of initiative and creativity, personal competence, cultural competence), which empower the self-realisation of the young generation, became a priority of Lithuanian education. Health education continued to be an integral part of the content of the taught subjects, which was executed in accordance with the aforementioned programmes. And only in 2012 the new General Health Education Programme [5] was prepared, reflecting modern education paradigm, which aims at establishing a positive attitude towards health and the factors promoting it, building up healthy living habits, improving knowledge of health protection and the ways to promote it. At the level of primary education, it is recommended that this programme would be integrated into the content of the taught subjects and non-formal education. At the level of lower and upper secondary education it should be integrated into all the taught subjects and non-formal education, as well as during at least one year of schooling at every level of education (5–8th, 9–10th and 11–12th forms) health education should be taught during special lessons, which encompass the entirety of the programme. The implementation of this General Health Education Programme [5] is based on the principle of continuity, ensuring a consistent development of competences in four areas: health and healthy living concept; physical health (physical activity, healthy nutrition, activity and rest, personal and environmental cleanliness); mental health (self-esteem, emotions and feelings, mind and positive thinking); social health (social awareness and culture, communication, the prevention of risky behaviour).

In response to the challenges to education arising in the contemporary society, the September of 2017 marks the start of the implementation of the newest General Health and Sexuality Education and Preparation for Family Life Programme [6] in the schools of general education. It was prepared taking into consideration the opinion of scientists and health specialists, the findings of scientific research, health statistics and the recommendations of the World Health Organization and as a result of merging of the Preparation for Family Life and Sexuality Education Programme [4] and the General Health Education Programme [5], which were being implemented in schools up to that time. The objective of the latest programme is to assist pupils in acquiring the holistic understanding of health and sexuality, developing health-enhancing skills and habits, promoting responsibility for their health and that of others, encouraging them to choose a healthy lifestyle, developing an ability to create and maintain harmonious and mature interpersonal relationships. The essential aspects of the content of the programme are presented in Table 1.

Table 1. Essential aspects of the content of General Health and Sexuality Education and Preparation for Family Life Programme.

PROGRAMME CONTENT		
1. Health, healthy living and family concepts		
2. Physical health	3. Mental health	4. Social health
Physical activity Healthy nutrition Harmonisation of activeness and rest Personal and environmental cleanliness Sexual maturation	Self-esteem Emotions and feelings Self-control Positive, constructive thinking and self-expression	Friendship and love resilience to risky behaviour
PROGRAMME OUTCOME		
Health education	Sexuality education	Preparation for family life
Healthy living (lifestyle, the entirety of daily habits, which assist in maintaining and promoting health)	Perception of innate sexual identity, responsible sexual behaviour, which means an ability, based on volition and motivation, to choose, for self and another person, the safest (in a physical and mental sense) sexual behaviour, encompassing respect for self and another person, the realisation of the consequences of a risky sexual behaviour	Ability to create and maintain interpersonal relationships, facilitating a mature friendship and love in order to consummate it in marriage and to create a harmonious family

It is recommended that the General Health and Sexuality Education and Preparation for Family Life Programme [6] would be implemented in a joined effort of a school and family (parents or guardians) through the collaboration of the teachers of different subjects and other education professionals and the participation of non-governmental organisations and healthcare specialists. The implementers of the programme are advised to follow the implementation guidelines of the General Health Education Programme [5], which had been executed up to that point. In addition to this compulsory programme, schools may optionally implement preventive programmes accredited by the National Centre for Special

Needs Education and the Psychology Centre as well: an early prevention programme "Zippy's Friends" and "Apple's Friends"; Child's Emotional Competences Training Programme; the programme "Second Step" to develop social-emotional skills; a preventive program "Unplugged"; Olweus bullying prevention programme; an alcohol, tobacco and other drug use prevention programme "Snowball"; LIONS QUEST Life Skills education programme "The Crossroads of Adolescence".

It should be noted that some Lithuanian schools of general education have been participating in the activities of the European Network of Health Promoting Schools or the national network since 1993. Scientific research [7, 8] has concluded that those are the schools, in which complex and coordinated health promotion is more often integrated into the system of general education. Compared with the schools in which targeted health education activity was not performed, the schools promoting health show positive changes towards healthy living among pupils: the obvious tendencies of healthier food consumption, increased levels of pupils' physical activity, lower levels of alcoholic beverages consumption, greater attention to oral hygiene. Recent scientific evidence [7, 8] showed that health promoting school concept is acceptable for the schools of general education; health promotion activity, which is based on it, is effective and can be used for the further development of the activity in order to engage more schools of general education into the national network of health promoting schools.

The review of health education programmes shows that, in Lithuanian education system, conditions have been established for the formation of healthy living of the young generation. It is corroborated by the experience of the schools, which participate in the network of health promoting schools. However, the findings of the scientific research indicate that the schools of general education still "lack coherent development of social skills, too little importance is still given to the building up of healthy living habits. The general view is that the priority of school activity is to prepare pupils for the final examination, whereas the implementation of health education by teachers is marked by fragmentation" [9, p. 147].

1.2 Health literacy as a multidimensional concept in health education

Since the end of the 20th century in different countries around the world, health literacy has become an important precondition for social and economic development of society. It is necessary for the daily ability of people to manage their health, healthy living and the quality of life, which depends on them. According to L. Kannas [10, p. 13], "improving people's health literacy across the lifespan has become an important health promotion approach in order to support healthy development in childhood and healthy ageing". High or low literacy levels have different impacts, notably on the individual's health. There is an increasing scientific evidence that the general health status of the people, whose health literacy is at a low level, is lower [11–14], their living is less healthy [15, 16], they have a lower preventive measures and services take-up [15, 17], their capacities to manage chronic diseases and to remember or to understand the information presented by the medical personnel are insufficient [18, 19]; such people are more prone to make mistakes when taking their medicines or to adhere to other instructions of the doctors [12, 20, 21].

One of the first health literacy concepts reflects traditional literacy and encompasses an individual's ability to perform tasks, requiring reading and arithmetic skills, related to information on health [22, 23]. However, over time, such narrow approach towards this concept was changing. One of the health literacy concepts, which is most frequently mentioned in scientific literature and recognised at international level, was proposed in 1998 by the World Health Organisation. The complemented Health Promotion Glossary points out that these are the "cognitive and social skills which determine the motivation and ability of individuals to gain access to, understand and use information in ways which promote and maintain good health" [24, p. 10].

Foreign scientific literature abounds with the diversity of concepts on health literacy. Having analysed them, the scientists [25–28] show that this concept is expanding from a narrow immutable understanding towards a renewing and multidimensional concept, which changes all lifelong. It is gradually changing from an individual towards communal level. One of the latest health literacy concepts is based on the systematic analysis and conceptualisation of the definitions presented in scientific literature: "health literacy is linked to literacy and entails people's knowledge, motivation and competences to access, understand, appraise, and apply health information in order to make judgments and take decisions in everyday life concerning healthcare, disease prevention and health promotion to maintain or improve quality of life during the life course" [28, p. 3]. In other words, the multidimensional concept of health literacy reflects a complex approach that combines four information processing stages (access, understand, appraise, apply) and three health relevant areas (health care, disease prevention, health promotion). In healthcare, health literacy is linked to a personal ability to extract information on medicine, to understand, to interpret and to appraise it and to make informed decisions and to adhere to the advice of healthcare specialists. Health literacy in disease prevention is an opportunity to receive and to understand, to interpret and to appraise information on health risk factors, to reach sound decisions regarding the management of health risk factors. In health promotion, health literacy is important as an ability to constantly renew, to realise, to interpret and to appraise information on social and environmental factors, which determine health, capacity to take sound decisions regarding the factors, which influence health. Consequently, health literacy is becoming an important contribution to the empowerment of people in the areas of healthcare, diseases prevention and health promotion.

There is no lack of research on health literacy abroad. However, there is a lack of sufficient research in Lithuania. Two more exhaustive studies published in the Lithuanian language can be mentioned: D. Zagurskiene "The Evaluation of Patients' Health Literacy" [18] and Z. Javtokas "Overview of the Research Data on Health Literacy of Lithuanian Population" [29]. Similar studies have not so far been conducted among Lithuanian young adult population. Therefore, it is becoming important to research and to analyse what level of health literacy the young adults, having graduated from schools of general education, possess.

The aim of the study is to identify the specific characteristics of health literacy of young adults and to reveal health literacy in health promotion as a challenge to health education.

2 Materials and methods

2.1 Research instrument

Subjective health literacy was measured using an HLS-EU-Q-47 questionnaire [30]. With the agreement of the HLS-EU Project coordinator K. Sørensen, the HLS-EU-Q-47 questionnaire was translated from English into the Lithuanian language using the translation–back-translation method. The questionnaire was pre-tested for comprehension and completeness. To explore internal consistency of the questionnaire, Cronbach's alpha was calculated. The analysis showed that Cronbach's alpha was satisfactory (Cronbach's $\alpha = 0.94$ respectively). In total, the HLS-EU-Q-47 questionnaire comprised 47 items which explored four dimensions of health literacy: access, understanding, appraisal and application of health information in three different domains: health care (16 items), disease prevention (16 items) and health promotion (15 items). The respondents were asked to report on 47 items using a four-point Likert scale (1 = *very difficult*, 2 = *difficult*, 3 = *easy*, and 4 = *very easy*). The "*I don't know*" answer option was used when stated spontaneously and coded as a missing value.

Demographic characteristics measured via the questionnaire included age (year), gender (male, female).

2.2 Characteristics of the research sample

According to the statistical information of Statistics Lithuania (`http://www.stat.gov.lt`), on January 1, 2014, 482,142 young adults aged 18–29 (247,578 males and 234,564 females) lived in Lithuania. The research sample was determined using sample size calculator (`http://www.surveysystem.com/sscalc.htm`). Having conducted calculations, it was determined that 798 respondents (399 males and 399 females) was a sufficient research sample that would represent the general population of 18–29 year old young adults.

The final study random sample was composed by 842 young adults, who voluntarily decided to participate. Data were collected during the period of 2014–2015. The paper-and-pencil questionnaire or online questionnaire was completed anonymously, and the participants were assured of data confidentiality. Table 2 presents the characteristics of the study sample.

Table 2. Distribution of sample by age group and gender.

Age group	Male (N = 414)		Female (N = 428)		Total (N = 842)	
	N	%	N	%	N	%
18–19	143	34.5	89	20.8	232	27.5
20–24	144	34.8	190	44.4	334	39.7
25–29	127	30.7	149	34.8	276	32.8

2.3 Data analysis

The data was analyzed using SPSS version 17.0. The following methods were applied to analyse the research data: descriptive analysis (frequencies, percentage), Pearson's Chi-square tests (nonparametric test was used to compare health literacy levels between two groups – young men and young women), Cronbach's alpha was used to test the reliability of the questionnaire scores and measure the internal consistency of propositions. The statistically significant value was set at $p < 0.05$.

Health literacy index scores for three scale (health care, disease prevention, and health promotion) and general health literacy index were standardized on a metric scale between 0 (lowest level of health literacy) and 50 (best level of health literacy) and constructed using the formula [30]:

$$I = (X - 1) \times \frac{50}{3}$$

In this formula, I – health literacy index, X – mean of all participating items for each individual.

The study participants were divided into 4 groups, which reflected their health literacy abilities. Group 1 (*inadequate health literacy*) included people who scored between 0–25, group 2 (*problematic health literacy*) people who scored between 25.01–33, group 3 (*sufficient health literacy*) people who scored between 33.01–42, and group 4 (*excellent health literacy*) people.

3 Results and discussion

3.1 Young adult health literacy by gender

As has already been mentioned, using HLS-EU-Q-47 questionnaire it is possible to find out general health literacy, health literacy on health care, health literacy on disease prevention, and health literacy on health promotion. Analysing the findings of the study, which are presented in Table 3, it is noted that, when it comes to the levels of health literacy, the percentage frequencies of the young men and the young women are similar, no statistically significant difference has been established.

Thus, having generalised the research results, it is possible to state that persons with sufficient or excellent levels of health literacy on health care made up 30.2% and 9.5% of the total, respectively. This study showed that 13.7% of the young adults had excellent health literacy on disease prevention, and 33.8% of the young adults had sufficient health literacy in this area. Analysing the respondents' replies about health literacy on health promotion, it is possible to claim that only two-fifths of the young adults had sufficient or excellent health literacy on health promotion.

Table 3. Distribution of young adults by levels of health literacy and gender (%).

	Health literacy levels	Male ($N = 414$)	Female ($N = 428$)	Total ($N = 842$)
Health literacy on health care	Inadequate	20.5	18	19.2
	Problematic	39.9	42.3	41.1
	Sufficient	30.2	30.1	30.2
	Excellent	9.4	9.6	9.5
Health literacy on disease prevention	Inadequate	18.3	15.2	16.7
	Problematic	35.3	36.2	35.8
	Sufficient	32.9	34.8	33.8
	Excellent	13.5	13.8	13.7
Health literacy on health promotion	Inadequate	23.2	25.2	24.2
	Problematic	37.7	34.6	36.2
	Sufficient	27.8	30.8	29.3
	Excellent	11.3	9.4	10.3
General health literacy	Inadequate	20.8	16.6	18.6
	Problematic	38.2	41.8	40
	Sufficient	34	33.9	34
	Excellent	7	7.7	7.4

Having calculated the index of general health literacy of young adults, it was determined that 18.6% respondents had inadequate, 40% – problematic, 34% – sufficient and 7.4% had excellent general health literacy. The results of this study can be compared with the results obtained by European Health Literacy Survey [30]. The obtained data were similar to the ones obtained in Austria, Bulgaria and Spain. However, the number of Lithuanian young adults that were characterised by sufficient or excellent levels of health literacy was lower than in Germany, Greece, Ireland, Netherlands and Poland [30]. However, the comparison of such results is only partly possible, because it must be noted that the age of the respondents in the study on health literacy, which was conducted in eight European countries, was 15-year olds and older.

3.2 Young adult's health literacy on health promotion

Our article provides a more in-depth analysis of young adult health literacy on health promotion. As already been mentioned, only 10.3% of the young adults have excellent health literacy on health promotion, whereas 29.3% – sufficient (Table 3).

Although information and theoretical knowledge are not a sufficient reason to change one's lifestyle or promote health, it is affirmed that the possession of theoretical knowledge provides an opportunity to achieve a better understanding of the influence of a lifestyle on health. Consequently, in order to promote one's own health, it is firstly necessary to acquire the knowledge necessary for that. The data of the study revealed that the majority of the young adults, who participated in the study, find it easy to extract information on a healthy lifestyle (exercising, healthy nutrition and the like), as well as ways to promote emotional health. However, only half of the respondents pointed out that they found it easy to extract information on possibilities for health promotion in their closest environment. Less than half of the respondents pointed out that they found it easy to extract information on how people in their community would be able to live healthier, information on health policy changes, which may influence their health. Summarising the data of the study it can be presumed that only 48.3% of the young adults possess sufficient or excellent ability to update themselves on determinants of health in the social and physical environment. The capacities of the rest of the young adults are either problematic or inadequate.

Unfortunately, the acquisition of information and knowledge is not a sufficient prerequisite for one's lifestyle to change. Only having realised the obtained information, having embraced the notion of the conveyed knowledge, it is possible to build up healthy living skills and attitude, which will determine the behaviour, which is favourable to health. On the basis of the findings of the study, it was established that the majority of the young adults found it easy to understand the advice of their family members or friends, or the information presented in the media how to promote health. However, only slightly more than half of the respondents agreed that they found it easy to understand the information given on food labels or the information on how to safeguard mental health. Consequently, summarising it can be presumed that only half of the young adults, who participated in the study, possess excellent or sufficient abilities to understand information.

The ability to critically analyse and appraise health-related information, to interpret it is becoming particularly important where contemporary youth is concerned. The media (the internet, the press, television) present expressive and abundant textual information and images, which, beyond reasonable doubt, influence the learning on health and healthy living of those, who live in the digital world. Therefore, the critical selection and appraisal of health-related information are becoming inextricably linked to health promotion. The majority of the young adults, who participated in the study, concurred that they find it easy to appraise how and what daily behaviour of theirs determines health and well-being and how their present living conditions assist in safeguarding health. Consequently, the study established that 74.8% of the young adults possess sufficient or excellent abilities to interpret and evaluate information on health determinants in the social and physical environment.

The critical appraisals of the information on health create prerequisites to change one's lifestyle and to reach appropriate decisions regarding health. 73.4% of the young adults, who participated in the study, pointed out that they find it easy make decisions how to promote their health. And although three-fifths of the young adults (more young men than young women, $\chi^2 = 21.050$; $p < 0.0001$) identified that they found it easy to go to a sports club, discotheque or take up another physical activity, however, only half of the young adults concurred that they found it easy to change their living conditions, in order to influence health and well-being (there were more young women, who pointed out that they find difficulty doing that, than young men, $\chi^2 = 11.973$; $p < 0.05$). Even fewer of the respondents pointed out that they found it easy to participate in the activity of their local community, which influences health. Based on such results it can be presumed that only 39% of the young adults, who participated in the study, possess an excellent or sufficient ability to make informed decisions. There were more young women, whose abilities are insufficient, than young men ($\chi^2 = 15.031$; $p < 0.01$).

Consequently, summarising the results of the study, it can be presumed that the health promotion, which is executed in schools, does not yet ensure the development of the abilities of a sufficient level. It poses a particular challenge to the educators and family of the generation, which is growing up.

4 Conclusions

The new conception of health literacy involves abilities to access, understand, appraise and apply information across the domains of health care, disease prevention and health promotion. Having generalised the research results, it is possible to claim that less than half of Lithuanian young adults have sufficient or excellent health literacy on health promotion, which is essential in maintaining good health and improving the quality of life. These findings are important evidence that health education should become a priority area at all levels of educational process at every stage of the life: from childhood through adulthood.

This study was funded by the Research Foundation of Lithuanian University of Educational Sciences.

References

1. A. Davidavičienė. Sveikatos ugdymo programa. *Universalios ugdymo programos*. 62–81 (1992). *(In Lithuanian)*
2. L. Bulotaitė, V. Gudžinskienė. *Gyvenimo įgūdžių ugdymo programa* (2003). Available at `http://www.smm.lt/web/lt/pedagogams/ugdymas/ugdymo_prog` *(In Lithuanian)*
3. Alkoholio, tabako ir kitų psichiką veikiančių medžiagų vartojimo prevencijos programa (2006). Available at `http://www.smm.lt/web/lt/pedagogams/ugdymas/ugdymo_prog` *(In Lithuanian)*
4. Rengimo šeimai ir lytiškumo ugdymo programa (2007). Available at `http://www.smm.lt/web/lt/pedagogams/ugdymas/ugdymo_prog` *(In Lithuanian)*
5. Sveikatos ugdymo bendroji programa (2012). Available at `http://www.smm.lt/web/lt/pedagogams/ugdymas/ugdymo_prog` *(In Lithuanian)*
6. Sveikatos ir lytiškumo ugdymo bei rengimo šeimai bendroji programa (2016). Available at `http://www.smm.lt/web/lt/pedagogams/ugdymas/ugdymo_prog` *(In Lithuanian)*
7. L. Stankevičienė, A. Zaborskis. Sveikatą stiprinančių Lietuvos mokyklų moksleivių gyvensena ir sveikata. Medicina, **35**, 780–787 (1999). *(In Lithuanian)*
8. A. Jociutė. *Sveikatos stiprinimo ir ugdymo veiklos veiksmingumo Lietuvos bendrojo lavinimo mokyklose įvertinimas*: daktaro disertacija (2002). *(In Lithuanian)*
9. A. Petrauskienė, S. Ustilaitė. *Mokyklos sveikatos ugdytojų vaidmenys* (2013). *(In Lithuanian)*

10. L. Kannas. Health Education as a stand-alone School Subject to improve Health Literacy of Future Generations: Lessons to learn from Finland. *8th Nordic Health Promotion Research Conference: abstract book.* (2016). Available at `https://www.jyu.fi/sport/laitokset/terveys/en/NHPRC`
11. Agency for Healthcare Research and Quality. *Health Literacy Universal Precautions Toolkit* (2010). Available at `http://www.ahrq.gov/qual/literacy/`
12. N.D. Berkman, S.L. Sheridan, K.E. Donahue, D.J. Halpern, & K. Crotty. Limited health literacy and health outcomes: An updated systematic review. Ann Intern Med **155**(2), 129–130 (2011)
13. A.E. Schnitzer, M. Rosenzweig, & B. Harris. Health Literacy: A Survey of the Issues and Solutions. J Consum Health Internet **15**(2), 164–179 (2011)
14. R. Mottus, C. Murray, J.M. Staff, W. Johnson, M.S. Wolf, I.J. Deary. Towards understanding the links between health literacy and physical health. Health Psychol **33**, 164–173 (2014)
15. S. White, J. Chen, & R. Atchison. Relationship of preventive health practices and health literacy: A national study. Am J Health Behav **32**(3), 227–242 (2008)
16. I.M. Bennett, J. Chen, J.S. Soroui, & S. White. The contribution of health literacy to disparities in self-rated health status and preventive health behaviors in older adults. Ann Fam Med **7**, 204–211 (2009)
17. Y.I. Cho, S.Y. Lee, A.M. Arozullah, & K.S. Crittenden. Effects of health literacy on health status and health service utilization amongst the elderly. Soc Sci Med **66**, 1809–1816 (2008)
18. D. Zagurskienė. Pacientų sveikatos raštingumo vertinimas: daktaro disertacija (2009). *(In Lithuanian)*
19. D. Zagurskienė, I. Misevičienė. Slaugytojų nuomonė apie pacientų sveikatos mokymą ir savo žinių sveikatos klausimais vertinimą. Sveikatos mokslai **21**(5), 191–195 (2011). *(In Lithuanian)*
20. S. Kripilani, M.E. Gatti, & T.A. Jacobsen. Association of age, health literacy, and medication management strategies with cardiovascular medication adherence. Patient EducCouns **81**, 177–181 (2010)
21. J.L. Lenahan, D.M. McCarthy, & T.C. Davis. A drug by any other name: patients' ability to identify medication regimens and its association with adherence and health outcomes. J Health Commun **18**, 31–39 (2013)
22. R.M. Parker, D.W. Baker, M.V. Williams, & J.R. Nurss. The Test of Functional Health Literacy in Adults (TOFHLA): a new instrument for measuring patient's literacy skills. J Gen Intern Med **10**, 537–542 (1995)
23. M.V. Williams, R.M. Parker, D.W. Baker, N.S. Parikh, K. Pitkin, W.C. Coates, & J.R. Nurss. Inadequate functional health literacy among patients at two public hospitals. JAMA **274**, 1677–1682 (1995)
24. World Health Organization. *Health Promotion Glossary* (1998). Available at `http://apps.who.int/iris/handle/10665/64546`
25. B. Kwan, J. Frankish, & I. Rootman. The development and validation of measures of "health literacy" in different populations (2006)
26. N.D. Berkman, T.C. Davis & L. McCormack. Health Literacy: What Is It? J Health Commun **15**(S2), 9–19 (2010)
27. R. Marks. *Health Literacy and School-Based Health Education* (2012)
28. K. Sørensen, S. van den Broucke, J. Fullam, G. Doyle, J. Pelikan, Z. Slonska, & H. Brand. Health literacy and public health: A systematic review and integration of definitions and models. BMC Public Health **12**(1), 80 (2012)

29. Z. Javtokas. Lietuvos gyventojų sveikatos raštingumo tyrimo duomenų apžvalga (2012). Available at http://www.smlpc.lt/media/file/SMLPC-informacija/Veikla/Tyrimai-analizes/Sveikatos%20rastingumas-2012.pdf *(In Lithuanian)*
30. HLS-EU Consortium. *Comparative report of health literacy in eight EU member states.* The European Health Literacy Survey HLS-EU (2012). Available at http://www.health-literacy.eu

The relationship between anomia and maladaptive personality traits

J. Levina*, V. Perejolkina, K. Martinsone, S. Mihailova, and J. Kolesnikova

Rīga Stradiņš University, Riga, Latvia

Abstract. Research of anomia is topical in the period when a society endures social and economic changes. On the other hand, the modern Western society is characterized by a rapid speed of development, by diversity, variability, multicultural life. In such conditions for some individuals it could be difficult to make life choices. Such individuals can get into "an anomic state". In a situation of rapid social and political changes as well as simply in the modern diverse and rapidly developing society pathological personality traits can serve as a factor that aggravates anomic feelings. Thus, the main purpose of this research is to investigate relations between anomia and pathological personality traits. The sample consisted of 83 Latvian inhabitants aged from 19 to 47 years (16.9% males, 83.1% females). The Anomia Questionnaire – AQ [1] and the Latvian Clinical Personality Inventory – LCPI v2.2 [2] were used. It is found that there are multiple relations between all six sub-dimensions of anomia and certain different maladaptive personality traits. Irresponsibility is the only personality trait that predicts deviation from prescribed rules or customs. Negative emotionality is the only significant predictor for social distrust. Suspiciousness that falls within a first-order domain Psychoticism and a second-order domain Schizotypy is the most significant predictor of estrangement to others as a dimension of anomia. Eccentricity that falls within a first-order domain Psychoticism and a second-order domain Schizotypy is the most significant predictor of cultural isolation. Separation insecurity that falls within a first-order domain Dependence and a second-order domain Neuroticism is the most significant predictor of a lack of goal clarity as a dimension of anomia. Depression as a pathological personality trait that falls within a first-order domain Negative emotionality and a second-order domain Neuroticism is the most significant predictor of a lack of goal clarity as a dimension of anomia.

1 Introduction

Research of anomia is topical in the period when a society endures social and economic changes. It is acknowledged that social, economic, political, cultural changes at the level of a society cause ***anomie*** – the condition or state, in which there is a breakdown of social norms and guidance for the citizens of a society [3] as well the discrepancy between common social goals and the legitimate means to attain those goals [4]. In its turn, at the

* Corresponding author: `jelena.levina@rsu.lv`

individual level such changes can cause ***anomia*** – "subjectively perceived anomie". While ***anomie*** refers to the state of a society characterized by the lack of norms and regulations, ***anomia*** characterizes the level of the individual state [5].

During the last decades both worldwide and in Latvia, rapid social and economic changes took place, such as the economic crisis, changes in the labour market, labour migration and others. All these transitions in the development of our society, social and economic changes could affect individuals and cause anomia.

On the other hand, the modern Western society is characterized by a rapid speed of development, by diversity, variability, multicultural life. In such conditions for some individuals it could be difficult to make life choices, to choose unambiguous and accurate norms and guidance for their behaviour. Such individuals can get into "an anomic state", namely, they could develop a deviant behaviour, lose social trust, feel social isolation, feel difficulties when it is necessary to choose life goals, feel a sense of generalized meaninglessness.

There is an empirical evidence of negative influences of anomia. It was found that anomia is negatively associated with subjective well-being – overall life satisfaction, satisfaction with separate life domains, and general sense of happiness [6, 7]. Anomia is negatively related with such significant values as politics and religion [8]. In a recent brief review of publications on anomia [9] it was shown that (1) anomic feelings are connected with higher suicidal tendencies [10, 11]; (2) anomia is associated with substance use [12]; (3) anomic individuals have low self-esteem [13]. Therefore, an important task is to investigate factors affecting anomia.

Pathological personality traits are recognized as one of important determinants of pathology and unsuccessful human functioning. For example, there is empirical evidence that pathological personality traits are associated with internalizing psychopathology [14], the aversive interpersonal behaviours [15], intimate partner violence [16], pathological gambling [17], the aggressive and self-defeating humour styles [18].

In a situation of rapid social and political changes as well as simply in the modern diverse and rapidly developing society pathological personality traits can serve as a factor that aggravates anomic feelings. Thus, the main purpose of this research is to investigate relations between anomia and pathological personality traits.

2 The concept and the integrative multidimensional model of anomia

The integrative multidimensional model of anomia was developed by Ļevina, Mārtinsone and co-authors [19–23]. According to this model *anomia* is defined as an individual psychological state when a person believes that there are no social norms, rules and prescriptions, which can regulate his/her own behaviour as well as the behaviour of other citizens of a society, when a person endures the psychological state of social isolation and meaninglessness.

According to this model, anomia is a multidimensional construct and includes three main dimensions: ***Normlessness***, ***Social isolation***, and ***Meaninglessness***. Normlessness is defined as a perceived breakdown of a social order in which norms no longer regulate behaviour. Social isolation is defined as a loss of a sense of community. At last, meaninglessness is defined as an absence or unclearness of terminal goals (life meanings) as prescriptions for an individual's behaviour.

Each major dimension of anomia is divided into two sub-dimensions:

- ***Normlessness*** includes an *individual's deviation from prescribed rules or customs* and *social distrust*;
- ***Social isolation*** includes *estrangement to others* and *cultural isolation*.

- ***Meaninglessness*** includes *lack of goal clarity* and *generalized sense of meaninglessness*.

3 Factors influencing anomia

Theorists of anomia [3, 24, 25] consider that both negative and positive social and economic changes, e.g. socially economic blossoming and crises, can cause *anomie* at the level of society and *anomia* at the level of an individual, however, more often exactly periods of sharp crises lead to this state.

From a psychological point of view, it is important to investigate which individual factors affect anomia. Up to now some studies were conducted in which such factors as socially demographic characteristics, namely, sex and age, and participation in social activities, including voluntary work were investigated [7, 9, 22].

Regarding socially demographic characteristics it was found that there was a significant difference in multidimensional anomia between different age groups. Middle-aged adults had higher scores for social isolation than respondents from the group of early adults. Late adolescents-youths had the lowest scores for meaninglessness comparing with all other age groups. Early adults had lower scores for meaninglessness than middle and late adults. Sex differences were no found [22].

Regarding participation in social activities it was found that such aspects of anomia as social distrust, social isolation and meaninglessness were negatively associated with social activities – interacting with others without doing a specific activity with them, doing an activity with others, helping others and contributing to society or civic activities [9].

While a great significance of maladaptive personality traits for health and psychosocial functioning is obvious, there were no studies in which relations between anomia and pathologic personality traits were explored.

4 Multidimensional models of pathologic personality traits

The DSM-5 Personality and Personality Disorders Work Group proposed a substantial shift to a dimensional conceptualization and diagnosis of personality pathology [26]. In the alternative DSM-5 model for personality disorders (PDs), PDs are characterized by impairments in personality functioning and particular constellations of *pathologic personality traits*. The recognition of the many limitations of the categorical model of personality disorder classification led to the development of quit a number of alternative proposals for a dimensional classification of personality disorders [27]. During past decades, different models of pathologic or maladaptive personality traits have been developed [27] and most of them more or less represent maladaptive variants of the five domains of the extensively validated and replicated personality model known as the "Big Five", or Five Factor Model of personality (FFM).

One of such models of pathologic personality traits was developed in Latvia as a theoretical frame for Latvian Clinical Personality Inventory (LCPI-v3) [2]. First, based on a theoretical analysis of existing models of pathologic personality traits, a list of 40 traits – a combination of traits listed in Samuel et al. [26], Wright et al. [28], DSM-5 Section III (PSY-5 model, APA, [29]) was prepared and used in deductively derived preliminary versions of 40 pathologic personality trait scales [30], and then, based on the results of a factor analytic study [31] a final model of 33 facet-level pathological personality traits that comprise seven first-order domains (namely, *Psychoticism*, *Introversion*, *Narcissism*, *Impulsivity*, *Dependence*, *Negative emotionalism* and *Compulsivity*), which in their turn, can be grouped in four second-order domains (*Schizotypy*, *Antagonism*, *Neuroticism* and *Compulsivity*) was developed.

According to this multidimensional hierarchical model six facet-level pathological personality traits – cognitive non-regulation, dissociation proneness, eccentricity, suspiciousness, unusual beliefs, and self-harm – fall within a higher-order domain *Psychoticism*. Three pathological personality traits – restricted affectivity, social withdrawal, close relationship avoidance – fall within a higher-order domain *Introversion*. In their turn, *Psychoticism* and *Introversion* form a second-order domain ***Schizotypy***.

Four facet-level pathological personality traits – aggression, irresponsibility, rashness, risk taking – fall within a higher-order domain *Impulsivity*. Six pathological personality traits – dominance, arrogance, attention seeking, manipulative behaviour, harshness, and deceitfulness – fall within a higher-order domain *Narcissism*. *Impulsivity* and *Narcissism* form a second-order domain ***Antagonism***.

Seven facet-level pathological personality traits – intemperance, emotional stability, emotional instability, depressiveness, anxiousness, impersistence, and distrustfulness – fall within a higher-order domain *Negative emotionality*. Four facet-level pathological personality traits – evaluation apprehension, submissiveness, indecisiveness, and separation insecurity – fall within a higher-order domain *Dependence*. *Negative emotionality* and *Dependence* form a second-order domain ***Neuroticism***.

Finally, three facet-level pathological personality traits – pedantry, perseveration, and perfectionism – fall within a higher-order domain *Compulsivity*. This domain remains as a second-order factor ***Compulsivity***.

5 The present study

Examination of the existing literature allows concluding that social, economic, *political*, *cultural* changes as well as overall such characteristic of modern society as a high speed of development and diversity cause anomia at the individual level. Anomia is a multidimensional construct and could be experienced as an individual's deviation from prescribed rules or customs, social distrust, estrangement to others, cultural isolation, a lack of goal clarity and generalized sense of meaninglessness.

In the current study we hypothesized that maladaptive personality traits (a tendency to feel, perceive, behave, and think in relatively consistent ways across time and across situations in which the trait may manifest (APA, [29]) can be factors that mediate the relations between a situation of rapid changes and transitions and anomia. Thus, the purpose of this research is to determine whether there are the relations between anomia and maladaptive personality traits.

We formulate the following research questions:

1) What are the relations between dimensions of anomia and pathological personality domains and facet-level traits?
2) What pathological personality domains and traits best predict specific dimensions of anomia?

6 Methods

6.1 Participants

The sample consisted of 83 respondents, aged from 19 to 47 years ($M = 24.29$, $SD = 6.65$), 16.9% were males and 83.1% were females. Majority (67.5%) of respondents live in Riga, 13.3% of respondents live in another large city in Latvia. The rest respondents live in a small town or in a rural settlement; 84.3% are Latvians, 12.0% are Russians, the rest part of respondents (2.4%) have another nationality or simply did not indicate it.

Regarding family status, 18.1% indicated that they are married, 26.5% live together with a boyfriend/girlfriend, but are not officially married, 1.2% is divorced, and 53.0% never had registered official marriage or lived with a partner unofficially. Other respondents did not indicate their family status.

Regarding education level, 18.1% of respondents have secondary education, 4.8% of respondents have special secondary education or vocational secondary education, the largest part (51.8%) of respondents have unfinished higher education, 4.8% of respondents have a higher education (I level) and 14.5% have a higher education (II level, Bachelor's degree), 6% of respondents have a Master's degree.

Income level distribution in the sample was as follows: 4.8% of respondents have level of income that is completely satisfying and 57.8% have level of income that was subjectively characterised as satisfying, 33.7% of respondents are unsatisfied with their level of income. The rest respondents did not specify their level of income.

6.2 Instruments

6.2.1 Anomia Questionnaire (AQ)

The Anomia Questionnaire – AQ [1] was used. The questionnaire includes 24 statements. Responses are made on a 5-point Likert scale (from 1 – strongly disagree till 5 – strongly agree). The questionnaire is in Latvian.

The questionnaire consists of six scales (Deviation from prescribed rules or customs, Social distrust, Estrangement to others, Cultural isolation, Lack of goal clarity, Generalized sense of meaninglessness) and, respectively, measures six sub-dimensions of anomia. All the AQ scales have high internal consistency – Cronbach's alpha vary from .73 to .86 (*Ļevina, Mārtinsone*, in press).

6.2.2 Latvian Clinical Personality Inventory (LCPI v2.2)

Participants filled in the second preliminary item pool of the Latvian Clinical Personality Inventory (LCPI v2.2, $k = 664$) [32], but for scale scoring the final version of this inventory (LCPI-v3; [2]) was used. Responses were made on a 4-point Likert scale (from 0 – strongly disagree till 3 – strongly agree).

The questionnaire is in Latvian. The LCPI-v3 includes 322 statements which could be scored in nine clinical scales, five functional scales, five additional scales and 33 facet-level pathological personality trait scales that form seven first-order domain-level scales and four second-order personality domain-level scales. Clinical scales of LCPI-v3 are: Depression symptoms scale (DP), Symptoms of posttraumatic stress disorder (PTSD), General anxiety symptoms (GA), Panic attack symptoms (PA), Symptoms of social anxiety (SA), Alcohol related problems (ALCO), Drug related problems (DRUG), Somatic symptoms (SM) and Psychotic symptoms (PSS); five functioning scales of LCPI-v3 are: Sleep problems (F1), Energy level (F2), Psychomotor retardation and diminished cognitive function (F3), Problems associated with attention functions (F4), and Communication problems (F5); Additional scales of LCPI-v3 are: Suicide ideation (SI), Low self-esteem (LSE), Stress symptoms (STRS) Perceived social support (PSS) and Unstable and intense interpersonal relationships (REL).

In this research only these 33 facet-level, seven first-order and four second-order domain-level scales of pathological personality traits were used. Used facet-level scales are as follows: Cognitive non-regulation, Dissociation proneness, Eccentricity, Suspiciousness, Unusual beliefs, Self-harm, Restricted affectivity, Social withdrawal, Close relationship avoidance, Aggression, Irresponsibility, Rashness, Risk taking, Dominance, Arrogance, Attention seeking, Manipulative behaviour, Harshness, Deceitfulness, Intemperance, Emotional stability, Emotional instability, Depressiveness, Anxiousness, Impersistence, Distrustfulness, Evaluation apprehension, Submissiveness, Indecisiveness, Separation insecurity, Pedantry, Perseveration, and Perfectionism. Seven personality domain-level scales (first-order factors) are as follows: Psychoticism, Introversion, Impulsivity, Narcissism, Negative emotionality, Dependence, Compulsivity. Finally, four major personality domain-level scales (second-order factors) are as follows: Schizotypy, Antagonism, Neuroticism, and Compulsivity. All used scales of LCPI-v3 have high internal consistency – Cronbach's alpha vary from .70 to .95 [2].

6.2.3 Procedure

Data was collected on a voluntary basis via internet from January 2016 to May 2016.

7 Results

7.1 Results of the research of relations between dimensions of anomia and maladaptive personality traits

To clarify the relations between different dimensions of anomia (i.e., deviation from prescribed rules or customs, social distrust, estrangement to others, cultural isolation, lack of goal clarity and generalized sense of meaninglessness) and maladaptive personality traits (on facet and domain level) the Pearson's correlation analysis was used (Table 1).

7.1.1 Correlations between measures of anomia and personality domains (first-order and second-order factors)

As it was shown in the theoretical part of this paper, in the model of personality developed by Perepjolkina et al. [2] there are seven personality domains that form first-orders factors and four personality domains that form second-orders factors. At the first stage of data analysis associations between dimensions of anomia and these personality domains (first-order and second-order factors) were analysed.

It was found that deviation from prescribed rules or customs as a dimension of anomia was not statistically significantly correlated with any of personality domains.

Social distrust was positively associated only with one first-order personality domain such as Negative emotionality ($p < .01$), and none of the second-order factors of personality domains, including Neuroticism showed statistically significant correlations with this dimension of anomia.

Such dimension of anomia as estrangement to others was positively associated with (1) Psychoticism ($p < .01$) and Introversion ($p < .01$) that refer to such second-order factor as Schizotypy and correlation with this second-order factor was also significant ($p < .01$); (2) Negative emotionality ($p < .01$) and Dependence ($p < .01$) that refer to such second-order factor as Neuroticism, and with Neuroticism itself ($p < .01$); (3) Compulsivity ($p < .01$).

Cultural isolation as a dimension of anomia was positively associated with such a first-order factor as Psychoticism ($p < .01$), and with corresponding second-order factor – Schizotypy ($p < .01$).

Such dimension of anomia as lack of goal clarity was positively associated with (1) Psychoticism ($p < .05$) and Introversion ($p < .05$), and with corresponding to them second-order factor Schizotypy ($p < .01$); and with (2) Negative emotionality ($p < .05$) and Dependence ($p < .01$) and with corresponding to them second-order factor Neuroticism ($p < .01$).

Finally, such dimension of anomia as generalized sense of meaninglessness was positively associated with (1) Psychoticism ($p < .05$) that refers to such second-order factor as Schizotypy; (2) Negative emotionality ($p < .01$) and Dependence ($p < .01$) that refer to such second-order factor as Neuroticism; (3) Compulsivity ($p < .05$). Moreover, generalized sense of meaninglessness was positively associated with both Schizotypy ($p < .05$) and Neuroticism ($p < .01$).

Table 1. Pearson's correlations coefficients for measures of anomia and maladaptive personality traits.

Correlations						
Facet- and domain-level scales of maladaptive personality traits	**Dimensions of anomia**					
	Deviation from prescribed rules or customs	Social distrust	Estrangement to others	Cultural isolation	Lack of goal clarity	Generalized sense of meaninglessness
Schizotypy	.023	-.051	.610**	.311**	.275*	.254*
Psychoticism	.059	-.035	.551**	.330**	.228*	.230*
Cognitive dysregulation	.074	-.018	.286**	.214	.080	.086
Dissociation proneness	.129	-.028	.419**	.162	.395**	.168
Eccentricity	-.002	-.142	.420**	.396**	.183	.176
Suspiciousness	-.029	.037	.577**	.233*	.237*	.274*
Unusual beliefs	.047	.068	.176	.092	-.161	.075
Self-harm	.008	.052	.261*	.117	.105	.133
Introversion	.018	-.152	.473**	.195	.246*	.135
Restricted affectivity	.139	-.198	.379**	.127	.211	.146
Social withdrawal	-.101	-.169	.375**	.230*	.228*	.087
Close relationship avoidance	.033	.013	.313**	.073	.117	.082
Antagonism	.128	.163	.175	-.087	.069	.057
Impusivity	.185	.166	.169	.080	.144	.064
Aggression	-.015	.216*	.284**	-.024	.063	.136
Irresponsibility	.254*	.128	.190	.161	.190	.105
Rashness	.078	.159	.038	-.002	.173	.095
Risk taking	.174	-.015	.032	.088	-.033	-.122
Narcissism	.094	.197	.111	-.192	.026	.043
Dominance	.013	.198	-.058	-.210	-.102	.039
Arrogance	.190	.025	.134	-.057	.051	.066
Attention seeking	.028	.078	.176	-.198	.096	.035
Manipulativeness	-.027	.206	.132	-.162	.040	.091
Harshness	.188	.177	.057	-.019	.050	-.089
Deceitfulness	.082	-.011	.258*	.112	.112	.140
Neuroticism	-.057	.179	.394**	.103	.405**	.323**
Negative emotionality	.012	.298**	.333**	.101	.252*	.301**
Intemperance	.014	.296**	.154	-.018	.089	.131
Emotional stability	-.067	-.264*	-.035	-.001	-.113	-.072
Emotional lability	.029	.355**	.243*	.141	.153	.245*
Depressivity	.000	.118	.546**	.112	.426**	.538**
Anxiousness	-.059	.081	.389**	.148	.275*	.270*
Impersistence	.006	.057	.281*	.146	.391**	.186
Distrustfulness	-.085	.135	.391**	.152	.170	.269*
Dependence	-.142	.009	.367**	.055	.456**	.281**
Evaluation apprehension	-.156	.075	.292**	-.060	.412**	.247*
Submissiveness	-.096	-.087	.331**	.178	.285**	.228*
Indecisiveness	-.117	-.032	.363**	.054	.423**	.230*
Separation insecurity	-.092	.046	.231*	.047	.353**	.221*
Compulsivity	-.062	.067	.376**	.194	.101	.226*
Pedantry	-.140	.023	.332**	.219*	.030	.243*
Perseveration	.094	.012	.124	-.024	-.079	.168
Perfectionism	-.078	.104	.352**	.211	.238*	.097

*$p < .05$, **$p < .01$

7.1.2 Correlations between measures of anomia and facet-level personality traits

As it was shown in the theoretical part of this paper, in the model of personality developed by Perepjolkina and colleagues [2] there are 33 facet-level personality traits. At the second stage of data analysis associations between dimensions of anomia and these personality traits were analysed.

Firstly, it was found that such dimension of anomia as deviation from prescribed rules or customs was positively associated with irresponsibility ($p < .05$) that refers to such domain of personality (first-order factor) as impulsivity.

Secondly, it was found that such dimension of anomia as social distrust was positively associated with such maladaptive personality traits as (1) aggression ($p < .05$) that refers to such domain of personality as impulsivity (first-order factor) and (2) with three facet-level traits from first-order factor Negative emotionality: intemperance ($p < .01$), emotional instability ($p < .01$), and negatively was associated with emotional stability ($p < .05$).

Thirdly, it was found that such dimension of anomia as estrangement to others was positively associated with such maladaptive personality traits as (1) cognitive non-regulation ($p < .01$), dissociation proneness ($p < .01$), eccentricity ($p < .01$), suspiciousness ($p < .01$), self-harm ($p < .05$) that all refer to such domain of personality (first-order factor) as Psychoticism; (2) restricted affectivity ($p < .01$), social withdrawal ($p < .01$) and close relationship avoidance ($p < .01$) that refer to Introversion (first-order factor); (3) aggression ($p < .01$) that refers to Impulsivity (first-order factor); (4) deceitfulness ($p < .05$) that refer to Narcissism (first-order factor); (5) emotional instability ($p < .05$), depressiveness ($p < .01$), anxiousness ($p < .01$), impersistence ($p < .05$), and distrustfulness ($p < .01$) that refer to Negative emotionality (first-order factor); (6) evaluation apprehension ($p < .01$), submissiveness ($p < .01$), indecisiveness ($p < .01$), separation insecurity ($p < .05$) that refer to Dependence (first-order factor), and (7) pedantry ($p < .01$) and perfectionism ($p < .01$) that refer to such domain of personality (first-order factor) as Compulsivity.

Such dimension of anomia as cultural isolation was positively associated with (1) eccentricity ($p < .01$) and suspiciousness ($p < .05$) that refer to Psychoticism (first-order factor); (2) social withdrawal ($p < .05$) that refers to Introversion (first-order factor), and (3) pedantry ($p < .05$) that refer to such domain of personality (first-order factor) as Compulsivity.

It was discovered that such dimension of anomia as lack of goal clarity was positively associated with (1) dissociation proneness ($p < .01$), suspiciousness ($p < .05$) that refer to such domain of personality (first-order factor) as Psychoticism; (2) social withdrawal ($p < .05$) that refers to Introversion (first-order factor); (3) depression ($p < .01$), anxiousness ($p < .05$), impersistence ($p < .01$) that refer to Negative emotionality (first-order factor); (4) evaluation apprehension ($p < .01$), submissiveness ($p < .01$), indecisiveness ($p < .01$), separation insecurity ($p < .01$) that refer to Dependence (first-order factor), and (5) perfectionism ($p < .05$) that refers to such domain of personality (first-order factor) as Compulsivity.

Finally, it was found that such dimension of personality as generalized sense of meaninglessness was positively associated with (1) suspiciousness ($p < .05$) that refers to such domain of personality (first-order factor) as Psychoticism; (2) emotional instability ($p < .05$), depressiveness ($p < .01$), anxiousness ($p < .05$), and distrustfulness ($p < .05$) that refers to such domain of personality (first-order factor) as Negative emotionality; (3) evaluation apprehension ($p < .05$), submissiveness ($p < .05$), indecisiveness ($p < .05$) and separation insecurity ($p < .05$) that refer to Dependence (first-order factor), and (4) pedantry ($p < .05$) that refers to such domain of personality (first-order factor) as Compulsivity.

7.2 Results of the research of prediction of levels of anomia from indicators of personality domains and personality traits

In order to determine which personality domains best predict dimensions of anomia regression analysis was performed. Regression analysis was performed three times, namely, using as predictors (1) second-order personality domains, (2) first-order personality domains, and (3) facet-level personality traits. This strategy will allow evaluation which personality domains with different level of generality (domains that are first-order factors and that are second order factors) as well lower level personality traits predict dimensions of anomia best of all.

7.2.1 Prediction of dimensions of anomia from indicators of personality domains (second-order factors)

As some dimensions of anomia, namely, deviation from prescribed rules or customs and social distrust were not related to any of personality domains that are second-order factors, for these dimensions of anomia regression analysis was not performed. In those cases when significant correlations between certain dimensions of anomia and some personality domains were found hierarchical multiple regression using Enter method was employed. In those cases when there was a significant correlation between one certain dimension of anomia and the only personality domain, simple linear regression was employed. Hierarchical multiple regression was performed to determine what personality domain that were second-order factors best of all predict estrangement to others, lack of goal clarity, and generalized sense of meaninglessness. Simple regression was employed to find whether and to what extent personality domain predicts the indicators of cultural isolation. Table 2 summarizes the results of regression analysis on dimensions of anomia as dependent variable.

Table 2. Regression analysis summary for personality domains (second-order factors), predicting dimensions of anomia.

	β	R^2	F
Estrangement to others			
Step 1		.37	48.04***
Schizotypy	.61***		
Step 2		.38	24.97***
Schizotypy	.55***		
Neuroticism	.13		
Step 3		.39	16.84***
Schizotypy	.52***		
Neuroticism	.10		
Compulsivity	.09		
Cultural isolation		.10	8.64**
Schizotypy	.31***		

Table 2 continuation. Regression analysis summary for personality domains (second-order factors), predicting dimensions of anomia.

Lack of goal clarity			
Step 1		.16	15.86***
Neuroticism	.41***		
Step 2		.17	8.28**
Neuroticism	.36**		
Schizotypy	.10		
Generalized sense of meaninglessness			
Step 1		.10	9.43**
Neuroticism	.32**		
Step 2		.12	5.27**
Neuroticism	.26*		
Schizotypy	.13		
Step 3		.12	3.58*
Neuroticism	.24		
Schizotypy	.11		
Compulsivity	.07		

Note: $*p < .05$, $** p < .01$, $*** p < .001$.

The regression analysis reveals that Schizotypy best of all predicts estrangement to others. Neuroticism in the second model does not significantly improve the prediction (R^2 change = .002, F (1, 80) = 1.57, $p > .05$) as well as Compulsivity in the third model also does not improve the prediction (R^2 change = .006, F (1, 79) = .75, $p > .05$).

Schizotypy is the only significant predictor for cultural isolation.

It was found that Neuroticism best of all predicts lack of goal clarity. Schizotypy in the second model does not significantly improve the prediction (R^2 change = .008, F (1, 80) = .76, $p > .05$).

The regression analysis reveals that Neuroticism best of all predicts generalized sense of meaninglessness. Schizotypy in the second model does not significantly improve the prediction (R^2 change = .012, F (1, 80) = 1.10, $p > .05$). Compulsivity in the third model also does not improve the prediction (R^2 change = .003, F (1, 79) = .30, $p > .05$).

7.2.2 Prediction of dimensions of anomia from indicators of personality domains (first-order factors)

As such dimension of anomia as deviation from prescribed rules or customs was not related to any of personality domains that are first-order factors, for this dimension of anomia regression analysis was not performed. In those cases when significant correlations between certain dimensions of anomia and some personality domains were found hierarchical multiple regression using Enter method was employed. In those cases when there was a significant correlation between one certain dimension of anomia and the only personality domain, simple linear regression was employed. Hierarchical multiple regression was performed to determine what personality domain that were first-order factors best of all predict estrangement to others, lack of goal clarity, and generalized sense of meaninglessness. Simple regression was employed to find whether and to what extent personality domain predicts the indicators of social distrust and cultural isolation. Table 3 summarizes the results of regression analysis on dimensions of anomia as dependent variable.

Table 3. Regression analysis summary for personality domains (first-order factors), predicting dimensions of anomia.

	β	R^2	F
Social distrust			
Negative emotionality	.30**	.09	7.90**
Estrangement to others			
Step 1		.30	35.40***
Psychoticism	.55***		
Step 2		.34	21.00***
Psychoticism	.42***		
Introversion	.24*		
Step 3		.37	15.15***
Psychoticism	.35**		
Introversion	.23*		
Compulsivity	.16		
Step 4		.37	11.50***
Psychoticism	.32**		
Introversion	.24**		
Compulsivity	.13		
Dependence	.09		
Step 5		.37	9.18***
Psychoticism	.30*		
Introversion	.24*		
Compulsivity	.13		
Dependence	.06		
Negative emotionality	.07		
Cultural isolation		.11	9.93**
Psychoticism	.33**		
Lack of goal clarity			
Step 1		.21	21.24***
Dependence	.46***		

Table 3 continuation. Regression analysis summary for personality domains (first-order factors), predicting dimensions of anomia.

Step 2		.21	10.54***
Dependence	.48***		
Negative emotionality	-.04		
Step 3		.23	7.83***
Dependence	.45**		
Negative emotionality	-.05		
Introversion	.15		
Step 4		.24	6.01***
Dependence	.48**		
Negative emotionality	-.02		
Introversion	.20		
Psychoticism	-.11		
Generalized sense of meaninglessness			
Step 1		.09	8.07**
Negative emotionality	.30**		
Step 2		.11	4.77*
Negative emotionality	.21		
Dependence	.16		
Step 3		.11	3.30*
Negative emotionality	.19		
Dependence	.13		
Psychoticism	.08		
Step 4		.12	2.57*
Negative emotionality	.18		
Dependence	.11		
Psychoticism	.06		
Compulsivity	.08		

Note: $^*p < .05$, $^{**}p < .01$, $^{***}p < .001$.

The regression analysis reveals that Negative emotionality is the only significant predictor for social distrust.

It was found that Psychoticism best of all predicts estrangement to others. Introversion in the second model significantly improves the prediction (R^2 change = .040, $F(1, 80) = 4.89$, $p < .05$). Compulsivity in the third model does not significantly improve the prediction (R^2 change = .021, $F(1, 79) = 2.61$, $p > .05$). Dependence in the fourth model also does not significantly improve the prediction (R^2 change = .006, $F(1, 78) = .70$, $p > .05$). Also in the fifth model Negative emotionality does not significantly improve the prediction (R^2 change = .003, $F(1, 77) = .33$, $p > .05$).

Psychoticism is the only significant predictor for cultural isolation.

The regression analysis reveals that Dependence best of all predicts lack of goal clarity. Negative emotionality in the second model does not significantly improve the prediction (R^2 change = .001, F (1, 80) = .078, p > .05). Introversion in the third model also significantly does not improve the prediction (R^2 change = .021, F (1, 79) = 2.13, p > .05). Psychoticism in the fourth model also does not significantly improve the prediction (R^2 change = .006, F (1, 78) = .64, p > .05).

Finally, it was found that Negative emotionality best of all predicts generalized sense of meaninglessness. Dependence in the second model does not significantly improve the prediction (R^2 change = .016, F (1, 80) = 1.42, p > .05). Psychoticism in the third model also does not significantly improve the prediction (R^2 change = .005, F (1, 79) = .42, p > .05) as well as Compulsivity in the fourth model also does not significantly improve the prediction (R^2 change = .005, F (1, 78) = .45, p > .05).

7.2.3 Prediction of dimensions of anomia from indicators of personality traits

As some dimensions of anomia were not related to certain personality domains (second-order and first-order), nevertheless these dimensions of anomia statistically significantly correlated with facet-level personality traits that referred to such personality domains. Therefore the additional question is which facet-level personality traits best of all predict dimensions of anomia. Taking into account a significant number of personality traits, namely, 33, in those cases when significant correlations between certain dimensions of anomia and some personality traits were found stepwise regression analysis was employed. The only exception is deviation from prescribed rules or customs as a dimension of anomia that correlated with the only facet-level personality trait – irresponsibility; in this case simple linear regression was employed. Table 4 summarizes the results of regression analysis on dimensions of anomia as dependent variable.

Table 4. Regression analysis summary for personality traits, predicting dimensions of anomia.

	β	R^2	F
Deviation from prescribed rules or customs		.06	5.58*
Irresponsibility	.25*		
Social distrust		.13	11.70**
Emotional lability	.36**		
Estrangement to others			
Model 1		.33	39.98***
Suspiciousness	.58***		
Model 2		.39	25.73***
Suspiciousness	.39**		
Depressivity	.31**		

Table 4 continuation. Regression analysis summary for personality traits, predicting dimensions of anomia.

Model 3		.43	19.40***
Suspiciousness	.31**		
Depressiveness	.31**		
Distrustfulness	.20*		
Cultural isolation		.16	15.08***
Eccentricity	.40***		
Lack of goal clarity			
Model 1		.18	17.91***
Depressiveness	.43***		
Model 2		.25	13.54***
Depressiveness	.37***		
Separation insecurity	.27**		
Generalized sense of meaninglessness		.29	33.05***
Depressiveness	.54***		

Note: $^{*}p < .05$, $^{**}p < .01$, $^{***}p < .001$.

Firstly, it was found that irresponsibility is the only facet-level personality trait that predicts deviation from prescribed rules or customs explaining 6% of variance.

The regression analysis reveals that emotional instability best of all predicts social distrust explaining 13% of variance.

Using Stepwise method three models were found for predicting estrangement to others. It was found that suspiciousness best of all predicts estrangement to others. Depressiveness statistically significantly improves the prediction in the second model (R^2 change = .06, F (1, 79) = 7.98, $p < .01$) as well as distrustfulness statistically significant improves the prediction in the third model (R^2 change = .03, F (1, 78) = 4.48, $p < .05$). Thus, the combination of three personality traits, namely, suspiciousness, depressiveness and distrustfulness best of all predicts estrangement to others explaining 43% of variance in total.

The regression analysis reveals that eccentricity best of all predicts cultural isolation explaining 16% of variance.

Using Stepwise method two models were found for predicting lack of goal clarity. It was found that depressiveness best of all predicts lack of goal clarity while separation insecurity statistically significantly improves the prediction in the second model (R^2 change = .07, F (1, 80) = 7.69, $p < .01$). Thus, the combination of two personality traits, namely, depressiveness and separation insecurity best of all predicts lack of goal clarity explaining 25% of variance in total.

Finally, the regression analysis reveals that depressiveness best of all predicts generalized sense of meaninglessness explaining 29% of variance.

8 Discussion

The present study was designed to investigate the relationship between anomia and pathological personality traits. In particular, two research questions were formulated: (1) What are the relations between dimensions of anomia and pathological personality domains and facet-level traits? (2) What pathological personality domains and traits best predict specific dimensions of anomia?

It was found that the first dimension of anomia – deviation from prescribed rules or customs – was positively associated with one personality trait – irresponsibility. However, associations between deviation from prescribed rules or customs and first- and second-order personality domains were not found. Irresponsibility is the only personality trait that predicts deviation from prescribed rules or customs. This finding is obvious, because a lack of readiness to be responsible for own behaviour, to respect and complete financial and other requirements, disrespect regarding own promises as an essence of such personality trait as irresponsibility will serve as a base for deviation from prescribed rules or customs for a person in an "anomic state".

It was found that the second dimension of anomia, namely, social distrust, was positively associated with such personality traits as aggression, intemperance, emotional instability, and was negatively associated with emotional stability. Social distrust was positively associated with a first-order personality domain Negative emotionality. However, social distrust was not associated with any of second-order personality domains. Emotional instability as a pathological personality trait best of all predicts social distrust. Respectively, it was found that Negative emotionality is the only significant predictor for social distrust. These obtained results allow to assume, that emotional imbalance, difficulties in emotional control, sharp changes in mood as personality characteristics primarily will allow to expect that in the anomic state a person will lose social trust, that he/she will feel a lack of trust to government and other social institutions as well as a lack of trust to other people, that is, will have doubts that others are generally fair, honest and respect presumed norms.

It was found that the third dimension of anomia – estrangement to others – was positively associated with such maladaptive personality traits as cognitive non-regulation, dissociation proneness, eccentricity, suspiciousness, self-harm, restricted affectivity, social withdrawal, close relationship avoidance, aggression, deceitfulness, emotional instability, depressiveness, anxiousness, impersistence, distrustfulness, evaluation apprehension, submissiveness, indecisiveness, separation insecurity, pedantry and perfectionism. Estrangement to others was positively associated with such first-order personality domains as Psychoticism, Introversion, Negative emotionality, Dependence and Compulsivity. Estrangement to others was positively associated also with second-order personality domains, namely, Schizotypy, Neuroticism, and Compulsivity. The combination of three facet-level personality traits, namely, suspiciousness, depressiveness and distrustfulness best of all predicts estrangement to others. Regarding first-order personality domains it was found that Psychoticism together with Introversion best of all predicts estrangement to others. Regarding second-order personality domains it was found that Schizotypy best of all predicts estrangement to others. These obtained results allow to assume that more likely suspiciousness that fall within a first-order domain Psychoticism and a second-order domain Schizotypy is the most significant predictor of estrangement to others as a dimension of anomia. Really, suspicious attitude towards others as a stable personality trait, tendency to have doubts regarding loyalty of the others should serve as a base for a sense of estrangement to others, that is, senses of a lack of social support, loneliness, alienation, inferiority when a person endures anomia.

It was found that the fourth dimension of anomia – cultural isolation – was positively associated with such maladaptive personality traits as eccentricity, suspiciousness, social withdrawal and social detachment, and pedantry. Cultural isolation was positively associated with a first-order domain Psychoticism and a second-order domain Schizotypy. Eccentricity as a pathological personality trait best of all predicts cultural isolation. Psychoticism as a first-order personality domain is the only significant predictor for cultural isolation and, respectively, Schizotypy is the only significant predictor for cultural isolation. Thus, exactly, eccentricity as unusual, strange behaviour, speech and beliefs will as stable personality characteristics will serve as a base for such an anomic sign as a sense of a loss of internalized social norms and values that is an indicator of cultural isolation in the anomic state.

It was found that the fifth dimension of anomia – lack of goal clarity – was positively associated with such pathological personality traits as dissociation proneness, suspiciousness, social withdrawal and social detachment, depressiveness, anxiousness, impersistence, evaluation apprehension, submissiveness, indecisiveness, separation insecurity, and perfectionism. Lack of goal clarity was positively associated with such first-order domains as Psychoticism, Introversion, Negative emotionality and Dependence as well as with such second-order domains as Schizotypy and Neuroticism. The combination of two personality traits, namely, depressiveness and separation insecurity best of all predicts lack of goal clarity. Dependence as a first-order domain best of all predicts lack of goal clarity and, respectively, Neuroticism as a second-order domain best of all predicts lack of goal clarity. These obtained results allow us to assume that more likely separation insecurity that fall within a first-order domain Dependence and a second-order domain Neuroticism is the most significant predictor of a lack of goal clarity as a dimension of anomia. Really, fear to stay alone, doubts that significant others "will stay with me", fear that nobody will promote necessary support and care, a lack of belief that "I'm able to take care about myself" can force a person to feel doubts in the anomic state that he/she is able to attain important life goals. In order to conceptualize this state Bandura [33] used a term of self-efficacy that means the extent or strength of one's belief in one's own ability to complete tasks and reach goals. Likely, that if a person feels lonely and is unsure that anyone will support his/her plans, intentions, and dreams, feels fear to lose significant others, he/she can refuse some important life goals and as a result feels uncertainty regarding own future.

Finally, the last sixth dimension of anomia, namely, generalized sense of meaninglessness, was positively associated with such pathological personality traits as suspiciousness, emotional instability, depressiveness, anxiousness, distrustfulness, evaluation apprehension, submissiveness, indecisiveness, separation insecurity, and pedantry. Generalized sense of meaninglessness was positively associated with such first-order domains as Psychoticism, Negative emotionality, Dependence, and Compulsivity. Generalized sense of meaninglessness was positively associated with such second-order domains as Schizotypy, Neuroticism, and Compulsivity. Depressiveness as a pathological personality trait best of all predict generalized sense of meaninglessness. Negative emotionality as a first-order domain best of all predicts generalized sense of meaninglessness and, respectively, Neuroticism as a second-order domain best of all predicts generalized sense of meaninglessness. Thus, exactly depressiveness can be viewed as the most important pathological personality trait in prediction of generalized sense of meaninglessness. If a person has a tendency to feel dejectedness, unhappiness, pessimism, hopelessness and these characteristics are typical for him/her, then in the anomic state more likely he/she will experience as a state of generalized meaninglessness.

Thus, this study contributes to the knowledge and understanding of such issue as personality factors predicting anomia.

Limitations and future directions. One of the main limitations of this research is the size of the sample. We can perceive this research only as a pilot study. Secondly, the majority of respondents were females. In spite of the fact that in the previous research [22] sex differences in levels of anomia have not been found, in future, regarding the relations between anomia and pathological personality traits it would be important to organize a sample in which females and males would be represented equally. In future it would be necessary to realize a similar study in a more representative sample.

The paper was supported by the National Research Program 5.2. "Economic Transformation, Smart Growth, Governance and Legal Framework for the State and Society for Sustainable Development – a New Approach to the Creation of a Sustainable Learning Community (EKOSOC-LV)" and by Latvian National Research Programme Biomedicine for Public Health (BIOMEDICINE) 2014–2017 (sub-project Nr.5.8.2.).

References

1. J. Ļevina, K. Mārtinsone. Psychometric properties of the anomia questionnaire (AQ) (in press)
2. V. Perepjolkina, J. Koļesņikova, K. Mārtinsone, A. Stepens. *Latvijas Klīniskais personības tests (LKPT-v3). Tehniskā rokasgrāma* (in press)
3. É. Durkheim. *The Division of Labour in Society*. NY: Free Press ([1893] 1964)
4. R.K. Merton. Anomie, Anomia, and Social Interaction: Contexts of Deviant Behavior. In M.B. Clinard (Ed.) *Anomie and Deviant Behavior: A Discussion and Critique* (pp. 213–242). NY: Free Press (1964)
5. E. Lytkina. From anomie to alienation: An approach towards the measurement of social well-being and deviance. Papers of the IV Russian Sociology Congress. Moscow: Institute of Sociology RAN, Russian Sociological Society, 8227–8231 (2012)
6. J. Levina, K. Martinsone, & D. Kamerade. Anomia as a factor predicting subjective well-being. Med. psihol. Ross, **3**(32), iegūts no: `http://www.mprj.ru/archiv_global/2015_3_32/nomer07.php` (2015)
7. J. Ļevina, K. Mārtinsone. Anomia, social participation and subjective well-being: A psychological perspective. *SOCIETY. INTEGRATION. EDUCATION. Proceedings of the International Scientific Conference* (in press)
8. J. Ļevina, K. Mārtinsone, & K. Klince. Relations between anomia and values of the inhabitants of the Baltic states. *SOCIETY. INTEGRATION. EDUCATION. Proceedings of the International Scientific Conference, Vol. 1* (pp. 431–446). Rēzekne: RA izdevniecība. doi:10.17770/sie2016vol1.1518 (2016)
9. J. Ļevina, K. Mārtinsone, & D. Kamerāde. The relationship between anomia and participation of Latvian inhabitants in social activities. *International Interdisciplinary Scientific Conference "Society. Health. Welfare." Proceedings* (2016)
10. A.F. De Man, C.P. Leduc. Suicidal ideation in high school students: Depression and other correlates. Journal of Clinical Psychology **51**(2), 173–181 (1995)
11. A. Heydari, A. Teymoori, & H. Nasiri. Development of suicidality within socioeconomic context: Mediation effect of parental control and anomie. Omega **68**(1), pp. 63–76 (2013–2014)
12. D.I. Lasky, J.T. Ziegenfuss (1979). Anomie and drug use in high school students. The International Journal of the Addiction **14**(6), 861–866 (1979)
13. A.F. De Man, L. Labrèche-Gauthier, & C.P. Leduc. Correlates of anomie in French-Canadian adolescents. The Journal of Social Psychology **133**(2), 141–145 (1993)

14. C.C. Conway, M.G. Craske, R.E. Zinbarg, S. Mineka. Pathological personality traits and the naturalistic course of internalizing disorders among high-risk young adults. Depression and Anxiety **33**(1), 84–93 (2016)
15. T.F. Williams, K.M. Thomas, M.B. Donnellan, C.J. Hopwood. The aversive interpersonal behaviours associated with pathological personality traits. Journal of Personality Disorders **28**(6), 824–840 (2014)
16. E.A. Dowgwillo, K.S. Ménard, R.F. Krueger, A.L. Pincus. DSM-5 Pathological Personality Traits and Intimate Partner Violence Among Male and Female College Students. Violence and Victims **31**(3), 416–437 (2016)
17. I. Ramos-Grille, M. Gomà-i-Freixanet, N. Aragay, S. Valero, V. Vallès. Predicting treatment failure in pathological gambling: the role of personality traits. Addictive Behaviors **43**, 54–59 (2015)
18. V. Zeigler-Hill, G.A. McCabe, J.K. Vrabel. The Dark side of humor: DSM-5 pathological personality traits and humour styles. Europe's Journal of Psychology **12**(3), 363–376, doi: 10.5964/ejop.v12i3.1109
19. J. Levina, K. Martinsone, J. Kolesnikova, & V. Perepjolkina. Possibilities of research of anomie in Latvian society. In A. Vētra, A.Vilks (Eds.) *5th International Interdisciplinary Scientific Conference "Society. Health. Welfare". 2nd Conference of Speech Therapists (Rīga, November 26–28, 2014): Abstracts* (p. 39). Rīga: RSU (2014)
20. J. Levina, K. Martinsone. A revised integrative hierarchical model of anomia: Towards the construction of anomia indices. The 14th European Congress of Psychology "Linking technology and psychology: feeding the mind, energy for life" (Milan, Italy, 7–10 July 2015): Abstract book (p. 2250). Downloaded from `http://www.ecp2015.it/wp-content/uploads/2015/07/ECP-2015-Abstract-Book upd27jul1.pdf` (Accessed 31 July 2015) (2015)
21. J. Ļevina, K. Mārtinsone, & D. Kamerāde. Individuālās anomijas modeļi Latvijas populācijā. *Rīgas Stradiņa universitāte. 2015.gada zinātniskā konference: Tēzes* (375.1 pp.). Rīga: RSU (2015)
22. J. Ļevina, K. Mārtinsone, & D. Kamerāde. Sex and Age Differences in Levels of Anomia of Latvian Inhabitants. *SOCIETY. INTEGRATION. EDUCATION. Proceedings of the International Scientific Conference, Vol. 3* (pp. 567–576). Rēzekne: RA izdevniecība. doi: 10.17770/sie2015vol3.475 (2015)
23. J. Ļevina, K. Mārtinsone, & D. Kamerāde. A Model of Anomia in the Baltic States. The International Journal of Interdisciplinary Civic and Political Studies, **11**(3), 1–12. doi: 10.18848/2327-0047/CGP (2016)
24. É. Durkheim. *Suicide: A Study in Sociology*. London, England: Routledge ([1897] 1951)
25. R.M. Maclver. *The Ramparts We Guard*. New York: The Macmillan Company (1950)
26. D.B. Samuel, D.R. Lynam, T.A. Widiger, & S.A. Ball. An expert consensus approach to relating the proposed DSM-5 types and traits. Department of Psychological Sciences Faculty Publications. ***Paper 13***. `http://dx.doi.org/10.1037/a0023787` (2012)
27. T.A. Widiger, E. Simonsen. Alternative dimensional models of personality disorder: Finding a common ground. Journal of Personality Disorders **19**(2), 110–130 (2005)
28. A.G.C. Wright, K.M. Thomas, C.J. Hopwood, K.E. Markon, A.L. Pincus, & R.F. Krueger. The hierarchical structure of DSM-5 pathological personality traits. Journal of Abnormal Psychology **121**(4), 951–995 (2012)
29. American Psychiatric Association. *Diagnostic and Statistical Manual of Mental Disorders: DSM-5* (5th ed.). Arlington VA: American Psychiatric Association (2013)

30. J. Kolesnikova, V. Perepjolkina, K. Martinsone, A. Stepens. The theoretical integrative model for the Latvian Clinical Personality Inventory. Society, integration, education. *Proceedings of the International Scientifical Conference, Volume **I***. Rēzekne: RA izdevniecība, pp. 388–398. `http://dx.doi.org/10.17770/sie2016vol1.1518` (2016)
31. V. Perepjolkina, J. Kolesnikova, K. Martinsone, A. Stepens, E. Rancans. *Development of the pathological personality trait scales for the Latvian Clinical Personality Inventory (LCPI v2.2)*. Report presented at the 10th Conference of the ITC, Vancouver, Canada, July 1–4, 2016 (2016)
32. V. Perepjolkina, J. Kolesnikova, K. Mārtinsone, & A. Stepens. Latvijas Klīniskā personības testa otrās sākotnējo apgalvojumu kopas izstrāde: integratīvās pieejas pielietošana. (Development of the Second Preliminary Item Pool of the Latvian Clinical Personality Inventory: an Integrative Approach). *Society, integration, education. Proceedings of the International Scientifical Conference (Sabiedrība, integrācija, izglītība. Starptautiskās zinātniskās konferences materiāli), Volume **I***, May 27th–28th, pp. 469–483. Rēzeknes Tehnoloģiju akadēmija, 2016. `http://journals.ru.lv/index.php/SIE/article/view/1521/1673` (2016)
33. A. Bandura. Self-efficacy: Toward a unifying theory of behavioral change. Psychological Review **84**, 191–215 (1977)

Health procrastination: The experience of 35–44 years old men

V. Silkane* and I. Austers

University of Latvia, Riga, Latvia

Abstract. The focal concept of this study is health procrastination. Delay in seeking help from medical practitioners is an increasing problem in Latvia, particularly with men between the ages of 35–44. Failures in primary and secondary prevention create a number of negative consequences, for instance, longer and more complex treatment; lower recovery prognosis as well as higher treatment costs. Nevertheless, the aforementioned group often avoids medical treatment. The aim of this study was to understand the experience of 35–44 years old men in terms of their health procrastination in qualitative terms. Semi-structured in-depth interviews with 48 men within the age range of 35–44 years were conducted. Data was analysed using inductive thematic analysis. The following dominant themes are: delay of health behaviour, time management, motivation emerged during the analysis: conscientiousness, task avoidance, effort, short-term vs. long-term thinking, gains and losses assessment, and time discounting. Conclusions followed: there are three main concepts – personality traits, discounting and task characteristics – that bring together the experience of health procrastination for 35–44 years old males.

1 Introduction

Failure to take care of one's health in a timely manner is a wide-spread problem that is often debated, but at the same time has shown itself lacking a clear solution. Taking care of own health means both leading a healthy life-style and seeking medical assistance in a timely manner. Very often, unless health problems are severe, like acute pain or other distressing symptoms, people simply neglect them. For example, a 2014 survey on health related habits of inhabitants of Latvia, carried out by The Centre for Disease Prevention and Control [1], showed that 68.9% of respondents had never visited a general practitioner to have a free medical check-up (75.3% men and 62.7% women) and 27% of respondents had not visited a general practitioner during the past year (35.9% men and 18.4% women). Those who most commonly do not visit general practitioners are men aged 25–34 (43.4%) [1] and men aged 35–44 (47.8%) [2]. A delayed or a postponed visit to a doctor, a failure to abide the treatment instructions and to take preventive measures cause several problems – illnesses that require a longer and more complicated treatment, worse recovery forecasts, as well as higher treatment costs.

* Corresponding author: vineta.silkane@gmail.com

Even though in the long term people regard health as an important asset, they often postpone actions in this respect – they not only miss preventive check-ups and visits to a doctor, they also do not follow practices related to a healthy life-style. Despite possessing knowledge associated to a healthy life-style, people postpone the cessation of smoking, starting physical activities, and following healthy nutrition habits [1].

Therefore it is necessary to examine factors related to postponing health behaviour and delay in taking care of one's health, in other words – health procrastination.

An important group that deserves additional research attention is men aged 35–44. At this age men often get their first health problems in addition to the unhealthy life-style they have, for example, smoking, eating unhealthy food, and suffering from overwork and stress. At the same time men in this age group rarely visit doctors to solve health-related issues [2]. Thus, men falling within this age group have a bigger risk of not noticing health problems. In order to gain a better understanding of men's health procrastination and the most common reasons why men neglect their health, it is important to apply the methods of a qualitative approach.

Men's health behaviour is studied fairly rarely and is mostly formed by quantitative research design dealing with masculinity and sexuality topics [3, 4]. In order to conduct a better examination of men's health behaviour experience and the most common reasons of not caring for their health, it is important to use a qualitative study approach. The aim of the study was to explore the experience of 35–44 years old men in terms of their health procrastination.

1.1 Procrastination

In the absence of a unified definition [5] the majority of researchers focus on three criteria to define procrastination: (1) activity is delayed or postponed, (2) postponement of activity is unnecessary and (3) activity is unproductive [6–8]. Since it is difficult to formulate non-necessity and non-productivity in an objective way, definitions focus more on delayed or postponed activity [5].

Procrastination is the voluntary [9] and irrational delay of intended activities or postponement of tasks [10–13], despite the expected aggravation of the situation [9]. Irrationality is characterised by the fact that such behaviour is chosen even though it is unsuitable and contrasts the individual's material (e.g., financial) or psychological (e.g., wellbeing) interests or objectives. An individual opts for worsening their situation although other solutions exist [15]. If hesitation is rational or has a sensible explanation, it is not procrastination [12].

Klingsieck [5], after analysis of the most popular definitions, clearly distinguishes between procrastination and functional forms of delay, like strategic delay. Taking into account everything written above, procrastination is "the voluntary delay of an intended and necessary and/or [personally] important activity, despite expecting potential negative consequences that outweigh the positive consequences of the delay" [5, 9].

The most common explanations as to why procrastination emerges are (1) time discounting, (2) task characteristics and (3) personality traits [9, 14, 15]. The majority of studies focus on procrastination as a personality trait. Some researchers suggest that procrastination can be reduced by changing behaviour [11, 16]. Task characteristics are analysed less profoundly, for example, when an individual perceives the task as insignificant, difficult, aversive or boring [17, 18] tasks with failure potential [19–21] and whether there is a deadline for the task [21, 22].

1.2 Procrastination and personality

In numerous studies procrastination is conceptualized as a personality trait [10]. Procrastination can be examined as a personality trait that is created by repeated procrastination behaviour [7]. If procrastination is constant over time and stable in various situations [9], procrastination is regarded as a personality trait.

Procrastination is related to a low conscientiousness level [9, 19, 23–25] as well as to a high neuroticism level [9, 19, 23–25]. For self-disciplined, accurate individuals it is much easier to meet deadlines and overcome resistance against aversive and tough tasks, therefore it is less likely that they would procrastinate. Impulsivity explains how neuroticism is related to procrastination. For instance, individuals would more often postpone fulfilment of tasks if there is strong drive for a more pleasant activity, and their self-control is too weak to be able to overcome desires, put forward by the moment, and to do what is necessary instead. For impulsive individuals it is difficult to get over their desires and they tend to succumb to various temptations and enticements [7]. Due to the lack of impulse control they are unable to postpone immediate satisfaction; at the same time they are unable to refuse pleasant short-term benefits for the sake of long-term results [19]. Studies that have used the Big Five personality model do not render significant results regarding the relationship between procrastination and extraversion, openness and agreeableness.

1.3 Procrastination and discounting

Some researchers argue that procrastinators avoid focusing on future tasks and that procrastination is related to time perception [26]. In studies that examine procrastination from the perspective of time, procrastination is most commonly described using time discounting [24, 27, 28]. Procrastinators frequently do not appreciate the importance of present efforts in relation to future achievements [24]. Even though the future benefit is very much welcome, or drawbacks are dangerous and terrible, it is difficult for an individual to get over discounting [28]. The longer one has to wait to get a reward for the particular behaviour, the less attractive that behaviour seems [24]. Future results are discounted or undervalued in comparison with immediate benefits [14].

Discounting relates to higher procrastination levels [27]. From the point of view of discounting, it is considered that procrastination develops when present costs seem higher than the future costs. Discounting of future costs may, at least partially, be explained by topicality of other activities that are to be done in the near future; people mostly believe that they will have more time to do it in the future [29] therefore procrastination often may be limited by setting deadlines or defining a limited time-frame [30].

1.4 Procrastination and task characteristics

Several studies have defined that one of the main causes of procrastination is averse, boring or uninteresting tasks [9, 15]. Procrastination studies usually define task avoidance by asking study participants to grade it [31] and it usually means boredom, frustration and offence. Task avoidance has a positive correlation with procrastination [18, 32].

People procrastinate more often if they have to do unpleasant tasks or the task is expected to be unpleasant [9, 14], it requires more effort, it is vaguely defined, less pleasing [26], tough and unattractive [18, 33, 34].

1.5 Procrastination and health behaviour

Relatively few studies have explored the relationship between procrastination and health behaviour. There are studies that examine the role of procrastination in the explanation of illness and each of them uses and analyses different mediators: (1) stress as a mediator between procrastination and illness [35, 36] and (2) treatment delay as a mediator between procrastination and illness [35, 36]. One of the studies has addressed the relationship between procrastination and health behaviour intention [37] and the mediators are self-efficiency and future consequences [37].

Procrastination is related to a higher stress level and worse health [38], anxiety and depression [31]. Procrastination is also related to more acute health problems, non-practicing of a healthy life-style, rarer visits to the doctor [36], and postponement of looking for help and treatment [36].

2 Method

Participants. Participants were recruited by using snowball sampling and participation in this study was voluntary. Forty-eight individual interviews were conducted among 35–44 year old men, the average age of the sample was $M = 40.44$, $SD = 3.96$. When quoted, study participants were identified by code number and age (for example, A15, 37).

Data collection. Semi-structured in-depth interviews were used as a data collection method. Interview questions were formulated in the way that would be possible to find out the instances of participants' health procrastination, as well as examining the following topics: taking care of health, healthy lifestyle motivators, reason for not taking care of health, delay or postponement of health behaviour or health checks. Participants were informed about the confidentiality of any information collected.

Data analysis. Inductive thematic analysis was used. First, repeated examination of each interview took place. Then, the search and definition of topics and sub-topics was performed.

3 Results

The aim of the study was to explore the experience of 35–44 years old men in terms of their health procrastination. The interviews allowed for the identification of the following behaviours related to procrastination: postponement, laziness, effort, lack of motivation, impulsivity, time management, task avoidance, assessment of gains and losses, discounting.

Several men mentioned *health behaviour postponement.* It was shown to manifest through postponement of a visit to a doctor because one "has to wait for a better blood test if too many sweets have been consumed" (the man has diabetes mellitus) (A33, 36). Or it may be a prevention postponement: "generally, the idea is good, but sometimes later (A38, 39). Another participant characterised men's behaviour: "Men tend to postpone solving problems as much as possible and finally get used to them. The reaction follows only after a problem has occurred" (A47, 35). It can be explained by short-term planning, assuming that the symptoms will vanish and no actions will be necessary. However, such an approach can be a problem solving model. An example mentioned by another man illustrates it: "One has to reach an edge of a catastrophe and then it happens" (A42, 44). One of the men postponed wisdom tooth extraction because the previous time he had done it, recovery took him 3–4 days and he could not find time for that. He postponed it "until an emergency situation occurred, there was sharp pain and nothing could be done about it anymore" (A48, 37). These examples show that postponement can make the situation worse.

Sometimes men evaluate the situation in terms of pros and cons, which can be regarded as a *strategic postponement* mentioned by Klingsieck [5]. For example, the postponement of an operation, since it affects working capacity in the long-term. "I have no spare six months in which I can afford to do nothing. Therefore I wait and hope that maybe some new technologies will be introduced. It does not affect my daily life; the only thing is that I can't play basketball" (A40, 43). Another man acted in a similar way, he postponed removal of a metallic element from his arm after a fracture because he had difficulties in finding a time for an operation. In both situations, postponement of surgery did not affect the daily routine and had no harmful consequences therefore one cannot speak of health procrastination in these particular situations.

Men often explain not doing something, for example, postponement of physical activities, by *laziness*. For example, "I am too lazy for running" (A3, 37). At the same time the man added that "as to health matters, I am not lazy," because "I value health very much" (A5, 43). When explaining the phenomenon of laziness, lack of motivation is often mentioned, as well as the effort that needs to be present to achieve something: "One has to have motivation in order to start doing things" (A38, 39). In this case attention has to be paid to the use of terms, as 'laziness' in its daily use does not always mean the same as the scientific definition of the term. Steel [39] is of an opinion that laziness and procrastination are contrasting concepts that only partly overlap, but for the purpose of this study it was important to take into account all kinds of explanations that are related to procrastination. A failure to do something or a postponement of doing things is something that laziness and procrastination have in common. The same applies to a lack of motivation. Both concepts differ in terms of feeling bad and guilty, which is present in case of procrastination, while it is not characteristic of laziness [39].

In order to follow different tasks, including engagement in health related activities, *effort* is required. It is known that procrastination is more widespread if the task seems aversive, difficult or boring. When explaining their behaviour, men often discussed activities that require effort, meanwhile acknowledging that they lack the willpower for that. Cessation of smoking and taking part in physical activities are among activities mentioned most often when speaking about effort, for example, "I have tried [to quit smoking], but I do not have enough willpower" (A2, 44), or "I am tired after work therefore it is difficult to do sports" (A31, 42) and "I would like to [engage in physical activities], but I lack determination" (A36, 44). In the example about wisdom tooth extraction given in the discussion previously, another source of procrastination related to the task characteristics appears – *task avoidance*.

In the interviews the men talk about *lack of motivation*, as well as mention that finding the right motivation is a key for success. For example, one man said that there are "a thousand reasons not to" (do sports), but "I decided to lose some waist weight" (A33, 36) and it helped to start. Another man explained that he will not choose a healthy life-style because he does not believe in it and the results do not double life expectancy anyway: "If I was sure that quitting smoking or eating carrots would allow me to live up to 150 years, maybe I would do that" (A14, 44). Such approach illustrates lack of motivation.

One of the interviews revealed *impulsivity*, which is a common companion of procrastination: "I instantly decide to do something and if it is not possible straight away most likely my instant decision will be gone" (A42, 44). In this particular case the man speaks about paying a visit to a doctor and about a preventive examination.

Frequently in their explanations men talk about *time management*. Lack of time may be connected with difficulties in time-planning "[...] time, lack of time has always been an obstacle" (A4, 38), as well as scrupulosity. One man stated that time management is "a matter of self-discipline" and "it is hard to be regular enough to make particular activity a tradition" (A48, 37).

Assessment of gains and losses was also present in the interviews and it may characterise short-term vs long-term thinking. For example, "If a dentist discovers cavities in teeth, it is cheaper and less painful. [...] If I don't visit a dentist, at the end the bad tooth has to be extracted and it costs thousands to replace it, or else I have to go on without a tooth" (A44, 35). The example about postponement of a surgery mentioned before portrays a similar situation. The man has already considered that maybe the technologies will improve in the future while shoulder trauma is bearable in his daily life. Sometimes such choice can be explained by *discounting*, since an immediate but smaller benefit is chosen. For example, "Men tend to postpone the solving of problems as much as possible and finally get used to them. Reaction follows only after problem occurrence" (A47, 35) or "I postpone going for a jog because I choose to watch a movie instead" (A44, 35). It can be seen also in the quotation given above where the man says that a healthy life-style will not extend his life expectancy up to 150 years, therefore it is not even worth starting because "[...] a man has as a pre-destined length of life either way" (A14, 44) and, at its best, it can be prolonged for one or two years.

4 Discussion and conclusions

The thematic analysis of interviews proves that stories told by the men allow for the identification of health procrastination and related concepts, thus, characterising the circumstances of procrastination occurrence. The categories mentioned in the results section can be divided into three groups which comply with the three most common procrastination explanations mentioned in the research literature – task characteristics (e.g., effort, unpleasantness), personal traits (e.g., impulsivity, conscientiousness) and discounting. However, further quantitatively designed studies are needed, in order to verify the role of all three concepts used in the explaining of procrastination as a single model.

The limitation of the study is gender difference between the research participants and the interviewer; it may have an impact on what and how answers are given about health behaviour and health procrastination. It is probable that other health-related nuances would have been revealed and slightly different behaviour explanations would have been given to a male interviewer.

The results of the study can be useful for the development of a relevant health policy. Currently, health policy in Latvia is using a variety of interventions, for example, informing individuals about healthy lifestyle and health-related regulations. However, the psychological aspects are not always taken into account, for example, when trying to understand an individual's decisions and the possibilities to influence those decisions. For example, in the case of restricting smoking both information and various regulations are used; yet, despite the fact that people know about the dangers of smoking, quitting is delayed. More severe regulations and restrictions do not necessarily have positive results, since the forced behaviour rather creates resistance than confirms the expected patterns. Therefore, there is a need for developing interventions that promote a healthy lifestyle and at the same time are based on understanding the psychological factors affecting a healthy lifestyle.

At the level of public health, it is difficult to create a health intervention that could change all aspects contributing to health procrastination, like personality traits, that the study illuminates. However, it is likely to influence the intentions related to a healthy lifestyle, as well as preferences, and motivation. It is also possible to reduce the systemic barriers related to health behaviour and thereby reduce effort to implement a healthy lifestyle, for example, to make preventive visits to the doctor more routine. Special attention should be paid to men's health-related campaigns. They should focus on the evaluation of benefits, losses, and risks in the long run.

References

1. The Centre for Disease Prevention and Control, *Health Behaviour among Latvian Adult Population* (2014)
2. The Centre for Disease Prevention and Control, *Health Behaviour among Latvian Adult Population* (2012)
3. J.R. Mahalik, M. Levi-Minzi, G. Walker, (2007). Psychol. Men Masculin. **8**, 240–249 (2007)
4. T. Calasanti, I. Pietilä, H. Ojala, N. King, N., Health Psychol, **32**, 15–23 (2013)
5. K.B. Klingsieck, *Eur. Psychol.* **18**, 24–34 (2013)
6. H.C. Schouwenburg, *Procrastination and Task Avoidance: Theory, Research, and Treatment*, 71–96 (1995)
7. H.C. Schouwenburg, C.H. Lay, *Pers. Individ. Dif.* **18**, 481–490 (1995)
8. G. Schraw, T. Wadkins, L. Olafson, *J. Educ. Psychol.* **99**, 12–25 (2007)
9. P. Steel, Psychol. Bull. **133**, 65–94 (2007)
10. C.H. Lay, *J. Res. Pers.* **20**, 474–495 (1986)
11. J.B., Burka, L.M. Yuen, *Procrastination: Why You Do It, What to Do About It* (2004)
12. M. Silver, J. Sabini, *J Theory Soc Behav*, **11**, 207–221 (1981)
13. C. Andreou, *J. Theory Soc. Behav.* **37**, 183–193 (2007)
14. J. Wistrich, *Wm. & Mary L. Rev.* **50**, 607–667 (2008)
15. Wilson, T.D. Nguyen, *Int. J. Psychol. Res. (Medellin)* **4**, 211–217 (2012)
16. W.J. Knaus, *J. Pers. Soc. Psychol.* **15**, 153–166 (2000)
17. D.S. Ackerman, B.L. Gross, *J. Mark. Educ.* **29**, 97–110 (2007)
18. K. Blunt, T.A. Pychyl, *Pers. Individ. Dif.* **28**, 153–167 (2000)
19. W. Van Eerde, *Pers. Individ. Dif.* **35**, 1401–1418 (2003)
20. C.H. Lay, *J. Pers. Soc. Psychol.* **5**, 91–103 (1990)
21. R. Gupta, D.A. Hershey, J. Gaur, *Curr. Psychol.* **31**, 195–211 (2012)
22. J.R. Ferrari, *Pers. Individ. Dif.* **13**, 315–319 (1992)
23. D.C. Watson, *Pers. Individ. Dif.* **30**, 149–158 (2001)
24. S. Dewitte, H.C. Schouwenburg, *Eur. J. Pers.* **489**, 469–489 (2002)
25. P. Steel, T. Brohen, V. Wambach, *Pers. Individ. Dif.*, **30**, 95–106 (2001)
26. J.R. Ferrari, C.P. Mason, C. Hammer, *Individ Differ Res* **4**, 28–37 (2006)
27. J. Howell, D.C. Watson, R.A. Powell, K. Buro, *Pers. Individ. Dif.* **40**, 1519–1530 (2006)
28. H.C. Schouwenburg, J. Groenewoud, *Pers. Individ. Dif.* **30**, 229–240 (2001)
29. S.B. Shu, A. Gneezy, *J. Mark. Res.* **47**, 933–944 (2010)
30. D. Ariely, K. Wertenbroch, *Psychol. Sci.* **13**, 219–224 (2002)
31. L.J. Solomon, E.D. Rothblum, *J. Couns. Psychol.* **31**, 503–509 (1984)
32. K. Krause, A.M. Freund, *Eur. Psychol.* **1**, 1–13 (2013)
33. D.S. Ackerman, B.L. Gross, *J. Mark. Educ.* **27**, 5–13 (2005)
34. C.H. Lay, *J. Pers. Soc. Psychol.* **7**, 483–494 (1992)
35. F. M Sirois, M.L. Melia-Gordon, T.A. Pychyl, *Pers. Individ. Dif.* **35**, 1167–1184 (2003)
36. F.M. Sirois, *Pers. Individ. Dif.* **43**, 15–26 (2007)
37. F.M. Sirois, *Pers. Individ. Dif.* **37**, 115–128 (2004)
38. D.M. Tice, R.F. Baumeister, *Psychol. Sci.* **8**, 454–458 (1997)
39. P. Steel, *The Difference Between a Procrastinator and a Lazy Person* [Video file], 2016

Anxiety indicators for women in the postpartum period

V. Sudraba[1] and L. Zemite[2]

[1]Rīga Stradiņš University, Riga, Latvia
[2]Riga Maternity Hospital, Riga, Latvia

Abstract. Situations, which are dangerous and sad, provoke state anxiety (S-anxiety), but trait anxiety (T-anxiety) shows anxiety like person's acquired behavioural disposition. The aim was to evaluate postpartum anxiety indicators of women that have had the vaginal birth and women that have had surgery in Riga regional medical institutions that provide obstetric services. Respondents – 100 women (50 – with vaginal birth; 50 – with caesarean section surgery) were surveyed by 2 instruments: the demographic survey and the state – trait anxiety inventory, STAI Form Y-1 (Spielberger et al., 1983), adapted in Latvian by Škuškovnika (2004). The results reflect the statistical difference between State anxiety indicators (S-anxiety) of women that have had the vaginal birth (M = 35.38; SD = 9.98) and women that have had the surgery (M = 37.20; SD = 9.41) was not found (p = 0.304). Indicators of Trait anxiety (T-anxiety) show that there was not statistically significant (p = 0.059) difference between women that have had the vaginal birth (M = 36.92; SD = 7.81) and women that have had the surgery (M = 39.24; SD = 7.56), however, it should be noted that the average anxiety scores were higher after surgery.

1 Introduction

Time from the birth of a child until the first year after the birth is a critical period to develop affective disorders [1]. Immediately after the childbirth the mother experiences stress due to the lack of energy caused by new responsibilities [2]. While pregnancy is like a defender against panic attacks, depression and development of anxiety, postpartum period increases their risk. During this period the inclination to the development of anxiety increases because demands to a woman grow, and she often has to face lack of sleep [3].

As aggravating factors which promote development of anxiety and postpartum depression are mentioned a history of mental illness, smoking, low socio-economic status, relationship or marriage problems, single mothers, loss of a previous child, lack of knowledge in the care of a new-born [2, 4, 5]. At the same time married women or physically fit women had a lower degree of anxiety [6].

Postpartum anxiety is stated to be one of the most common psychotic disorders in Canada [7]. In Israel about 13% of women suffer from symptoms of depression in postpartum period [3]. In Georgia anxiety in the postpartum period is found in 17–20% cases [8], but in Greece it ranges from 6.1 to 27.9% in the postpartum period [1]. In the study [9] in Germany prevalence of postpartum anxiety was found in 11.1% cases but in

18.4% of anxiety cases depressive disorders were diagnosed. Studies [6] show that 1.5% of puerperal women can develop any mental disorder (mood disorders, anxiety, tendency to alcoholism, etc.).

Comparing to vaginal birth, Caesarean section surgery is related to a higher risk of complications, longer recovery period, lower satisfaction and greater risk of postpartum mood disorders. Depressed mood was more often observed in women who had unplanned Caesarean section surgery. Study results show that women, who give birth themselves, control the process and outcome, have a reduced possibility to postpartum mood disorders and an increased feeling of satisfaction [10].

Postpartum depression is a widely studied subject, however, there is no sufficient attention given to panic and anxiety studies in the postpartum period, although they are often faced on a daily basis [3].

Spielberger (1985a), as mentioned [11] distinguishes two anxiety aspects: (1) state anxiety (S-anxiety) – it is characterized by subjective, deliberate perception of threat and tension which are connected with activation and irritation of autonomic nervous system. Situations which are considered to be dangerous and threatening provoke this state in a human. (2) Anxiety as a personality trait (trait anxiety/T-anxiety) – means motive or acquired behavioural disposition which encourages an individual to perceive broad, objectively safe object range as containing threat and encourages to react to them.

It was assumed that T-anxiety could be measured in real time and that it changes over time, indicating to what extent an individual perceives the environment as dangerous or threatening. S-anxiety is defined as anxiety differences between individuals; it means differences in tendency to perceive stress situations as dangerous or threatening reacting to them with more or less explicit situational anxiety changes [11]. Average indicators of S-anxiety in women from Latvia [11] in reproductive age group 19–39 are 38.34, but for age group 40–49 it is 36.74, while for T-anxiety it is accordingly 43.00 and 40.73. If anxiety indicators of a respondent correspond to the average indicators in appropriate age group it is predictable that an individual mostly is able to assess the situation adequately and react accordingly. If indicators exceed the average indicators of a particular age group there are possible rapid changes in mood, tears, and nervousness about expected difficulties not concentrating on the current situation, too long and non-effective analyzing of own mistakes as well as insecurity, uneasiness and disbelief in themselves [11].

The goal of this work was to state the differences of anxiety indicators (S-anxiety and T-anxiety) between women that have had vaginal birth and women that have had the surgery.

2 Materials and methods

2.1 Object of study

Research data were collected during the time period from 31.03.2016 to 01.05.2016. In total 100 respondents took part in the research: 50 women who had vaginal birth and 50 women who had Caesarean section surgery. Criteria for participant inclusion in the research: puerperal women 2–4 days after labor with a child born alive and able to fill in questionnaires independently (in Latvian). The research was done obtaining permission from Riga Stradiņš University Ethics Committee and considering all ethical standards.

2.2 Study instruments

Respondents filled in socio-demographic data survey (age, place of residence, education, employment, marital status, the number of previous births, types of birth) and The state –

trait anxiety inventory, STAI Form Y-1 (*Spilberger, Gorsuch, Lushene, Vagg, Jacobs, 1983*), which was adapted in Latvian by Škuškovnika [11].

STAI Form Y-1 consists of 40 closed questions which are divided into two sub-groups. First 20 questions (S-anxiety) evaluate current anxiety state asking respondents to describe their feelings at present moment using subjective questions about sadness, tension, nervousness, worry, etc. Next 20 questions (T-anxiety) evaluate stable aspects which are long-lasting – peace, trust and security [12]. Answers are evaluated according to the Likert scale 4 point gradation starting from "no, not at all" ending with "yes, definitely".

2.3 Statistical analysis

Data processing was done using SPSS version 20. Statistically processing research data with Kolmogorov–Smirnov test help, it was possible to determine whether obtained data correspond to standard breakdown. Further statistical data processing was done using Mann–Whitney *U* test.

3 Results

100 puerperal women took part in the research – 50 women who had vaginal birth and 50 women who had the surgery in the age group 19–43 (see Table 1). It is obvious that the biggest group consists of women in the age group 26–37. Most respondents live in registered relationships, have higher education, are employed and live in a city. Half of respondents had previously given birth.

Table 1. Socio-demographic data of respondents.

Socio-demographic data		All (N = 100)	Vaginal birth (n = 50)		Caesarean section surgery (n = 50)	
			n	%	n	%
Age	19–25	22	14	28	8	16
	26–31	33	17	34	16	32
	32–37	32	14	28	18	36
	38–43	13	5	10	8	16
Family status	Married	61	30	60	31	62
	Unregistered	37	19	38	18	36
	Live alone	2	1	2	1	2
Education	University	67	33	66	34	68
	High school	30	15	30	15	30
	Elementary	3	2	4	1	2
Employed	Yes	84	43	86	41	82
	No	16	7	14	9	18
Place of residence	City	94	48	96	46	92
	Country	6	2	4	4	8
Previous birth	Yes	50	26	52	24	48
	No	50	24	48	26	52

Comparing socio-demographic indicators with state or trait anxiety indicators (see Table 2), it is clear that statistically significant differences (p = 0.036) appear only in group of women that had the surgery regarding age and trait anxiety. The lowest Trait anxiety indicators were in the age group 26–31. The highest Trait anxiety indicators were in the age group 38–43. The rest socio-demographic data between respondent groups regarding anxiety do not differ.

Table 2. Socio-demographic data comparison by groups.

	Vaginal birth (*n* = 50)		**Caesarean section surgery (*n* = 50)**	
	sig		**sig**	
	S-anxiety	**T-anxiety**	**S-anxiety**	**T-anxiety**
Age	0.424	0.291	0.341	**0.036**
Family status	0.115	0.058	0.603	0.167
Education	0.219	0.700	0.743	0.330
Employed	0.123	0.559	0.397	0.785
Place of residence	0.343	0.691	0.649	0.155

Looking at the indicators of descriptive statistics for women that had vaginal birth (see Table 3) and women that had the surgery (see Table 4), it is clear that both state anxiety and trait anxiety average indicators are bigger for women that had the surgery, although maximum indicators are bigger for women that had vaginal birth. However, comparing data with conclusive statistical method (see Table 5), it is clear that statistically significant differences appear neither in state anxiety indicators (p = 0.304) nor in trait anxiety indicators (p = 0.059), although close to the significant border.

Table 3. Anxiety indicators of descriptive statistics for women that had vaginal birth (n = 50).

	Min	**Max**	**Mean**	**SD**	**95% CI**
S-anxiety	21	66	35.88	9.975	33.05–38.71
T-anxiety	25	58	36.92	7.809	34.70–39.14

Table 4. Anxiety indicators of descriptive statistics for women that had the surgery (n = 50).

	Min	**Max**	**Mean**	**SD**	**95% CI**
S-anxiety	22	58	37.20	9.413	34.52–39.88
T-anxiety	24	55	39.24	7.561	37.09–41.39

Table 5. Comparison of anxiety indicators by groups.

Kind of anxiety	**Vaginal birth (*n* = 50)**		**Caesarean section surgery (*n* = 50)**		***p***
	M	**SD**	***M***	**SD**	
S-anxiety	35.88	9.98	37.20	9.41	0.304
T-anxiety	36.92	7.81	39.24	7.56	0.059

4 Discussion

Looking at trait anxiety and state anxiety average indicators it is clear that for women who had the surgery they are higher than for women who had vaginal birth, although women who had vaginal birth got more maximum number of points. It was found [13] that using epidural anaesthesia or performing Caesarean section surgery, anxiety indicators were higher than giving birth naturally.

Comparing with vaginal birth, Caesarean section surgery is connected with higher risk of complications, longer recovery period and higher risk of postpartum mood disorders. Depressed mood was more often observed in women who had unplanned Caesarean section surgery. Study [10] results show that women, who give birth themselves, control the process and outcome, have a reduced possibility to postpartum mood disorders and an increased feeling of satisfaction.

Looking more specifically at trait anxiety indicators and age relationship of women who had the surgery, which is statistically significant, it was observed that the lowest trait anxiety indicators were in the age group 26–31, but the highest trait anxiety indicators were in the age group 38–43. Such result could be related to woman's greater concern for health and bigger responsibility for being able to deal with new-born baby and herself in the post-surgery period.

Looking more specifically at relationship between trait anxiety for women who had vaginal birth and marital status, which is close to critical value, it was observed that in case of registered marriage anxiety indicators were lower than in cases of not registered relationship or non-being in relationship. With reference to the research [4] it was found that single women have a greater risk to face depression and anxiety in the perinatal period.

The research gives general information about the tendency of anxiety indicators in specific respondent group. Such evaluation can be used as a basis for further research of factors affecting anxiety indicators, with the aim to reduce them ultimately and thus providing high-quality care in Latvian medical institutions. Repeated survey would be necessary to improve and specify data.

5 Conclusions

Statistically significant difference in anxiety indicators between women who had vaginal birth and women who had the surgery was not found. Although the range between the lowest and the highest anxiety indicators is bigger for women who had vaginal birth, both average state anxiety indicator and average trait anxiety indicator are higher for women who had the surgery. Between demographic data and anxiety indicators for women who had vaginal birth and women who had the surgery statistically significant differences have not been found, except connection between trait anxiety and age for women who had the surgery. As the average anxiety indicators are not too high, it is possible to conclude that the environment in Riga postpartum units corresponds to the required because slightly increased indicators are natural for such an emotional experience as the birth of a child or the surgery.

References

1. O. Giakoumaki, K. Vasilaki, L. Lili, M. Skouroliakou, G. Liosis. J. Psychosom. Obstet. Gynecol. **30**, 21–28 (2008)
2. S. Shrestha, K. Adachi, M.A. Petrini, S. Shrestha. Int. Nurs. Rev. **61**, 427–434 (2014)
3. I.S. Polachek, L.H. Harari, M. Baum, R.D. Strous. Isr. J. Psychiatry Relat. Sci. **51**, 128–134 (2014)
4. L.E. Sockol, C.N. Epperson, J.P. Barber. Arch. Womens Ment. Health. **17**, 199–212 (2014)
5. J. Yelland, G. Sutherland, S.J. Brown. BMC Public Health. **10**, 771 (2010) `http://www.ncbi.nlm.nih.gov/pmc/articles/PMC3022850/`

6. V. Sagayadevan, S. Pheng Lee, E. Abdin, J. Vaingankar, H. Chen, S.A. Chong, M. BMC Women's Health. **15**, 1–9 (2015)
7. W. Simpson, M. Glazer, N. Michalski, M. Steiner, B.N. Frey. Can. J. Psychiatry. **59**, 434–440 (2014)
8. S.L. Farr, P.M. Dietz, M.W. O'Hara, K. Burley, J. Ko. J. Womens Health. **23**, 120–128 (2014)
9. C. Reck, K. Struben, M. Backenstrass, U. Stefenelli, K. Reinig, T. Fuchs, C. Sohn, C. Mundt. Acta Psychiatr. Scand. **118**, 459–468 (2008)
10. R.S. DeLuca, M. Lobel. Anal. Soc. Issues Public Policy. **14**, 183–204 (2014)
11. D. Škuškovnika. Trauksme latviešiem un Latvijā dzīvojošiem krieviem. Promocijas darbs. Rīga: Latvijas Universitāte [In Latvian: Anxiety for Latvian and Russion, who live in Latvia. Dissertation.] (2004)
12. L.J. Julian. Arthritis Care Res. **63**, 467–472 (2011)
13. J. Henderson, M. Redshaw. J. Reprod. Infant Psychol. **31**, 465–478 (2013)

Risks to health of the population of the nonindustrial city from quality of atmospheric air

I. Abdikadirova[1], B. Zasorin[1], L. Yermukhanova[1], V. Sabyrakhmetova[1], A. Zhexenova[1], and U. Urazova[2]

[1]Marat Ospanov West Kazakhstan State Medical University, The Republic of Kazakhstan
[2]Astana Medical University, The Republic of Kazakhstan

Abstract. Risk assessment to public health living in the environmental problems is of great importance for humans today. However, the main attention is paid to major industrial centers, while for cities not having developed industries such studies are rare. The study of the intensity and structure of chemical air pollution was conducted to determine the risk levels for human health from effects of the chemical air pollution in residential areas of Aktau city (West Kazakhstan). Method of retrospective analysis of emissions into the atmosphere was used, and damage to health was determined in accordance with the risk assessment procedure. We take into account the major pollutants, the city being in the air basin – sulfur dioxide, hydrogen sulfide, carbon monoxide, nitrogen oxides, ammonia and total hydrocarbons. The main threat to children's health is caused by common hydrocarbons (88%), nitrogen oxides (5%), and ammonia (3%). The estimated cancer risk is defined at the level of 10^{-3} to 10^{-4}, it corresponds to the limits and requires regular monitoring.
Key words: risk evaluation, population morbidity, air pollution.

1 Introduction

Health risk assessment system allows us to obtain not only qualitative, but also quantitative characterization of the influence of the environment on population's health before the onset of the effects of this influence. It distinguishes risk assessment of epidemiological methods of analysis and makes it possible to predict the result and work on the basis of management decisions to minimize adverse effects, based on existing chemical environmental factors, monitoring of the environment and the population's health. In this regard, the health risk assessment system can be easily integrated into the overall management system and decision-making in administrative practice. The ability to assess health of the population or the extent of its loss by financial categories is important in today's economic environment, both for health and the management system. Health risk assessment system allows us to determine the health risk from a variety of factors [1, 2].

At present, atmospheric air is the leading object of the environment, which is connected with the greatest part of all health risks [3]. This problem is significantly worsened at the border of the XX–XXI centuries due to the increase of capacity of thermal power plant, petrochemical metallurgy, and as a result of the growth of units of vehicles in urban areas [4, 5].

Considerable development of the Caspian region of Kazakhstan which is characterized by hot and humid climate that leads to regional specificity in the evaluation of risks for health of the population living here is being continued.

Aktau is a port on the Caspian Sea, through which the main maritime transit of Kazakhstan goods is realized. There are not industrial plants polluting the air in the city but the network of transport companies is well developed.

The aim of this work is assessment of levels of the carcinogenic and non-carcinogenic risks from chemical factors of air pollution for population's health in Aktau.

2 Materials and methods

Risk assessment of public health is carried out by the usual method [6]. The risks were calculated for the following age groups: children (0–6 years) [7], adolescents (14–16 years) and adults (over 20 years).

Monitoring data of the air quality of the city for a five-year period (2011–2015) was used as the information source. The reference concentration for chronic inhalation exposure was used for risk calculation. Comprehensive assessment of the carcinogenic risk was performed using the computer program "RISK * ASSISTANT", in accordance with the methodology set out in the documents: Risk Assessment Guidance for Supperfund. Vol. I: Human Health Evaluation Manual No. 540/189/002. (US EPA, 1989); Guidance for Risk Characterization (US EPA, 1995). Quantitative parameters for the evaluation of the carcinogenic risks were taken from the documents U.S. Environmental Protection Agency, Integrated Risk Information System, Environmental Criteria and Assessment Office, Office of Health and Environmental Assessment, Office of Research and Development (Cincinnati, 1993) [8].

Information on actual pollution of the atmospheric air of Aktau city (for a 5-year period) was received at the Department on consumers' rights protection of Mangystau region.

Calculation of non-carcinogenic risks taking into account vital organs and systems is made for the following pollutants of the atmospheric air as: sulphur dioxide, hydrogen sulphide, carbon monoxide, nitric oxides, ammonia, volatile organic compounds (VOC), and general hydrocarbons.

3 Results and discussion

During the first stage of the research the calculation of average annual and average daily concentration of contaminants was made. On the basis of given calculations, concentration of the substance, entering the body with the atmospheric air, taking into consideration mg/kg of the body weight (for the mentioned period of time) was determined.

Hazard quotient of carcinogenic effects and hazard index for all the mentioned above substances was calculated. Besides, hazard index for every vital organ and system was calculated. Calculation of the complex air pollution index (API_5) for five substances with the largest values is made by the formula: $API_5 = \Sigma$(g cp.i/MAC (maximum allowable concentration)) (i) Ci where g cp.i is the average concentration of the pollutant;

MAC_i is the average daily MAC of the pollutant;

Ci is a coefficient that depends on the hazard class of the pollutant, taken to be 1.7, 1.3, 0.1 and 0.9, respectively, for the 1, 2, 3 and 4 hazard classes of the contaminant.

The multiplicity of MAC excess (K), including values less than 1, is established by dividing the existing concentration of the substance in the air by the amount of daily average MAC according to the formula:

$$K = \frac{C}{MAC}$$

where:
C – the existing concentration of this substance in the air in different periods averagely over time (average daily or average annual);
Characteristics of the risk of development of non-carcinogenic effects for individual substances are based on the calculation of the hazard factor using the formula:

HQ = AD / RfD or HQ = AC / RfC, where

HQ – hazard ratio;
AD – average dose, mg/kg;
AC – average concentration, mg/m^3;
RfD – reference (safe) dose, mg/kg;
RfC – reference concentration (safe), mg/m^3.

The hazard ratio is calculated separately for the conditions of short-term (acute), subacute and long-term exposure to chemicals. In this case, the averaging period of the exposures and the corresponding safe exposure levels should be similar.

Characteristics of the risk of developing non-carcinogenic effects in the combined and complex effects of chemical compounds are based on the calculation of the hazard index (HI). The hazard index for the conditions of simultaneous intake of several substances in the same way (for example, inhalation or oral) is calculated by the formula:

HI = Sum of HQ, where
I

HQ – hazard ratios for individual components of the mixture *i* of the active substances. Received results are presented in Table 1.

Table 1. Hazard quotient (HQ) of non-carcinogenic effects for population of Aktau.

Substance	Children	Adolescents	Adults	Vital organs
Sulphur dioxide	0.039	0.039	0.036	Respiratory organs
Hydrogen sulphide	1.0	0.36	0.23	Respiratory organs
Carbon monoxide	0.389	0.139	0.091	Blood, cardio-vascular system, physical development, central nervous system
Nitric oxides	2.059	0.736	0.478	Respiratory organs blood (MetHb formation)
Ammonia	1.313	0.471	0.872	Respiratory organs
General hydrocarb ons	37.023	13.227	8.581	Eyes, respiratory organs, liver, kidneys, central nervous system

Conducted calculations of non-carcinogenic risks indicate that children population of Aktau city are exposed to maximal risks for their health. In particular maximal ratings of HQ are established for general hydrocarbons and comprise 37.023. Hazard quotient indicators more than 1.0 are revealed for nitric oxides (2.059) and ammonia (1.313). Hazard quotient of hydrogen sulphide is equal to one.

Risk level for the health of adolescents living in Aktau city due to the pollution of the atmospheric air is slightly lower. The same tendency as in the group of children from 0 to 6 years is detected in this group. However risk level for the health of adolescents is 2.3–3 times lower. So, Hazard quotient for general hydrocarbons is 13.227 and it is the only chemical substance with Hazard quotient >1.

Similar results have been received while evaluating Hazard quotient for the adult population.

Results on hazard index evaluation for every vital organ and system are presented in Table 2.

General hazard index for children (from 0 to 6 years) is 41.895. Hazard index for respiratory organs takes the first place and is equal to 41.506. Diseases of central nervous system take the second place (hazard index = 37.412), liver and kidney impairments (hazard index = 37.023) take the third place. Excess of hazard index for the diseases of blood and cardio-vascular system should be mentioned as well (2.448).

For adolescents, hazard index comprised 14.971. Among vital organs the respiratory organs (hazard index = 14.832) are in the first place, the second place is for the diseases of central nervous system (hazard index = 13.366), the third place – the kidney and liver diseases (hazard index = 13.227).

In the group of adult population hazard index is 4 times lower comparing with children (0–6) and 1.45 times lower comparing with adolescents. Among the most vulnerable organs are respiration organs, central nervous system, kidneys and liver.

Table 2. Hazard index (HI) for different age groups of Aktau population.

HI	Children	Adolescence	Adults
Common	41.895	14.971	10.288
Physical development	0.389	0.139	0.091
Kidney diseases	37.023	13.227	8.581
Diseases of the blood, cardiovascular system	2.448	0.875	0.569
Diseases of the respiratory system	41.506	14.832	10.197
Disease of the CNS	37.412	13.366	8.672
Liver diseases	37.023	13.227	8.581

Summarizing the obtained results, it can be argued that the maximum non-carcinogenic risk for exposure to chemical substances in the air pool of the residential territories is set for the children population (0–6 years) Aktau. High risk of developing diseases of the respiratory organs, central nervous system, liver and kidney and blood and cardiovascular system is noted. It should be emphasized that to the risk of developing these diseases total hydrocarbons, nitrogen oxides, ammonia and hydrogen sulphide, contained in the atmospheric air, also made their contribution.

In the second stage of calculation of risks to public health was the assessed risk of carcinogenic effects from exposure to chemical substances emitted into the atmosphere of city of Aktau. The calculation of individual carcinogenic risk was carried out using data on the amount of exposure and the values of the factors of the carcinogenic potential (slope factor, unit risk). As a rule, for carcinogenic chemicals, the additional probability of developing cancer of the individual throughout life (CR) is assessed taking into account the average daily dose over lifetime (LADD) – Table 3.

Table 3. LADD (life average daily dose) for non-carcinogenic effects mg/(kg*day).

	Children	**Adolescence**	**Adults**
Sulfur dioxide	0.471	0.413	0.264
Hydrogen sulfide	0.013	0.011	0.007
Carbon monoxide	7.51	6.56	4.21
Nitrogen oxides	6.23	5.46	3.49
Ammonia	0.844	0.738	0.473
Total hydrocarbons	16.91	14 79	9.46
Volatile organic compounds	2.11	1.83	1.17
Σ	34.088	29.802	19.074

The obtained data on the lifetime average daily dose of admission of harmful substances from the air (subject to continuing levels of air pollution) indicate that a group of children (0–6 years) have the maximum average daily concentration for the entire period of life. This is followed by adolescence and adults.

The results obtained were used to calculate individual and population carcinogenic risk for the population of Aktau.

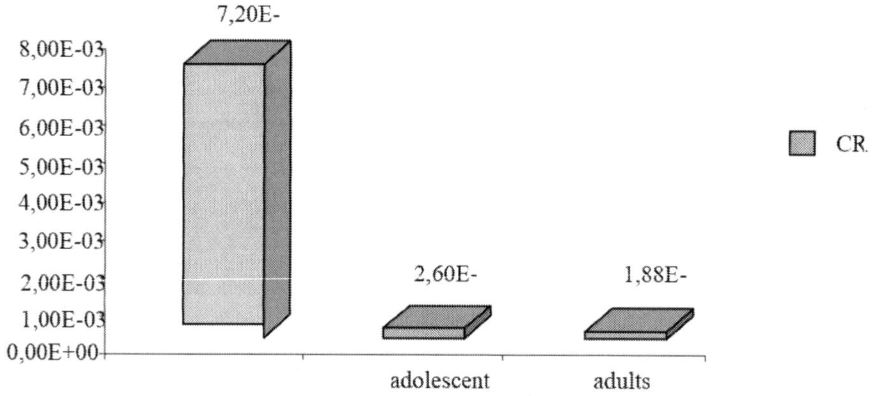

Fig. 1. Individual carcinogenic risk for the population of Aktau taking into account age.

The levels of carcinogenic risk for population's health of Aktau city in Fig. 1 show that the maximum individual risk of cancer is noted in the group of children population (0–6 years). The risk of malignancy among adolescent and adult population of the city on average is 30 times lower. Minimum individual cancer risk is determined among adults.

In a comparative risk characterization we used the value of the annual population risk (PCRa) – the estimated number of additional cancer cases during the year by sex and age (Fig. 2).

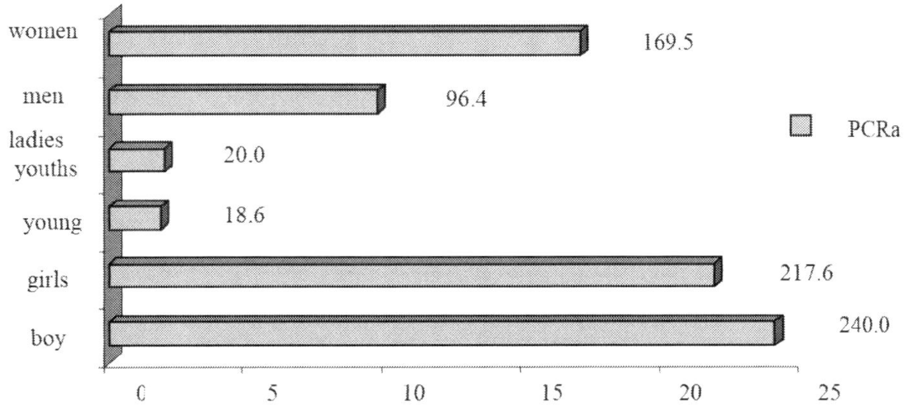

Fig. 2. Carcinogenic risks for the population of Aktau by sex and age.

Analysis of the results of the evaluation of carcinogenic risk population with regard to gender and age showed that the maximum cancer risk is for children population of Aktau (0–6 years) whose risk of occurrence of malignancies is by 1.5–2.5 times higher compared with the same categories of adults. Low population carcinogenic risk for adolescences may be explained by their relatively small number in the population.

To assess the representativeness of the obtained results, we carried out a correlation analysis between the dynamics of changes in atmospheric air of the studied pollutants and changes in the levels of morbidity in all age groups. The list of diseases was compiled on the basis of the "Guide to the International Statistical Classification of Diseases, Injuries and Causes of Death of tenth revision." "Coefficient of Spearman rank correlation" was used to assess the level of morbidity due to air pollution. Calculations were performed using SAS-9 statistical processing of the modular system. The data for the period from 2011 to 2015 were used for calculating.

Analysis of the data shows that there are correlations of moderate strength level of primary morbidity of children living in Aktau with air pollution. In particular, it should be noted that there is a correlation of 0.43–0.60 between the identified diseases of the urinary system and the content of hydrogen sulfide in the atmosphere of the residential area, carbon monoxide, nitrogen oxides, ammonia, and VOC. Neoplasms link frequency and the presence of hydrogen sulfide in air, carbon monoxide, nitrogen oxides and VOC (r between 0.43 and 0.6); respiratory diseases and the level of hydrogen sulfide, carbon monoxide and oxides of nitrogen (r = 0.50)is also noted. In addition, the frequency of congenital anomalies correlates with the content in the atmosphere of nitrogen oxides, ammonia, hydrocarbons, and volatile organic compounds (r between 0.43 and 0.64).

Among the chemicals for which connection of incidence level is often identified, carbon monoxide r from 0.49 to 0.60 (infectious and parasitic diseases, neoplasms, diseases of the endocrine, nutritional, metabolic and immunity systems, disease of the nervous system, eye and its appendages, respiratory diseases, diseases of the genitourinary system) rank first. In second place – r hydrogen sulfide from 0.43 to 0.66 (neoplasms, diseases of the eye and its appendages, diseases of the ear and mastoid process, cardiovascular diseases, respiratory diseases, diseases of the genitourinary system), on the third place – r nitrogen oxides from 0.47 to 0.64 (tumors, cardiovascular diseases, respiratory diseases, congenital anomalies, injuries and poisoning).

The low level of correlation (within statistical error) dependence is marked for diseases of the digestive system, diseases of the skin and subcutaneous tissue and diseases of the musculoskeletal system.

Analysis of the prevalence of morbidity among children shows that there are correlations of moderate strength with a number of air pollutants. In particular, a connection between the eye and its appendages diseases with the content in the ambient air of sulfur dioxide, hydrogen sulfide, carbon monoxide and oxides of nitrogen (*r* from 0.43 to 0.6) is revealed. The following rank place is taken by the diseases of the genitourinary system, *r* from 0.43 to 0.60 (sulfur dioxide, carbon monoxide, nitrogen oxides). In third place there are diseases of the blood and blood-forming organs, *r* from 0.41 to 0.6 (hydrogen sulfide, nitrogen oxides).

Among the chemicals, most frequently identified due to the incidence level, first place is occupied by *r* nitrogen oxides from 0.43 to 0.60 (diseases of the blood and blood-forming organs, cardiovascular diseases, diseases of the eye and its appendages, diseases of the urinary system). The second place is occupied by sulfur dioxide *r* from 0.49 to 0.60 (endocrine, nutritional, metabolic, and immune system, diseases of the eye and its appendages, ear and mastoid disease, diseases of the genitourinary system, diseases of the musculoskeletal system. In third place there are *r* hydrogen sulfide from 0.41 to 0.66 (infectious and parasitic diseases, diseases of the blood and blood-forming organs, eye diseases and its appendages).

There is a low level of correlation (within statistical error) for diseases of the nervous system, diseases of the digestive system.

Analysis of the data shows that there are correlations of medium and high power levels of primary morbidity of adolescents living in Aktau with a number of chemical substances, as defined in the atmosphere of residential areas. In particular, the maximum values of correlative links marked for diseases of the endocrine, nutritional, metabolic and immunity (hydrogen sulphide, carbon monoxide, nitrogen oxides, ammonia) *r* from 0.54 to 0.82, and diseases of the genitourinary system (hydrogen sulfide, ammonia, hydrocarbons, VOC) *r* from 0.71 to 0.89 is noted. Then come: *r* circulatory disease from 0.49 to 0.73, eye diseases and its appendages *r* from 0.47 to 0.77, diseases of the ear and mastoid *r* from 0.49 to 0.77 (ammonia, hydrocarbons, VOC).

Among the chemicals for which connection of incidence level is often identified, the first place is ammonia (diseases of blood and blood-forming organs, endocrine, nutritional, metabolic and immunity disorders of eye and its appendages, ear disease and the mastoid process, cardiovascular diseases, diseases of the genitourinary system) *r* from 0.48 to 0.88.In second place – hydrocarbons (diseases of the eye and its appendages, diseases of the ear and mastoid process, diseases of the circulatory system, diseases of the digestive system, diseases of the genitourinary system) *r* from 0.47 to 0.77 followed by: VOC (cardiovascular diseases, diseases of the eye and its appendages, ear and mastoid diseases, respiratory diseases, diseases of the genitourinary system) *r* from 0.51 to 0.89; hydrogen sulfide (diseases of the blood and blood-forming organs, endocrine, nutritional, metabolic and immunity, respiratory diseases, diseases of the genitourinary system) *r* from 0.51 to 0.82.

For infectious and parasitic diseases, diseases of the nervous system, diseases of the skin and subcutaneous tissue, diseases of the musculoskeletal system there is a low level of correlation (within statistical error).

Analysis of the prevalence of morbidity among adolescents indicates that there are correlations of medium and high strength with a number of air pollutants.

In particular, the diseases of the genitourinary system are associated with the content of hydrogen sulphide, ammonia, hydrocarbons and VOC (volatile organic compounds) (*r* 0.71–0.88) in the atmosphere followed by: neoplasms (sulphur dioxide, carbon monoxide, and hydrocarbons) r of 0.50 to 0.66; diseases of the blood and blood-forming organs (hydrogen sulphide, carbon monoxide, ammonia) r from 0.49 to 0.60; diseases of the eye and its appendages (ammonia, hydrocarbons, VOC) r from 0.46 to 0.73.

Among the chemicals, most frequently identified due to the incidence level, leading place are taken by: hydrocarbons (neoplasms, diseases of the ear and mastoid process, diseases of the eye and its appendages, diseases of the genitourinary system) r from 0.46 to 0.77; ammonia (diseases of the blood and blood-forming organs, diseases of the genitourinary system, diseases of the eye and its appendages, diseases of the ear and mastoid process) r from 0.47 to 0.70; VOC (diseases of the eye and its appendages, diseases of the genitourinary system, injury and poisoning) r from 0.73 to 0.87.

For diseases of the nervous system, respiratory diseases, diseases of the digestive system, diseases of the skin and subcutaneous tissue, diseases of the musculoskeletal system there is a low level of correlation (within statistical error).

Analysis of the data shows that there are correlations of medium and high power levels of primary morbidity of the adult population of Aktau, with a number of chemical substances, as defined in the atmosphere of residential areas. In particular, there is a high level of correlations complications of pregnancy and childbirth with the content of hydrogen sulphide in the air, nitrogen oxides, ammonia, hydrocarbons and VOC (r from 0.49 to 0.81). The next rank place is referred to the diseases of the blood and blood-forming organs level which is associated with the presence in the air of hydrogen sulphide, ammonia, hydrocarbons, VOC (r from 0.71 to 0.78), followed by infectious and parasitic diseases (hydrogen sulphide, ammonia, hydrocarbons, VOC) r from 0.77 to 0.84), and diseases of the circulatory system (ammonia, hydrocarbons, VOCs) r from 0.43 to 0.73.

Among the chemicals, most frequently identified due to the leading position of morbidity level, are: hydrogen sulphide (infectious and parasitic diseases, diseases of blood and blood-forming organs, nervous system diseases, disorders of ear and mastoid process, respiratory diseases, digestive diseases, pregnancy complications and childbirth) r from 0.43 to 0.77; ammonia (infectious and parasitic diseases, diseases of the blood and blood-forming organs, nervous system diseases, diseases of the ear and mastoid process, diseases of the circulatory system, digestive diseases, complications of pregnancy and childbirth) r from 0.43 to 0.89; hydrocarbons (infectious and parasitic diseases, diseases of blood and blood-forming organs, nervous system diseases, circulatory system diseases, complications of pregnancy and childbirth) r from 0.61 to 0.84; VOC (infectious and parasitic diseases, diseases of blood and blood-forming organs, nervous system diseases, circulatory system diseases, complications of pregnancy and childbirth) r from 0.61 to 0.84.

Analysis of the data shows that there are correlations of medium and high power levels of prevalence of morbidity of the adult population of Aktau, with a number of chemical substances, as defined in the atmosphere of residential areas. In particular, there is a high level of correlation between the incidence of diseases of the blood and blood-forming organs and the air containing hydrogen sulphide, ammonia, hydrocarbons and VOC (r from 0.43 to 0.94). This is followed by neoplasms, diseases of the endocrine, nutritional, metabolic and immune diseases of the circulatory system (hydrogen sulphide, ammonia, hydrocarbons and VOC) r from 0.43 to 0.88.

Among the chemicals, most frequently identified due to the level of morbidity, the leading place is taken by: VOC (neoplasms, diseases of the blood and blood-forming organs, endocrine, nutritional, metabolic, and immune system, nervous system diseases, diseases of the eye and its appendages, disease the circulatory system, diseases of the musculoskeletal system, complications of pregnancy and childbirth) r from 0.43 to 0.94; ammonia (neoplasms, diseases of the blood and blood-forming organs, endocrine, nutritional, metabolic, and immune system, nervous system diseases, diseases of the eye and its appendages, cardiovascular disease, complications of pregnancy and childbirth) r

from 0.43 to 0.71; hydrocarbons (neoplasms, diseases of the blood and blood-forming organs, endocrine, nutritional, metabolic, and immune system, nervous system diseases, circulatory system diseases, complications of pregnancy and childbirth) r from 0.43 to 0.70.

There is a low level of correlation (within the margin of error) only for the skin and subcutaneous tissue diseases.

Thus according to the results of the study a number of peculiarities in the age groups among people with newly diagnosed diseases are found. In particular in children a high level of connection of neoplasms, diseases of genitourinary system and congenital malformations with air pollutants is revealed. In the group of adolescents maximum connection is established between environmental factors and diseases of the genitourinary system, endocrine, nutritional and metabolic diseases, and immune disorders. In adults the maximum connection is established between atmospheric pollutants and complications of pregnancy and childbirth, blood disorders, diseases of the ear and mastoid process.

In the age groups, the prevalence of disease is characterized by the following features: in children the maximum connection is set for diseases of the eye and its appendages, diseases of the genitourinary system; in adolescents – diseases of the genitourinary system, neoplasms, diseases of blood and blood-forming organs; in adults – diseases of the circulatory system, digestive system diseases and complications of pregnancy and childbirth.

Chemicals that are mostly connected with the level of primary morbidity in the age groups are as follows: in children – hydrogen sulfide, carbon monoxide and nitrogen oxides; in adolescents and adults – ammonia, hydrocarbons, VOCs.

Chemicals that are connected with the level of disease prevalence in the age groups: in children – nitrogen oxides, sulfur dioxide, and hydrogen sulfide; in adolescents and adults – ammonia, hydrocarbons, VOCs.

4 Conclusions

Thus according to the results of evaluation of individual and population carcinogenic risk to health of people in the city of Aktau and correlation analysis of the population the individual risk of cancer among all age and sex groups can be assessed as very high. As in case of individual as well as population cancer risk, the most vulnerable is children population (0–6 years) of the city. However, it should be noted that the amount of the annual risk of population, as a rule, should not be used for any direct analogies between the levels of actual cancer incidence or mortality rates and the values of these risks.

The results obtained were presented to Aktau city administration to make managerial decisions on development of measures to reduce the risk to public health.

References

1. Brauer M, Freedman G, Frostad J, van Donkelaar A, et al. Environ Sci Technol. 2016 Jan 5; **50**(1), 79–88. doi: 10.1021/acs.est.5b03709. Epub 2015 Dec 4
2. Li Y, Henze DK, Jack D, Henderson BH, Kinney PL. Sci Total Environ. 2016 Jan 1; **539**, 515–25. doi: 10.1016/j.scitotenv.2015.08.129. Epub 2015 Sep 15
3. Raaschou-Nielsen O, Beelen R, Wang M, Hoek G, Andersen ZJ, et al. Environ Int. 2016 Feb; **87**, 66–73. doi: 10.1016/j.envint.2015.11.007. Epub 2015 Nov 28
4. Zasorin BV, Yermukhanova LS. Gigiyena i sanitariya. Vliyaniye faktorov okruzhayushchey sredy na immunologicheskuyu rezistentnost' organizma. **3**, 8–9 (2012)

5. Zasorin BV, Kurmangaliyev OM, Yermukhanova LS. Osobennosti immunnogo statusa u naseleniya urbanizirovannykh territoriy s povyshennym soderzhaniyem tyazhelykh metallov. J. Gigiyena i sanitariya. **3**, 17–19 (2012)
6. Zhang H, Zhou ZQ, Zhao HY, Xiong ZQ. Huan Jing Ke Xue. 15; **37**(1), 28–34. Chinese. (2016)
7. Wang C, Qian X, Li HM, Sun YX, Wang JH. Huan Jing Ke Xue. 15; **37**(5), 1662–9. Chinese. (2016)
8. Zasorin BV, Sabyr KK, Iskakov AZH, Kokzhiyek J. Riski zdorov'yu naseleniyu ot vozdeystviya faktorov sredy obitaniya urbanizirovannykh territoriy. Aktobe, (2009)
9. Zasorin BV. J. Meditsinskiy Zapadnogo Kazakhstana. **2**(26), 89–94 (2010)
10. Iskakov AZH, Boyev VM, Zasorin BV. J. Gigiyena i sanitariya. **3**, 52–54 (2009)

Study of dispensary patients of working age with circulatory system diseases through the automated control systems

I. Abdikadirova[1], A.S. Chukmaitov[2], L. Yermukhanova[1], and A. Bekbauova[1]

[1]West Kazakhstan Marat Ospanov State Medical University, The Republic of Kazakhstan
[2]Virginia Commonwealth University School of Medicine, USA

Abstract. Diseases of the circulatory system occupy a leading position among the diseases prevalence of which is increasing with each passing year and is sustainable. The leading reasons for the circulatory diseases as revealed by prophylactic medical examination are arterial hypertension, chronic heart failure, arrhythmia and coronary heart disease. The growth in the number of circulatory system diseases causes medical and social problem – disability of the working-age population. In order to determine the role of Automated Control System the study of morbidity, disability and mortality due to circulatory system diseases was conducted in the municipal polyclinic No. 4 of Aktobe city (West Kazakhstan). The method of retrospective data analysis of annual statistical reports and data from the portal "Electronic Register of dispensary patients", "Automated Information System", "Drug Supply Information System" was used. The incidence of circulatory system diseases in dispensary patients of working age had risen from 1.94‰ (2011) to 2.8‰ (2015). In the main classes of primary disability the circulatory system diseases in terms of growth steadily occupy the first place ranking.
Key words: automated control system, circulatory system diseases, morbidity, disability.

1 Introduction

In modern conditions the most urgent problem of public health and health care is the health of the working age population.

The second half of the 20th century and the beginning of the 21st century in many countries around the world along with real positive achievements in the prevention and treatment of many diseases are marked with increase in morbidity and disability in the population (including that caused by the circulatory system diseases).

Cardio-vascular diseases are a major problem of public health worldwide [1], as the number of patients with cardio-vascular diseases is constantly increasing and increases with age [2].

The territorial computerized register of all contingents of dispensary for diseases was established in the Republic of Kazakhstan [3]. In all medical institutions, patients with circulatory system diseases are a subject to dispensary observation. Prophylactic medical examination is the method of active dynamic hospital observation of the health of population groups with chronic illnesses [4]. The study of morbidity of patients under the dispensary observation will improve health outcomes and make it possible to evaluate the effectiveness of prophylactic medical examination. Improving the efficiency of dispensary observation remains one of the topical problems of modern health care of the Republic of Kazakhstan primarily in connection with prevalence of chronic non-communicable diseases among the working-age population. Organization of prophylactic clinical examination requires much time for completing the documentation, accounting and reporting. Currently information technologies are widely used in health care as it allows automation of physicians' activities and improves the quality and efficiency of their work; they help to move from paper to modern technologies of data storage and processing.

State programme of reforming and development of health care of Kazakhstan from September 13, 2004 No. 1438 covers the establishment of the Unified Health Management Information System of Kazakhstan which resulted in the introduction of information technologies in health care and the establishment of the National Information and Analytical Center with branches in all regions.

Implementation of automated control systems in the health care of Kazakhstan is a necessary process, as it provides computer support for medical decisions in diagnosis, clinical examination of the various population groups, as well as providing management decision-making in health care.

Automated control system is aimed at improving the management healthcare and improving the quality and accessibility of health aid. Automated control system will ease conducting of a complete medical electronic document management, it will help to get rid of the chore of keeping documentation and focus on the patient's problems [5].

According to the Order No. 71 from February 6, 2014 "Electronic Register of dispensary patients" was introduced.

Portal "Electronic Register of Dispensary Patients" is the system of dispensary patients' records. Daily work of medical clinics in conducting dispensary card is fully automated in the programme with the formation of statistical and operational reports.

To improve the system of drug provision for population in the framework of guaranteed free medical care, the Order No. 446 from September, 4 2009 "On approval of the list of drugs and medical products for free preferential provision of population within the guaranteed volume of free medical care on dispensary basis" was issued. In accordance with the above mentioned orders, dispensary patients having circulatory system diseases with only four nosologies: hypertension, coronary heart disease, arrhythmia, chronic heart failure are provided by free drugs.

Nowadays cardio-vascular diseases are one of the most common cause of mortality all over the world [6, 7], they are also the main reason of disability which requires socio-economic resources [8]. One of the indicators characterizing the medical and social significance of the problem is a high rate of primary disability of patients with these diseases. Cardio-vascular diseases such as coronary heart disease and hypertension are major causes of people's disability in regions.

The purpose of this work – to study morbidity, primary disability and efficacy of rehabilitation in dispensary patients of working age with cardiovascular disease using the automated control system programmes.

2 Materials and methods

We conducted a retrospective analysis of morbidity and disability due to the most common diseases of the circulatory system in the structure of dispensary patients of Aktobe municipal polyclinic No. 4 in the period of 2011–2015 years. Sources of research have been the analysis of the data of annual statistical reports and data from the portal "Electronic

Register of dispensary patients", "Automated Information System", "Information system of drug supply". Dynamics of morbidity due to circulatory system diseases was studied over the past 5 years (2011–2015), the incidence trends are determined by least squares. In conventional methods of biomedical statistics, extensive EI and rough indicators RI of circulatory system diseases are calculated. The average age of the patients (M), confidence interval (95% CI) and the average annual growth rate (TPA%) are determined.

Indicators of general and primary morbidity are calculated for 100,000 of the adult population, disability is for 100,000 of the adult population.

3 Results and discussion

Total number of patients 36,643 within 5 years (2011–2015) were registered in municipal policlinic No. 4 of Aktobe Republic of Kazakhstan, among them patients with the first-ever diagnosis of circulatory system diseases – 4974 and dispensary patients – 20,646 (Table 1).

Table 1. Structure of morbidity due to systemic blood circulation diseases in municipal polyclinic No. 4 Aktobe in 2011–2015.

Years	Total (women)		Newly diagnosed (including women)		Dispensary registered (including women)		Average age (95% CI)	MP- (95% CI)
2011	6399	3570 (55.8%)	765	408 (53.3%)	2846	1719 (60.4%)	55.5 (55.1 – 55.8)	8.72 (8.52 – 8.93)
2012	6502	3610 (55.5%)	804	457 (56.8%)	3221	1889 (58.6%)	52.7 (52.4 – 53.1)	8.49 (8.30-8.7)
2013	6799	3878 (57%)	846	504 (59.6%)	3289	1921 (58.4%)	52.5 (52.2 – 52.8)	8.81 (8.62 – 9.02)
2014	7732	5123 (66.2%)	1356	625 (46.1%)	4761	2682 (56.3%)	51.9 (51.6 – 52.2)	9.86 (9.65 – 10.07)
2015	9211	5932 (64.4%)	1203	725 (60.3%)	6529	3945 (60.4%)	52.1 (51.8 – 52.4)	11.95 (11.72 – 12.18)
Total	36643	22113	4974	2719	20646	12156	M=52.7	M = 9.56

Distribution of patients with circulatory system diseases by sex in 2011was the following: women – 55.8%, 2012 – 55.5%, 2013 – 57%, 2014 – 66.2% and in 2015 – 64.4% with increasing prevalence of female population (Table 1). The average age of patients was 52.7 ± 7.1 and dynamics tended to decrease with 56.6 ± 7.6 in 2011 and 52.6 ± 7.8 in 2015 when levelling these indicators noted above trend (Tub = -1.4%), i.e. there is a shift in the direction of "rejuvenation". The average rough indicator (R.I.) incidence of systemic blood circulation disease was 9.56 ± 1.43o/oooo. In rough indicator (R.I.) dynamics tended to grow: 8.72 in 2011 year to 11.95 in 2015 year.

In the study of the incidence of systemic blood circulation disease in dispensary patients of working age, this figure has tended to rise from 8.72‰ (2011) to 11.95‰ (2015) (Fig. 1).

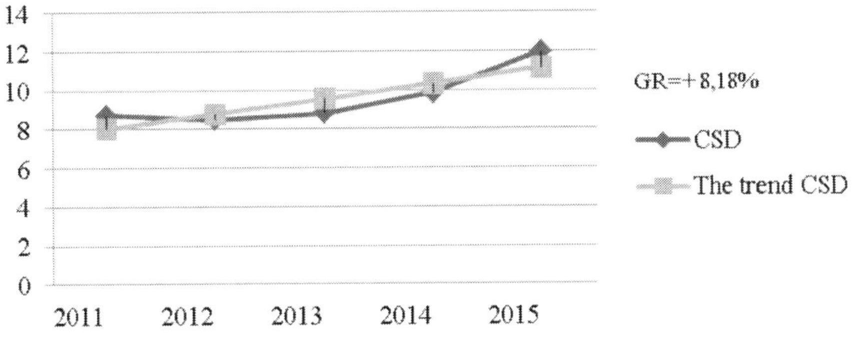

Fig. 1. The trend of overall morbidity of systemic blood circulation diseases in patients of municipal polyclinic No. 4 in Aktobe for 5 years.

When aligning number of indicators of the dynamic series, there is a tendency towards increase in the incidence of systemic blood circulation disease levels by an average of 8.18% annually. The maximum incidence is set in the age group 59–63 years (Fig. 2) with an increase in the growth rate GR = +4.7%, which resulted in a general increase in the incidence of systemic blood circulation disease in all categories of the population.

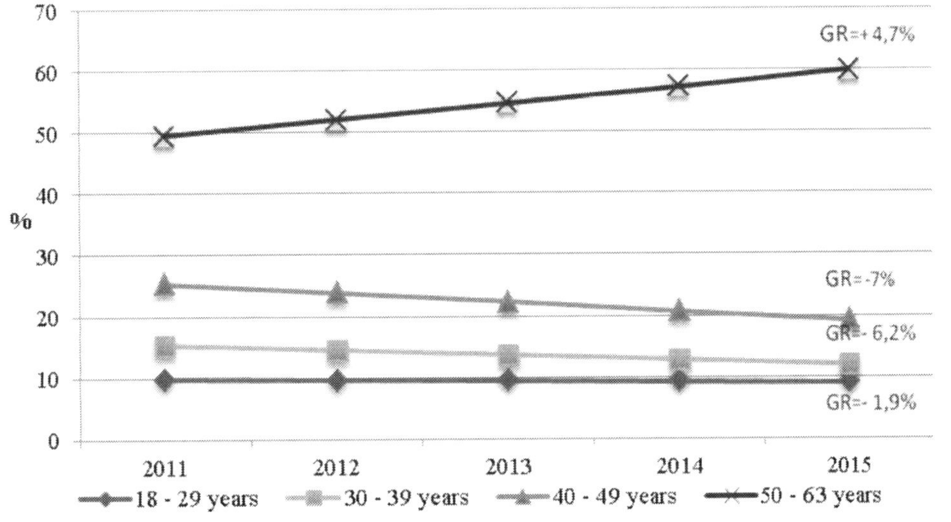

Fig. 2. The trend morbidity of systemic blood circulation disease incidence by age groups in patients of municipal polyclinic No. 4 in Aktobe for 5 years.

At the same time, trends in other age groups declined. But in general, the highest incidence rate for 5 years accounts for 2014. The average age of the patients remains almost unchanged 56.6 ± 7.6 in 2011 and 52.6 ± 7.8 in 2015.

Received results are presented in Table 2.

In the structure of patients in the dispensary of the circulatory system diseases, the largest number of patients is patients with high blood pressure and coronary heart disease. The number of patients with these nosological forms is growing, both among the indicators of general morbidity and primary morbidity. The prevalence of hypertension was 1595.3 in 2011 and 6716.2 in 2015 (100,000 people), which means increase by 32.3%, in connection with which the pace of growth took first place ranking; in second place – ischemic heart

Table 3. Dynamics in prevalence of morbidity and rate increase in the most frequent diseases among dispensary patients with circulatory system diseases for 2011–2015 years, municipal polyclinic No. 4 of Aktobe (for 100,000 people).

Years	**2011**	**2012**	**2013**	**2014**	**2015**	**Temp growth %**
Circulatory system diseases (total)	3880.5	4209.8	4264.6	6070.4	8470.2	20.5
Hypertension	1595.3	2704.1	2689.2	3709	6716.2	32.3
Ischemic heart disease	909.5	973.7	995.8	1236.8	1254.5	8.9

disease. The growth of the given nosology form during the study period reached 8.9% (909.5 and 1254.5, respectively) – Table 3.

Dynamics study of newly identified diseases in the nosological forms found an increase by 13.2% (from 1043 in 2011 to 1721.2 in 2014 on 100,000 of adult population). So on the first rank place are diseases characterized by high blood pressure (385.9 in 2011 and 793.9 in 2015), the second ranking place is ischemic heart disease. Indicators of coronary heart disease had wave like character of growth and decline. The maximum value was observed in 2015 (of 258.2 per 100,000 people), the minimum in 2014 (154.3 per 100,000) (Fig. 3).

One of the important indicators, characterizing the health status of the population is the indicator of disability.

We have studied the dynamics of primary disability due to cardiovascular diseases among the dispensary patients of working age per 100,000 population municipal polyclinic No. 4 of Aktobe city, for the period from 2011 to 2015. In the structure of primary disability

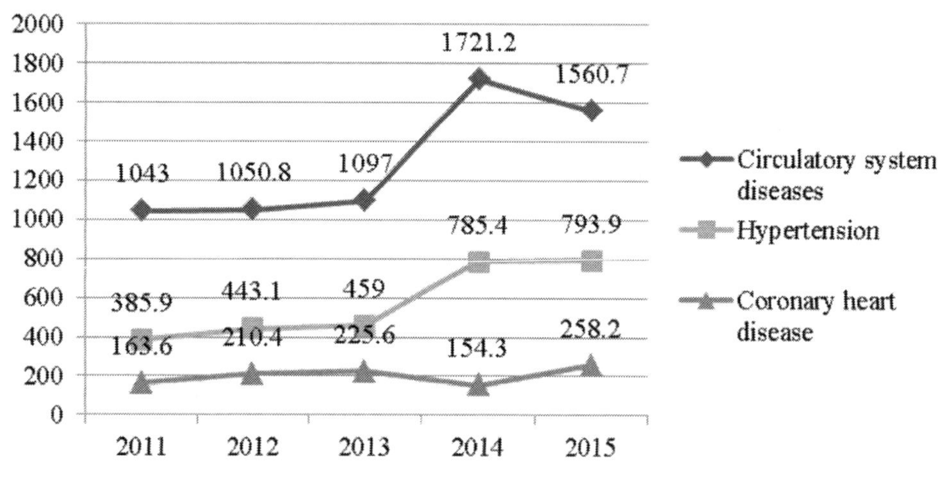

Figure 3. Dynamics of primary morbidity of the circulatory system diseases of the adult population in municipal polyclinic No. 4 of Aktobe according to nosological forms for 2011–2015 (per 100,000 people).

by main classes of diseases, growth rate of circulatory system diseases consistently holds a leading place, on the second rank place – neoplasm, and the third – diseases of the nervous system. So during the studied period the increase of diseases of the circulatory system accounted for 15.1% of neoplasm – 8.0%, and diseases of the nervous system – 2.9%.

Received results are presented in Table 4.

Table 4. The structure of primary disability by main categories of diseases among adult population of municipal polyclinic No. 4 of Aktobe for 2011–2015 (%).

Classes of diseases	**Years**					**Temp growth in %**
	2011	**2012**	**2013**	**2014**	**2015**	
Total on the main classes of diseases	100.0	100.0	100.0	100.0	100.0	
Diseases of the circulatory system	9.3	13.5	20.5	12.7	21.4	15.1
Neoplasms	18.6	32.4	31.3	18.3	36.7	8.0
Diseases of the nervous system	13.9	21.6	12	19.7	17.3	2.9

Out of initially recognized disabled in 2011, 11.2% are of working age, the number of which in 2012 amounted to 8.9%, that increased this figure by 2.3% from the previous year. And in the next 2 years, reduction in the yield on the primary disability from 6.5% in 2014 to 4.2% in 2015 is marked (Fig. 4).

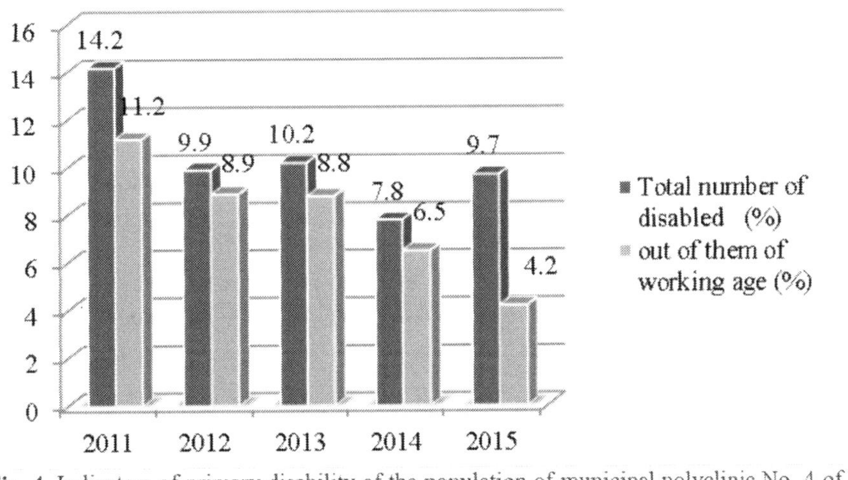

Fig. 4. Indicators of primary disability of the population of municipal polyclinic No. 4 of Aktobe for 2011–2015.

Significance of distribution of working-aged disabled people with diseases of circulatory system by disability groups are presented in Table 5.

Table 5. Distribution of working-aged disabled people with diseases of circulatory system by disability group in municipal polyclinic No. 4.

	2011		**2012**		**2013**		**2014**		**2015**	
	abs	**%**	**abs**	**%**	**abs**	**%**	**abs**	**%**	**abs**	**%**
I group	10	14.7	9	13.6	10	13.7	8	13.5	2	4.7
II group	25	36.7	25	37.9	31	42.4	28	47.4	25	58.1
III group	33	48.5	32	48.5	32	43.9	23	39.1	16	37.2
Total number	68		66		73		59		43	

Among recognized as disabled for the first time in 2011–2015 dominated the group III with a maximum value in 2011 – 48.5%.

The decline in output in the primary disability is firstly associated with the indicator of the effectiveness of rehabilitation. Received results are presented in Table 6.

Table 6. The effectiveness of rehabilitation of disabled of municipal polyclinic No. 4 for 2011–2015.

	2011		2012		2013		2014		2015	
	abs	%	abs	%	abs	%	abs	%	abs	%
Transferred from group I to group II	5	45.5	8	44.4	2	6.9	7	29.1	2	6.2
Transferred from group II to group III	2	18.1	3	16.7	15	51.8	8	33.3	15	46.9
Recognized as able-bodied	4	36.4	7	38.9	12	41.3	9	37.6	15	46.9
Total number rehabilitated	11		18		29		24		32	

The data presented in Table 6 of the effectiveness of rehabilitation of working-aged people with disability indicates that the rate of recognized capacity to work has grown from 36.4% in 2012 to 46.9% in 2015 while the number of the rehabilitated grows by every year.

Thus, we can draw the following conclusions:

1. Among the patients who are on the dispensary due to the diseases of the circulatory system, women are dominated and their number is growing every year (from 55.8% in 2012 to 64.4% in 2015 year).
2. Annually, among the population a tendency for increasing of diseases of the circulatory system by 8.18% is observed, which is also confirmed by the growth rate of TPR = +4.7%.
3. In the structure of morbidity of diseases of the circulatory system the leading place is occupied by patients with high blood pressure and coronary heart disease.
4. In the structure of primary disability for the main classes of diseases, diseases of the circulatory system show a stable growth rate.
5. The indicator growth of people, recognized as able-bodied contributed to the decline in output in the primary disability (from 36.4% to 46.9%), indicating the efficiency index of rehabilitation of patients with diseases of the circulatory system.

The obtained results once again confirm the significance of the preventive and therapeutic measures among patients with diseases of the circulatory system with the objective of reducing rates of disability and mortality among this category of patients.

A study conducted by Heidenreich et al. [9] also highlighted the prevalence of the cardiovascular diseases in the USA indicating the growing proportion in the health expenditure of the country. Moreover, the study forecasted the substantial increase in the prevalence and costs of care for cardiovascular diseases, particularly hypertension and coronary heart disease, until 2030. Also it has been suggested that prevention and early intervention are the key factors in decreasing both the prevalence and mortality from diseases of the circulatory system which have been confirmed in this study.

According to Townsend et al. [10] cardiovascular diseases are responsible for 45% of deaths in Europe and the majority of them are coronary heart diseases. These numbers justify the need for continuous study of the dispensary patients. Also they reported the 56.5% decrease in mortality from cardiovascular diseases in Kazakhstan in ten years (from 2002 to 2012). Despite the fact that the mortality and prevalence are different indicators, to some extent it can be viewed as consistent with the results of this study.

Another finding provided by Finegold et al. [11] implies that diseases of the circulatory system in the working age population are much higher in developing countries than in developed ones. This causes an additional challenge for the countries like Kazakhstan taking into account the increasing costs in the future projected by Heidenreich et al.

In conclusion, the results of this study are in line with other studies that have investigated morbidity, disability and mortality from circulatory system diseases. As the information systems that have been implemented in recent years may have had an impact on the statistics, as well as on the healthcare system in general, the comparison between before and after intervention periods of the abovementioned electronic portals and registers could be explored in future studies.

References

1. Ekkert N.V., Mihajlovskij V.J. Sibirskoe medicinskoe obozrenie Sovremennye problemy i puti povysheniya ehffektivnosti mediko-social'noj reabilitacii invalidov vsledstvie boleznej sistemy krovoobrashcheniya (na primere Tul'skoj oblasti) **2**, 77–81 (2015)
2. Gohlke H. Is it worth offering cardiovascular disease prevention to the elderly? Eur J Pre Cardiol. **20**, 121–6 (2013)
3. Abdikadirova I.T., Ermukhanova L.S., Bekbauova A.U., Umbetov M.U. The use of automated control systems in dispensary observation of patients in Aktobe city. J. MEDICINE (Almaty). **4**(166), 6–8 (2016)
4. Lisitsyn Y.P., Ulumbekova G.E. *Obshchestvennoye zdorov'ye i zdravookhraneniye* GEOTAR-Media (2014)
5. Abdikadirova I.T., Ermukhanova L.S., Ibrasheva A.Z., Umbetov M.U. Evaluation of automated control systems at the primary health care. J. Vestnik KazNMU. **1**, 650–653 (2016)
6. The top 10 causes of death [Electronic resource] // Mediacentre of WHO. — Fact sheet **310** (2014) Available at: `http://www.who.int/mediacentre/factsheets/fs310/en`
7. Piccolo R., Giustino G., Mehran R., Windecker S. Stable coronary artery disease: revascularization and invasive strategies. The Lancet. **386**(9994), 702–713. `http://dx.doi.org/10.1016/S0140-6736(15)61220-X` (2015)
8. Izet Masic and Jasmin Alajbegovic. The Significance of the Psychosocial Factors Influence in Pathogenesis of Cardiovascular Disease. Int J Med. **4**(11), 1323–1330 (2013)
9. Heidenreich P.A., Trogdon J.G., Khavjou O.A., Butler J., Dracup K., Ezekowitz M.D., ... & Lloyd-Jones D.M. Forecasting the future of cardiovascular disease in the United States. Circulation. **123**(8), 933–944 (2011)
10. Nick Townsend, Lauren Wilson, Prachi Bhatnagar, Kremlin Wickramasinghe, Mike Rayner, Melanie Nichols; Cardiovascular disease in Europe: epidemiological update 2016. European Heart Journal. **37**(42), 3232–3245 (2016)
11. Finegold J.A., Asaria P., & Francis, D.P. Mortality from ischaemic heart disease by country, region, and age: statistics from World Health Organisation and United Nations. International journal of cardiology. **168**(2), 934–945 (2013)

Adherence to treatment after CABG in terms of patients' personality

M. Iakovleva[1], O. Shchelkova[1], E. Lubinskaya[2], and O. Nikolaeva[2]

[1]Saint Petersburg State University, Saint Petersburg, Russian Federation
[2]Federal Almazov North-West Medical Research Centre, Saint Petersburg, Russian Federation

Abstract. The relevance of the problem is caused by the prevalence of cardiovascular diseases among the population, including coronary heart disease which requires surgical treatment, on the one hand, and low rates of treatment adherence among patients with chronic diseases, which reduce the effectiveness of the treatment and survival rates of patients, on the other hand. Patients after CABG were divided into groups according to the degree of their adherence to treatment and were examined by using the Big Five Questionnaire. Differences between the groups on the scales "conscientiousness" ($p < 0.001$), "openness" ($p < 0.001$), "agreeableness" ($p < 0.05$), and "neuroticism" ($p < 0.05$) were revealed. The study showed that the psychological factors are relevant in the definition of adherence to treatment and patients' personality traits are potentially significant for the prediction of its degree among patients after CABG.
Key words: clinical psychology, CABG, adherence to treatment, personality, Big Five Questionnaire.

1 Introduction

In 2014, the mortality from coronary heart disease (CHD) in the Russian Federation amounted to 492,303 people [1]. According to the European Society of Cardiology statistics, the mortality from CHD in Russia is one of the highest in Europe [2].

In this regard, nowadays the problem of treatment, rehabilitation and prevention of CHD (a disease which not only affects the quality of life of patients, but also leads to disability and death), is one of the most important and urgent problems in modern medicine [3, 4].

The data indicate that in patients with chronic diseases adherence to treatment does not exceed 50% [5, 6]. Poor adherence to treatment was also found in 1/3 of patients with psychosomatic diseases, more often among patients suffering from ulcer disease and coronary heart disease (40%) [7].

Despite the presence of numerous data on the socio-demographic and clinical factors affecting the degree of patients' adherence to prescribed therapeutic measures (e.g., the impact of the disease severity [8, 9], the dosing regimen [10], patients' sex and age [11–13], as well as their social status and education level [14]), the question of the influence of patients' personality on their therapeutic behaviour, and the question of influence of certain psychological characteristics on patients' degree of adherence, remain open.

Some researchers believe that this influence is insignificant and do not acknowledge the importance of patients' personality in determining adherence to treatment [15, 16], but physicians' clinical experience and the increasing interest of researchers in this issue [17] indicate the opposite.

Many studies have been carried out, trying to propose a unified classification of factors affecting the adherence to treatment [18, 19], as well as many meta-analyzes, considering the available data and making an attempt to systematize the accumulated material [20–22].

There is no "psychological factors of adherence" group in these classifications. A group named "factors related to the patient" usually corresponds to the patients' age and race, but not their personality features, characteristics of coping behaviour, cognitive functioning, mindsets or their beliefs. Taking into account all the aforesaid it seems relevant to investigate the psychological characteristics of patients in relation to their adherence to long-term treatment.

Among patients who underwent CABG, poor adherence to restorative treatment and supportive care is particularly dangerous, as it reduces the efficiency of surgery and the health outcomes, leading to disease relapses and increasing the treatment costs [5, 23].

The social consequences (impacts) of this phenomenon are also significant: the patient creates an idea of the physician's incompetence and the senselessness of further treatment, and the doctor creates an image of a "hopeless patient" [24].

Thus, the objective of this study was formulated. The aim of the research was to study the psychological factors of adherence to treatment (more specifically – features of the personality structure) among patients suffering from coronary heart disease, who underwent CABG.

2 Materials and methods

2.1 Materials

A total of 100 patients suffering from coronary heart disease (CHD) were examined in this study. The sample consisted of 82 men (82%, their average age was 58.96 years (SD = 8.68), at the age between 44 and 79 years), and 18 women (18%, their average age was 63.94 years (SD = 8.54), with age between 51 and 83 years).

About 40% of patients suffered CHD for more than 6 years, 23% – 3–6 years, 17% – 1–3 years, and about 20% of patients were diagnosed with CHD recently, less than one year before the necessity of surgery emerged. Of the patients, 68% suffered one or more myocardial infarction by the time of the surgery. All the studied patients underwent coronary bypass grafting (CABG) at the Federal Almazov North-West Medical Research Centre (Saint Petersburg, Russian Federation).

The following concomitant diseases were observed more often (without CVD, in descending order) among patients in the studied group: diseases of the gastrointestinal tract (chronic gastritis, gastric ulcer, duodenal ulcer), diseases of the pulmonary system (chronic bronchitis, chronic obstructive pulmonary disease), diseases of the urinary system (urolithiasis, cystitis), diseases of the locomotor system (arthritis, scoliosis of different regions of the spine, degenerative-dystrophic damage of the spine), endocrine diseases.

2.2 Methods

We conducted a complex study of patients' demographic, socioeconomic, clinical and psychological characteristics. The following set of methods was used: a detailed study of medical records of each patient, a structured interview with patients which consisted of a wide range of questions about their family, job, social activity, habits and mindsets, and a psychometric technique – the "Big Five Questionnaire" [25], adapted and standardized on a Russian sample [26].

Patients' adherence to treatment was assessed using an expert method, performed by an interdisciplinary team of experts (cardiologists and clinical psychologists). Several aspects were considered when determining the degree of patients' adherence: their adherence to the drug therapy and adherence to the recommended lifestyle changes, such as keeping to a diet, following the recommended level of physical activity, controlling blood pressure indicators and weight, reducing alcohol consumption and quitting smoking. Assessment of adherence to treatment was performed not earlier than 3 months after CABG, to insure patients' return to normal independent life and adaptation to their new health state and lifestyle.

3 Results and discussion

3.1 Results

Initially, the psychological characteristics of the studied patients were analyzed in accordance with the division into two groups – adherent and non-adherent to treatment (Table 1).

Table 1. Average scores on the scales of the Big Five Questionnaire in two groups of patients.

Factors of personality traits (scales of the Big Five Questionnaire)	Adherent to therapy (n = 52) M (SD)	Non-adherent to therapy (n = 48) M (SD)	*p*
Extraversion	26.33 (0.67)	26.29 (0.82)	0.973
Conscientiousness	31.92 (0.62)	27.31 (0.81)	0.000
Agreeableness	32.29 (0.64)	29.65 (0.92)	0.019
Neuroticism	24.48 (0.87)	27.54 (0.90)	0.016
Openness	27.46 (0.80)	21.00 (0.95)	0.000

It has already been shown that such a division is justified and allows determining significant differences between the patients on the characteristics of their personality [27]. The study of the patients' structure of personality using the technique "Big Five Questionnaire" revealed the individual psychological characteristics which are typical for the studied groups of patients.

Thus, it was found that patients who do not follow health recommendations showed significantly lower values on several scales of the questionnaire than adherent patients. These patients proved to have statistically significant lower values on the scale of "Conscientiousness" (meaning they are more careless, forgetful and irresponsible) ($p < 0.001$), as well as on the scale of "Openness" (meaning they have a lack of imagination, artistry and personal resources) ($p < 0.001$). These patients also showed statistically significant differences from the adherent to treatment group on the scale "Agreeableness": patients with poor adherence to treatment are characterized by lower values on this scale, which means they are more unfriendly and less trusting and cooperative ($p < 0.05$).

On the other hand, adherent to treatment patients have lower values on the scale "Neuroticism" ($p < 0.05$) in comparison with the second group, it means that patients with a higher degree of adherence are more anxious, moody, emotional and unstable.

The issue of the process of division of patients into groups itself, the possibility of pole division of patients suffering from CHD, into two polar groups – adherent and non-adherent to treatment – remains relevant. This brings up the question of the impact that may cause a more differentiated division into groups based on the degree of adherence to the treatment on the results obtained with respect to the psychological characteristics of patients.

Upon further study, the sample was divided more differentially into the following three groups:

1. Group I: included those patients who received the highest values in all aspects of adherence, i.e. complying with doctors' recommendations on the admission of drugs (dosage, frequency and so on) (adherent to pharmacological therapy), performing control (and, if required, decrease) of their weight, dieting (following the designated by the doctor restrictions on various foods), following the recommended level of physical activity, performing the recommended exercise and giving up harmful habits completely or largely – giving up smoking, reducing alcohol consumption (adherent to non-pharmacological therapy). It can be concluded that patients from this group are truly adherent to the treatment and have the most favourable prognosis with respect to following the recommendations of doctors. For brevity, this group will be henceforth designated "adherent to treatment patients".
2. Group II: consisted of patients, non-adherent to treatment in all aspects of adherence; they do not follow (or follow at a minimal degree) the doctors' recommendations on pharmacological treatment, do not perform monitoring of blood pressure, diet and exercise regime. This group has the least favorable indicators in terms of adherence to treatment. The second group of patients can be named "non-adherent to treatment patients".
3. Group III: included the patients who implemented the recommendations only for drug therapy (regularly taking prescribed medications) and were monitored regularly by a cardiologist, but who did not follow the recommendations on diet and refusal of harmful habits, or performing lifestyle changes. This group can be described as "relatively (partly) adherent patients" because, despite the fact that adherence to pharmacotherapy is an essential aspect of the therapy effectiveness, in isolation from other prescribed measures it is insufficient. Despite this circumstance, the prognosis in respect of adherence to treatment in this group of patients is significantly more favorable than that of patients from the second group.

In the course of the study the three groups of patients described above were compared once more on their psychological characteristics using analysis of variance (ANOVA) (Table 2).

Table 2. Average scores on the scales of the Big Five Questionnaire in three groups of patients.

Factors of personality traits (scales of the Big Five Questionnaire)	Group I ($n = 45$) M (SD)	Group II ($n = 27$) M (SD)	Group III ($n = 28$) M (SD)	Differences between groups	p
Extraversion	26.42 (5.20)	26.13 (5.40)	26.64 (5.36)	–	$p > 0.1$
Conscientiousness	31.53 (0.73)	27.87 (0.87)	28.80 (1.34)	I*II	0.030
Agreeableness	32.62 (0.72)	29.48 (1.20)	30.92 (1.16)	I*II	0.077
Neuroticism	24.56 (1.00)	28.96 (1.34)	25.64 (1.06)	I*II	0.028
Openness	25.76 (6.28)	22.22 (7.01)	24.28 (7.23)	–	$p > 0.1$

The analysis of average values on the questionnaire scales showed that there were statistically significant difference in several personality features among adherent, non-adherent and relatively (partly) adherent to treatment patients who underwent coronary bypass surgery.

Adherent to treatment patients (group I) have higher scores on the scale "Conscientiousness" ($p < 0.05$) and lower scores on the scale "Neuroticism" ($p < 0.05$) compared with patients with a poor level of adherence (group II). Difference between adherent and non-adherent patients on a statistical tendency level has also been identified on the scale "Agreeableness" ($p < 0.1$) – patients from group I tend to be more trusting and cooperative.

That is, the division of patients into three groups according to their adherence to therapy (that also seems more logical from the clinical point of view) is also productive in regard of research. However, it should be noted that, as data in Table 2 indicates, the major difference has been found between the groups I and II, i.e. between groups "adherent to treatment patients" and "non-adherent to treatment patients". This fact testifies the validity and sufficiency of the initial division into two groups, carried out at the first stage of the research.

3.2 Discussion

Patients with different degrees of adherence to treatment were found to be significantly different on several personality characteristics: adherent patients are characterized by a higher willingness to agreement and cooperation. These data are confirmed by other researchers in other medical fields. For example, it was determined that the lack of readiness to cooperate and oppositional behaviours have a negative effect on adherence to treatment also in patients with cystic fibrosis [28].

On the other hand, the examined in this study no adherent to treatment patients significantly differ by lower personal resources and lower responsibility, conscientiousness and self-discipline. Similar tendencies are marked in other studies: for example, Stilley et al. found that higher scores on the factor "conscientiousness" in patients are significant predictors of high adherence to hypolipidemic therapy [29]. Later, these data was confirmed also on a sample of patients suffering from other diseases. Karamanidou et al. in their study showed the negative effect of hostility and low levels of self-consciousness on adherence to drug therapy in patients with renal disease [30]. Carelessness and poor skills of self-organization in conjunction with laziness are considered to be a negative factor affecting adherence to therapy also in patients suffering from tuberculosis [31].

All the above mentioned confirms the researchers' high interest in regard to the patients' personality in the context of their adherence to treatment, and allows to assert that patients' personality features play an important role in determining the level of adherence to both – pharmacological and non-drug – treatment among patients with different diseases, including patients with CHD after CABG surgery.

Other psychological characteristics of patients that are related to their personality features and may affect the therapeutic behaviour, such as their coping behaviour (coping strategies and personality coping resources), system of values, beliefs (including beliefs about health and the disease), and the patients' attitude towards the disease, also require further studies.

4 Conclusions

Patients' characteristics of personality have a prognostic value in relation to the degree of adherence to treatment after CABG surgery: good self-organisation skills, responsibility, willingness to cooperate and the aspiration to self-development and personal growth have a positive effect, calmness and emotional stability – have negative effect. Thus, it was found that both strategies of dividing patients into either two or three groups according to their adherence to therapy is promising in terms of scientific investigation and reveals significant difference in the psychological characteristics (personality features in particular) of patients with different degrees of adherence to treatment.

Summarizing all the above mentioned, it is essential to indicate the importance of the psychological factors in the definition of adherence to treatment. The obtained empirical data on the patients' psychological characteristics contribute to increase in the efficacy of therapy by providing professional psychological support to patients who underwent CABG and making psycho corrective programs for no adherent to treatment patients.

References

1. *Estestvennoe dvizhenie naseleniya Rossiyskoy Federatsii v 2014 godu: byulleten'* [Vital Statistics of the Russian Federation in 2014: Bulletin] (Statisticheskiy byulleten' Publ., Moscow, 2015) (In Russ.)
2. N. Townsend, L. Wilson, P. Bhatnagar, K. Wickramasinghe, M. Rayner, M. Nichols, Eur Heart J. **37**(42), 3232–3245 (2016)
3. World Health Organization, *Prevention of cardiovascular disease: guidelines for assessment and management of total cardiovascular risk* (WHO, Geneva, 2007)
4. L.A. Bockeria, M.M. Alshibaya, N.G. Bendeliani, S.Ph. Nikonov, K.V. Krymov, The Bulletin of A.N. Bakoulev Scientific Center for Cardiovascular Surgery of Russian Academy of Medical Sciences. Cardiovascular diseases. **13**(1), 38–47 (2012) (In Russ.)
5. World Health Organization, *Adherence to long-term therapies, evidence for action* (WHO, Geneva, 2003)
6. B. Hagström, B. Mattsson, I.M. Rost, R.K. Gunnarsson, Fam Pract. **21**(1), 46–50 (2004)
7. I.A. Novikova, V.V. Popov, Med. psihol. Ross. **6**(35), 9 (2015) (In Russ.)
8. J. Leong, A. Molassiotis, H. Marsh, Clin Effect in Nurs. **8**(1), 26–38 (2004)
9. S. Choi-Kwon, S.U. Kwon, J.S. Kim, Eur Neurol. **54**(4), 204–211 (2005)
10. A.J. Claxton, J.A. Cramer, C. Pierce, Clin Ther. **23**, 1296–1310 (2001)
11. M.C. Wong, W.W. Tam, C.S. Cheung, E.L. Tong, A.C. Sek, N.T. Cheung, S. Leeder, S. Griffiths, Int J Cardiol. **167**(4), 1438–1442 (2013)
12. K. Krueger, L. Botermann, S.G. Schorr, N. Griese-Mammen, U. Laufs, M. Schulz, Int J Cardiol. **184**, 728–735 (2015)
13. M.W. Schoen, R.G. Tabak, J. Salas, J.F. Scherrer, F.R. Buckhold, Am J Cardiol. **117**(1), 48–53 (2016)
14. P. Theofilou, H. Panagiotaki, Oncol Rev. **6**(1), e5 (2012)
15. D. Hevey, H.M. McGee, J. Horgan, Psychosom Med. **69**, 793–797 (2007)
16. V. Fuster, Rev Esp Cardiol. **65**(Suppl 2), 10–16 (2012)
17. L.A. Phillips, H. Leventhal, E.A. Leventhal, Psychol Health. **28**(10), 1135–1151 (2013)
18. T. Mathes, T. Jaschinski, D. Pieper, Arch Public Health. **72**(1), 37 (2014)
19. A.F. Yap, T. Thirumoorthy, Y.H. Kwan, Geriatr Gerontol Int. **16**(10), 1093–1101 (2015)

20. J. Jin, G.E. Sklar, V. Min Sen Oh, S. Chuen Li, Ther Clin Risk Manag. **4**(1), 269–286 (2008)
21. F. Alsolami, X.-Yu. Hou, I. Correa-Velez, Clin Med Diagnostic. **2**(4), 27–32 (2012)
22. P. Kardas, P. Lewek, M. Matyjaszczyk, Front. Pharmacol. **4**, 91 (2013)
23. L. Osterberg, T. Blaschke, N Engl J Med. **353**(5), 487–497 (2005)
24. D.S. Danilov, Psychiatry and Psychopharmacotherapy (P.B. Gannushkin Journal), **10**(1), 13–20 (2008) (In Russ.)
25. L.A. Pervin, O.P. John, *Personality: Theory and research* (7th ed.) (John Wiley & Sons, New York, 1997)
26. D.P. Yanichev, *Cognitive aspects of self-perception of personality traits in patients with neurotic and neurosis-like symptoms* (Saint-Petersburg State University, Saint-Petersburg, 2006)
27. M.V. Iakovleva, O.Y. Shchelkova, C.I. Lubinskaya, J Cardiol Clin Res. **4**(1), 1055 (2016)
28. J.R. Weiner, E.L. Toy, P. Sacco, M.S. Duh, Expert Opin. Pharmacother. **9**(5), 751–766 (2008)
29. C.S. Stilley, S. Sereika, M.F. Muldoon, C.M. Ryan, J. Dunbar-Jacob, Ann Behav Med. **27**(2), 117–124 (2004)
30. C. Karamanidou, J. Clatworthy, J. Weinman, R. Horne, BMC Nephrol. **9**, 2 (2008)
31. S.A. Munro, S.A. Lewin, H.J. Smith, M.E. Engel, A. Fretheim, J. Volmink, PLoS Med. **4**(7), e238 (2007)

Physiotherapy following cardiac surgery: Program comparison

D. Pulmane[1,2], A. Vetra[1], R. Lacis[2], and D. Driba[1,2]

[1]Rīga Stradiņš University, Riga, Latvia
[2]Stradiņš Clinical University Hospital, Latvia

Abstract. The objective is to examine and compare the usability of two physiotherapy programmes, analyzing respiratory function in patients before and after cardiac surgery in hospital during seven postoperative days (POD).
Quantitative randomized prospective study of 157 patients before and after the valve replacement surgery, coronary artery bypass graft and combined surgeries, who moved independently. Participants were randomized into two groups (1 and 2) with different physiotherapy programmes. The routine physiotherapy of breath-enhancing techniques, micro-circulation improvement were used for the first group, for the second group – modified physiotherapy – the improvement of inspiration muscles, mm. quadriceps. gluteus max strength. Anthropometric measurements were defined for both groups on the day before surgery and during spirography – dynamic indicators – forced vital capacity (FVC), forced expiratory volume in the 2nd second (FEV1), Tiffeneau index (FEV1/VC (%)), peak expiratory flow (PEF), operation parameters.
By comparing the postoperative respiratory parameters between the two groups and using independent samples t test, it was found out that the difference in FVC between groups is 1.71 [95% CI: –8.25 to 4.8] and it is not statistically significant ($t\,(155) = -0.52$; $p = 0.60$). Based on the Leuven test results FVC the distribution variance is not statistically notably different for group 1 and group 2 ($F = 0.27, p = 0.60$).
Using routine and modified physiotherapy and comparing postoperative dynamic respiratory performance of the two groups, there was no statistically significant difference, proving that the two models are equally effective.

1 Introduction

During recent years the lifestyle of people has changed and the population is aging and cardiovascular diseases are an important problem, as evidenced by the high morbidity and hospitalization rates.

Various heart surgeries are one kind of the treatments of heart diseases. A variety of cardiac physiotherapy techniques are offered for reduction of complications during postoperative period in hospital. The most frequently mentioned techniques in literature and studies were chosen for this research – the promotion of expiration, coughing up, diaphragmatic breathing, the promotion of inhalation with inhalation spirometer [1–4].

Active movements of the limbs from the periphery to the centre, in lying, sitting, standing positions. Active movements of the body, strength stimulating techniques for major muscle groups, mm. gluteus max, mm. quadriceps, walking, climbing stairs. These techniques were applied in two different physiotherapy programmes – in programme 1 coughing up with holding the scar was done, diaphragmatic breathing, breath improvement, active movements of body and limbs from the periphery to the centre in lying, sitting, standing positions, walking and climbing stairs during seven post operation days. In the 2nd programme – the promotion of inhalation with inspiratory spirometers, improvement of strength in the large muscle groups in lying, sitting, standing positions, walking and climbing stairs was carried out during seven postoperative days.

The most common complications in the postoperative period are atelectasis, infectious problems in the airways and the operation wound, pleural effusion, pain in surgery scar area and intercostal nerve localization sites, fatigue, dizziness, rhythm disorders. Physiotherapy programmes were developed in order to reduce postoperative complications [5–7].

The aim of the research is to study and compare the usability of two physiotherapy programmes, to analyze respiratory functions of patients before and after cardiac surgery in hospital during seven postoperative days (POD).

2 Methods

The research was carried out in Cardiac Surgery Centre of Pauls Stradins Clinical University Hospital from 2015 until the end of 2016. Ethical permissions from Riga Stradins University and Paul Stradins Clinical University Hospital Ethics Committees were obtained for carrying out the research. The participants of this study were informed about the progress of research and they have signed the informed consent form.

2.1 Participants of study

Patients with cardiovascular and valvular diseases, with a planned cardiac surgery – aortocoronary bypass graft, valve replacement surgery and combined, median sternotomy, with artificial blood circulation, who move independently.

Exclusion criteria – acutely hospitalized patients, and patients who could not move independently. Participants were randomized into two groups, they were offered two distinct programmes under the guidance of a physiotherapist.

On the day before the planned heart surgery, the anthropometric data were defined – pulse rate, blood pressure, respiratory rate, body mass index, weight, waist/hip circumference. Biochemical data – total cholesterol, high-density, low-density cholesterol, triglycerides, blood glucose levels. During spirography such dynamic indicators as forced vital capacity (FVC), forced expiratory volume in the 1st second (FEV1), Tiffeneau index (FEV1/VC (%)), peak expiratory flow (PEF) were defined, conclusions were made about the functionality of respiratory system. Spirographic data were redefined on the 6th postoperative day.

On the seventh postoperative day a 6-minute walking test [8] was done, determining the distance travelled, heart failure functional class, pulse rate, blood pressure, oxygen saturation prior to the test, after the test, and after 5 minutes of recovery time. The patient's well-being was evaluated subjectively, using the Borg scale and Borg dyspnoea scale [9]. The test was carried out in 80-metre corridor, the corridor ends were marked which the participant went around, the walking took place as soon as it was possible, it was allowed to stop and rest during walking, if it was necessary, then to continue the test. The test could also be taken with the high walking frame, if it was necessary.

The following data of heart surgery were analyzed: the duration of surgery, artificial blood circulation time, aortic occlusion time, body temperature, heart rate recovery mechanism.

2.2 The evaluation of participants during the programme

Before the exercises for every participant was measured pulse rate, blood pressure, oxygen saturation, the number of repetitions of exercises, the time of sleeping, seating, standing in minutes, the walking distance in metres, the number of climbed steps, the iteration number of breathing techniques, breath spirometer display in millilitres on each postoperative day, the duration of exercises in minutes. The exercises were carried out once a day, every day.

2.3 Statistical analysis

The results obtained were analyzed using descriptive statistics (mean values, standard deviation, percentage, minimum and maximum values), independent sample t test and Leuven test.

3 Results

3.1 Basic data

The study (Table 1) included 157 participants who had undergone heart surgery via longitudinal sternotomy, valvular prosthesis 77 (49.04%), bypass graft 57 (36.3%), combined (valvular prosthesis and bypass graft) 23 (14.65%). 72 (45.86%) women and 85 (54.14%) men participated in research. Participants were randomly split into two groups. Group 1 carried out a routine physical therapy programme, and the second group took a modified physiotherapy programme. There were 41 (51.9%) women and 38 (48.1%) men in the first group, the average age was M = 65.84 (SD = 9.71). In this group there were 36 (45.57%) valve replacement surgeries, 25 (31.64%) grafts and 18 (22.78%) combined surgeries. The average Body Mass Index (BMI) was M = 29.54 (SD = 6.34), average respiratory rate before surgery was M = 20.18 (SD = 3.96). Other tests, which were carried out before the operation, are illustrated in Table 1. There were 31 (39.74%) women and 47 (39.74%) men in the second group. The average age was M = 67.26 (SD = 8.73). In this group there were 31 (39.73%) valve replacement, 32 (41.02%) bypass and 15 (19.23%) combined surgeries. The average BMI was M = 28.22 (SD = 5.39) and average respiratory rate before surgery was M = 20.81 (SD = 5.39). Both groups had 14 smokers.

3.2 Data about surgeries

The surgery was carried out in artificial blood circulation, one of the members from the first group, who had repeated surgery, was not tested and in the second group two patients, who had the operation without artificial blood circulation, were not tested. In the first group the artificial blood circulation time in minutes was M = 96.51 (SD = 33.10), aortic clamping time in minutes was M = 64.96 (SD = 22.10). The results are illustrated in Table 2. In the second group artificial blood circulation time in minutes was M = 92.33 (SD = 24.71), Aortic clamping time in minutes was M = 62.20 (SD = 17.41).

Table 1. Clinical characteristics data between first and second group participants (n = 157).

	1. group, n = 79		2. group, n = 78	
	Mean	SD	Mean	SD
Women:men	41:38		31:47	
Age, years	65.84	9.71	67.26	8.73
Valve replacement surgery, n	36		31	
Coronary artery bypass grafting, n	25		32	
Combined heart surgery, n	18		15	
Body mass index, kg/m^2	29.54	6.34	28.22	5.39
Breathing frequency prior to surgery, times per minute	20.18	3.96	20.81	7.88
Smoking yes, n:no, n	14:65		14:64	
Total cholesterol, mmol/l	4.60	1.09	4.43	1.18
Glucose, mmol/l	6.34	2.11	5.86	1.53
6 min walking test on 7th postoperative day	274.38	104.88	266.18	84.49

Table 2. Description of intraoperative data for first and second group participants (n = 157).

	1. Group, n = 78		2. Group, n = 76	
	Mean	SD	Mean	SD
Cardiopulmonary bypass, min	96.51	33.10	92.33	24.71
Aortic clamping, min	64.96	22.10	62.20	17.41
Reperfusion, min	23.49	10.08	23.40	9.53
Lowest body temperature, °C	35.85	0.39	35.78	0.54

3.3 Spirographic data

Before and after surgery during spirography the dynamic breathing ratios were defined – forced vital capacity (FVC), forced expiratory volume in the 1st second (FEV1), Tiffeneau index (FEV1/VC (%)), peak expiratory flow (PEF), maximal (mid-) expiratory flow (MEF 75, MEF 25). The results see in Table 3. In the first group the evaluation of postoperative respiratory dynamic performance was not done for 2 (2.53%) participants with complications – nosocomial infections (Acinetobacter baumannii) and flu, in the second group 2 (2.56%) participants were not investigated, who had complications – neurological dysfunction un prolonged ICU (8 days). Repeated respiratory dynamic indicators were defined on the sixth postoperative day and they were compared between groups. The data were compared using independent sample t test, it was concluded that the difference between group members of groups 1 and 2 – FVC difference is 1.71 [95% CI: –8.25 to 4.8] and it is not a statistically significant difference (t (155) = –0.52, p = 0.60). Based on Leuven test results, it was concluded that FVC distribution variance is not statistically significantly different (F = 0.27; p = 0.60) in the first and second groups. By analyzing other respiratory parameters with Leuven and t tests, it was concluded that the result was similar.

Table 3. Spirographic dynamical indicators for first and second group participants.

		1. Group, $n = 79$		2. Group, $n = 78$	
		Mean	SD	Mean	SD
Forced vital capacity (FVC), l	Norm	3.30	0.99	3.48	0.87
	Preoperative	3.01	1.00	3.28	0.90
	Postoperative	2.32	0.80	2.52	0.75
Forced expiratory volume in the 1st second (FEV1), l	Norm	2.64	0.76	2.77	0.65
	Preoperative	2.34	0.79	2.50	0.77
	Postoperative	1.78	0.64	1.91	0.63
Tiffeneau index (FEV1/VC(%)), %	Norm	76.22	1.67	75.78	1.58
	Preoperative	77.26	8.74	75.69	10.03
	Postoperative	76.31	7.68	75.33	8.74
Peak expiratory flow(PEF), l/s	Norm	6.93	1.46	7.23	1.27
	Preoperative	5.87	2.02	6.16	1.99
	Postoperative	4.66	1.73	5.02	1.70
Maximal (mid-) expiratory flow (MEF75), l/s	Norm	6.14	1.27	6.42	1.11
	Preoperative	5.04	1.87	5.25	2.01
	Postoperative	4.04	1.54	4.36	1.74
Maximal (mid-) expiratory flow (MEF50), l/s	Norm	3.85	0.73	3.96	0.56
	Preoperative	2.72	1.32	2.92	1.43
	Postoperative	2.03	0.97	2.17	1.11
Maximal (mid-) expiratory flow (MEF25), l/s	Norm	1.36	0.42	1.38	0.37
	Preoperative	0.89	0.55	0.89	0.52
	Postoperative	0.62	0.44	0.62	0.34

3.4 Programmes

During physiotherapy programmes for both groups of participants the vital parameters – pulse, blood pressure, oxygen saturation was controlled. In Table 4 we see that physiotherapy programme starts on the first postoperative day (POD1), In the 1st group 75 (94.9%) participants start the programme and 4 participants do not start it, because of complications – 3 intubated, 1 with neurological dysfunction. 74 (94.87%) participants from group 2 start the programme,4 patients do not start, complications – 2 atrial fibrillation paroxysm, 2 had repeated operations. Physiotherapist guided exercises are held on seven POD. Participants are involved in the physiotherapy sessions from POD1 to POD4. From the 5th POD the amount of activities decreases with starting independent activities of daily living.

Supervised coughing up technique was launched in two groups on the 1st POD and it was continued till the 4th POD for group 1, but was reduced for group 2, which did inhalation spirometry technique. In the 2nd group inhalation technique ($n = 37$) and muscle strength improvement ($n = 28$) starts on the 2nd POD and is intensively used till the 5th POD. Inhalation promotion technique ($n = 24$) for the 1st group starts on the 2nd POD and continues till 5th POD. The time of activities in minutes is growing up till the 4th POD, then remains constant. The intensity of activities (meters walked and the amount of climbed steps) is rising till the 6th POD for both groups.

3.5 6-minute walk test

On the 7th postoperative day the members of groups 1 and 2 carried out a 6-minute walking test to compare the distance walked. Two participants from group 1 did not take part in 6-minute walk test, one refused, the other had nosocomial infection (Acinetobacter baumannii). One patient from group 2 died. For the 1st group the walked distance in metres was $M = 274.38$ (SD = 104.88). For the 2nd group the walked distance in metres was 266.18 (SD = 84.49). Using the independent sample t test, it was concluded that the difference of the distance in metres of 6-minute walking test for groups 1 and 2, was 11.86 meters [95% CI: –16.23 to 39.95] and has no statistically significant difference ($t\,(152) = 0.83$; $p = 0.15$). On the basis of results of Leuven test a conclusion was made that for groups 1 and 2 walking distance variance does not have significant difference ($F = 1.82$, $p = 0.18$).

4 Discussion

The aim of cardiac rehabilitation is to return the patients' daily activities and social environment as quickly as possible, at the same time preserving human functioning in high physical and psychological level. The study used two different physiotherapy programmes. For the first programme expiration stimulating techniques were applied, the second programme used inhalation stimulating techniques and strength in the large muscle groups was improved. In both groups early mobilization was applied. Inhalation spirometry and the improvement of inspiratory muscle strength is one of the methods described in many studies [10–13]. It has been proved that this method is effective, but there are more effective programmes, that offer variety of techniques such as expiratory promoting, early mobilization, aerobic exercises [10–15]. The study with healthy people demonstrated that inhalation spirometry training shows the best results [16]. The aim of research was to find out which of the applications of this technique are more effective. Basic data for both groups as age, breathing frequency before the operation, the number of patients who smoke were similar. Spirographic results before the operation for two groups of patients were similar. Duration of artificial blood circulation during surgery for patients from group 1 was 96.51 minutes, for patients from group 2 it was 92.33 minutes which is not significantly different. On the sixth postoperative day spirography was repeatedly carried out, concluding that results in both groups are not significantly different. On the seventh postoperative day a 6-minute walk test was done to check the functional indicators for both groups. 6-minute walk test results did not differ significantly between the two groups, Group 1 on average walked 274.38 metres, the second group walked on average 266.18 metres. 6-metre walking test is recommended as a good tool for people with moderate disabilities [8, 17]. Different physiotherapy methods applied for the first and second groups show similar results for both – assessing spirographic data and analysis of the 6-meter walking test results after surgery. Both applied physiotherapy programmes were equivalent and applicable to patients in the period after cardiac surgery in hospital. Different physiotherapy techniques applied in programmes could be combined and successfully applied for faster return of patients to daily activities and social environments. Physiotherapy programs after heart operations should be standardized, they can be carried

Table 4. Comparison of Physiotherapy programs between first and second group participants.

	POD1		POD2		POD3		POD4		POD5		POD6		POD7	
	1. group	2. group	1. group	2. group	1. group	2. group	1. group	2. group	1. group	2. group	1. group	2. group	1. group	2. group
Excercises of wrists and feet, n	75	73	74	69	71	73	76	68	57	43	26	23	9	7
Excercises of extremities, n	38	8	70	60	70	73	75	68	57	43	26	23	9	7
Exercises of body, n	0	0	13	13	49	59	66	60	53	42	25	23	9	7
Exercises in lying position, n	75	74	67	62	26	26	11	13	5	7	3	4	0	2
Exercises in sitting position, n	56	54	73	63	68	73	75	68	57	44	25	23	9	7
Excercises in standing position, n	0	0	22	10	57	63	73	68	56	47	25	27	9	7
Drains, n	76	74	74	71	N/A	N/A	N/A	N/A	N/A	N/A	N/A	N/A	N/A	N/A
Directed coughing, n	79	78	28	55	73	71	76	40	56	26	26	11	9	5
Aspiratory excercises with incentive spirometry, n	N/A	N/A	N/A	37	N/A	73	N/A	72	N/A	52	N/A	26	N/A	7
Strength improvement exercises, n	N/A	N/A	N/A	28	N/A	63	N/A	66	N/A	51	N/A	27	N/A	7
Expiratory incentive excercises, n	N/A	N/A	24	N/A	57	N/A	75	N/A	57	N/A	25	N/A	9	N/A
Diaphragmatic breathing, n	N/A	N/A	37	N/A	66	N/A	75	N/A	53	N/A	23	N/A	9	N/A
Walking, m (n)	N/A	N/A	N/A	N/A	55,51 (73)	56,19 (52)	124,45 (66)	117,72 (66)	143,84 (54)	134,96 (49)	151,20 (25)	167,78 (27)	110,56 (9)	161,43 (7)
Walking upstairs, steps (n)	N/A	N/A	N/A	N/A	N/A	N/A	5,69 (24)	6,91 (25)	8,39 (25)	9,15 (26)	8,00 (13)	9,25 (13)	3,33 (2)	10,00 (3)
Duration of session, min	6.77	7.00	10.09	11.26	13.94	15.59	16.39	17.68	17.40	18.59	17.58	18.04	17.22	17.38
Complications, n	5	5	8	10	14	5	8	7	6	4	5	4	1	1
Participants, n (%)	75 (94,9)	74 (94,9)	74 (93,7)	69 (88,5)	71 (89,9)	74 (94,9)	76 (96,2)	72 (92,3)	57 (72,2)	51 (65,4)	26 (32,9)	27 (34,6)	9 (11,4)	8 (10,3)

*POD – Post operative day. N/A – not applicable.

out only on weekdays, or on all days of the week [5, 18]. Screening techniques are being searched that would anticipate the length of stay in hospital and a 5-metre walking test has been offered [19]. Good results and reduced complications are demonstrated by programs that started before the operation and continue in the postoperative period [20, 21]. Similar to our study, the highest number of complications is on the first three POD, the most common complications – respiratory, hemodynamic and cardiac conduction disorders [5, 6, 22]. In our study the patients performed independent physical activities from the 4th POD. If twenty years ago it was considered as unethical model [22], then currently secure, independent activity models are being studied under telemetry monitoring [23].

5 Conclusion

In this study, it is concluded that a variety of physiotherapy techniques for improvement of respiratory functions, as well as the preservation and improvement of mobility, and the promotion of strength during postoperative period are equivalent and can be included in the total, standardized model of physiotherapy for reducing complications after cardiac surgery in hospital period.

References

1. R.D. Restrepo, R. Wettstein, L. Wittnebel, M. Tracy, *Incentive Spirometry: 2011 Clinical Practice Guideline*, Respiratory Care **56**, 1600–1604 (2011)
2. C.M. Mans, J.C. Reeve, M.R. Elkins, *Postoperative outcomes following preoperative inspiratory muscle training in patients undergoing cardiothoracic or upper abdominal surgery : a systematic review and meta analysis*, Clinical Rehabilitation **29**(5), 426–438 (2015)
3. V.K. Ravaliya, *Functional Status in Post-CABG Patients Following Physiotherapy – A Retrospective Analysis*, Indian Journal of Physiotherapy and Occupational Therapy **7**(3), 270–274 (2013)
4. E. Crisafulli, E. Venturelli, G. Siscaro, *Respiratory muscle training in patients recovering recent open cardiothoracic surgery: a randomized controlled trial*, BioMed Research International, **Article ID 354276**, 1–7 (2013)
5. E.R. Winkelmann, F. Dallazen, A.B.S. Bronzatti, I.C.W. Lorenzoni, P. Windmöller, *Analysis of steps adapted protocol in cardiac rehabilitation in the hospital phase*, Brazil J Cardiovasc Surg **30**, 40–48 (2015)
6. V.O. Carvalho, *Phase 1 cardiovascular rehabilitation: be aggressive?*, Journal of Cardiothoracic Surgery **6**, 140 (2011)
7. R.M. Macedo, J.R. Faria-Neto, C. Constantini, *Phase I of cardiac rehabilitation: a new challenge for evidence based physiotherapy*, WJ of cardiology **3**(7), 248–255 (2011)
8. Approved by the ATS Board of Directors, *ATS Statement: Guidelines for the Six-Minute Walk Test*, Am J Respir Crit Care Med **166**, 111–117 (2002)
9. Borg GA, *Psychophysical bases of perceived exertion*, Med Sci Sports Exerc, **14**, 377–381 (1982)
10. S. Buurena, E.H.J. Hulzebosc, K. Valkenetd, R. Lindemand, *Reference chart of inspiratory muscle strenght: a new tool to monitor the effect of pre-operative training*, Physiotherapy **100**(2), 128–133 (2014)
11. A.L. Narayanan, J. Bahru, S.R. Hamid, E. Supriyanto, *Evidence regarding patient compliance with incentive spirometry interventions after cardiac, thoracic and abdominal surgeries: A systematic literature review*, CJRT **52**(1), 17–26 (2016)

12. P. Palau, E. Dominguez, E. Nunez, *Effects of inspiratory muscle training in patients with heart failure with preserved ejection fraction*, EJ of Preventive Cardiology **21**(12), 1465–1473 (2014)
13. C.R. Carvalho, D. Paisani, A. Lunardi, *Incentive spirometry in major surgeries: a systematic review*, Brazilian Journal of Physical Therapy **15**(5), 343–50 (2011)
14. H.S. García, L.E. Prendes, J.M. Oquendo, A. Jose, E.R. Estany, L.E. Prendes, *Hospitalization phase of cardiac rehabilitation. Protocol for cardiac surgery*, CorSalud **6**(3), 246–256 (2014)
15. D. Karaszewski, *Comparison of two models of hospital rehabilitation in patients after coronary artery bypass grafting*, Termedia Country of Publication **11**, 86–95 (2015)
16. D.N. Paiva, L.B. Lindstrand, D.F. Bordin, R. Gass, Revista Portugesa de pneumonia, Inspiratory muscle
17. D.L. Borges, M.G. Silva, Silva, L. Nascimento, J.V. Fortes, E.T. Costa, R.P. Assunção, C.M. Lima, N.V.J. da Silva, M. Bernardo-Filho, D.S. Caputo, *Effects of Aerobic Exercise Applied Early After Coronary Artery Bypass Grafting on Pulmonary Function, Respiratory Muscle Strength, and Functional Capacity: A Randomized Controlled Trial*, Index Medicus **13**(9), 946–951 (2016)
18. E. Westerdahl, M. Möller, *Physiotherapy – supervised mobilization and exercise following cardiac surgery: a national questionnaire survey in Sweden*, Journal of Cardiothoracic Surgery **5**, 67 (2010)
19. C.M. Wilson, S.R. Kostsuca, S.M. McPherson, M.C. Warren, J.W. Seidell, R. Colombo, *Preoperative 5 – Meter Walk Test as a Predictor of Length of Stay After Open Heart Surgery*, Cardiopulmonary Physical Therapy Journal **26**(1), 2–7 (2015)
20. A. Herdy, *Pre and postoperative cardiopulmonary rehabilitation in hospitalized patients undergoing coronary artery bypass surgery*, American journal of Physical Medicine & Rehabilitation **87**(9), 714–719 (2008)
21. C. Urell, E. Westerdahl, H. Hendenstrom, *Lung function before and two days after open heart surgery*, Critical Care Research and Practice, **Article ID 291628**, 1–7 (2012)
22. D. Johnson, C. Kelm, D. Thomson, B. Burbridge, I. Mayers, *The Effect of Physical Therapy on Respiratory Complications Following Cardiac Valve Surgery*, Chest **109**, 638–44 (1996)
23. S.F. Mungovan, P. Singh, G.C. Gass, N.A. Smart, A.D. Hirschhorn, *Effect of physical activity in the first days after cardiac surgery*, J Rhabil Med **49**, 71–77 (2017)

Perception of ethical issues by students of health professions

A. Kralova[1], V. Littva[2], and K. Fliegova[2]

[1]Central Military Hospital – Faculty Hospital in Ružomberok, Slovakia
[2]Catholic University in Ruzomberok, Slovakia

Abstract. In recent years in education in health professions in Slovakia big changes have taken place. New insights regarding approximation to the European education system and contextual understanding of health professionals and patients encourage need of change in approaches. The transition from deep-rooted paternalism to individual responsibility of the patients created a gap which was needed to fulfil in education. In our research we analyzed using ANOVA statistics with $p < 0.01$ a non-standardized questionnaire looking what impact had study of ethical issues on their approach to marginalization, labelling, empathy and holistic approach. Significant difference between both groups of students (125 students at the beginning of the study and 119 students at the end of the study) has been shown related to the study of professional ethics on one side and approach to marginalization, labelling, empathy and holistic approach at other side.
Key words: health professions, ethics, labeling, marginalization, holistic approach.

1 Introduction

In Slovakia, during recent years in education of health professions big changes have taken place. Qualifying studies in most fields have been transferred from secondary to higher education, and new insights regarding approximation to the European education system and contextual understanding of health professionals and patients encourage need of change in approaches. This requires above all a change in attitudes towards the patient in the course of medical practice. The transition from deep-rooted paternalism to individual responsibility of patients created a gap which is needed to be fulfilled in education.

2 Survey methods

The aim of this study was to assess the changes in perception of ethical issues by students of selected health professions in order to point out the importance of teaching Professional Ethics, Law and Legislation in Health Care as well as Psychology. We have set the following research tasks:

- Determination of the hypotheses,

- Evaluation of perception of ethical issues by students at the beginning of the study program,
- Evaluation of perception of ethical issues by students at the at the end of the study – after the completion of selected courses,
- Comparison and assessment of research outcomes.

In our survey we used a non-standardized questionnaire which we analyzed by ANOVA statistics with $p < 0.01$. Questionnaire included 20 closed questions. At the beginning of the study (first year of the study) 125 questionnaires were distributed and 119 questionnaires at the end of the study (by completion of third year of study). Respondents consisted of 125 students of the first year (age 19.2 ± 2.6) and 119 students of the third year of BSc studies (age 22.4 ± 2.7) in Nursing, Paramedics, Public Health, and Physiotherapy at the Faculty of Health in Ruzomberok. Survey was conducted in **2015** and 2016. We established 4 hypotheses:

H_1 We assume significant difference between both groups of students in approach to the issue of marginalization of patients.

H_2 We assume significant difference between both groups of students in approach to labelling of patients.

H_3 We assume significant difference between both groups of students in approach to empathy.

H_4 We assume significant difference between both groups of students in holistic approach.

3 Results

All results were analyzed in relation to the students after completing the courses: professional ethics, law and legislation in health and psychology. In all these study subjects they had opportunity to learn about the mentioned topics.

Multi cultural society gives more prominence to the fact of understanding other ethnicities and cultural groups. Students should be prepared to work with patients from different environment, cultural background, and religious group to eliminate the social, economic, health and other factors. They must focus on it comprehensively, because residents of poor communities are territorial, physically, educationally and culturally marginalized. The primary cause of marginalization is language and social barriers [1]. To decrease level of marginalization in nurses' work was introduced trans-cultural nursing into practice. Trans-cultural nursing helps overcome marginalization, because its aim is to provide cultural identical and differentiated care to healthy or sick persons and help them cope with the adverse conditions, disease, or dying culturally appropriate manner [2].

Regarding to marginalization in our study a significant difference between both groups of students has been shown related to the study of professional ethics as well as to law and legislation in health care on one side and approach to marginalization at other side. Significant difference between both groups was not confirmed in relation to psychology. Hypothesis H_1 was confirmed partially as is shown in Table 1.

The next issue in the work of medical professionals is labeling – describing someone or something in a word or short but simplified phrase. Often it is a false attribution of behaviour or assignation to a group that bears certain characteristics (mostly negative) without a deep examination of the truth. It features in plotting of prompt consideration and opinion. We meet it in the context of generalization for example, in patients – seniors. It is claimed that they are grumpy, irritable, or they do not understand something because they are already old [3].

Table 1. Approach of students to marginalization.

ANOVA					
	Sum of Squares	df	Mean Square	F	Sig.
Professional Ethics					
Between Groups	4.847	2	2.424	3.991	**.005**
Law and Legislation in Health Care					
Between Groups	7.653	2	3.827	4.495	**.000**
Psychology					
Between Groups	10.301	2	5.151	10.515	.025

Pearson Correlation * significant correlations at the level 1% ($p < 0.01$).

The cause of the simplification of their attitudes is often excessive stress at work. It is therefore essential that students are prepared for this work by building a mental state achieved by focusing one's awareness on the present moment, while calmly acknowledging and accepting one's feelings, thoughts, and bodily sensations, used as a therapeutic technique – mindfulness. Incorporating this coping strategy into the curriculum ensures that students will have the tools to deal with the stressors of the course, which in turn will enhance the student experience [4].

Results of our study confirmed significant difference before and after completion of all three study subjects – Professional Ethics, Law and Legislation in Health Care and also Psychology. Hypothesis H_2 has been fully confirmed as shown in Table 2.

Table 2. Approach of students to labeling.

ANOVA					
	Sum of Squares	df	Mean Square	F	Sig.
Professional Ethics					
Between Groups	7.733	2	3.377	6.339	**.000**
Law and Legislation in Health Care					
Between Groups	9.876	2	4.938	5.485	**.000**
Psychology					
Between Groups	8.197	2	4.099	9.031	**.000**

Pearson Correlation * significant correlations at the level 1% ($p < 0.01$).

Need to help yourself and others as well as to solve complicated life situations on the advice of others is a certain way to achieve peace. Empathy manifests itself in every single person, whether we like to admit it or not. It is a kind of feature in a personal "I," but not everyone uses it in equal measure. Even if we have it, it is extremely important to develop and support it. In the nursing literature we met statements that the ability to respond empathically is among the factors that improve the relationship between nurse and patient. According to them, empathy is considered as one of the most important and also the most complex factors affecting communication between nurse and patient. Empathy is considered as an essential prerequisite for a nurse to effectively care for a patient and for a holistic understanding of a patient's perspective in a student population [5]. Authors of the study titled "The Effect of Empathy Training of the Empathic Skills of Nurses" stated that profound impact of empathy training on quality of nursing care has been recognized and relevant training will lead to an improvement in nurses' empathic skills, which in turn, will enable them to understand their patients better, establish positive interpersonal relationships with them, and boost their professional satisfaction, but when they held an experiment with training, after the training, no statistically significant difference was found ($p = 0.886$) [6]. In our experiment, which was similar we get results showing statistical significance at the level $p = 0.01$ after successful completion of the course of Professional Ethics ($p = 0.008$), Law and Legislation in Health Care ($p = 0.003$) and psychology ($p = 0.007$). Results are indicated in Table 3. Hypothesis H_3 in our study has been confirmed.

Table 3. Approach of students to empathy.

ANOVA					
	Sum of Squares	df	Mean Square	F	Sig.
Professional Ethics					
Between Groups	4.745	2	2.373	2.625	**.008**
Law and Legislation in Health Care					
Between Groups	7.694	2	3.847	3.804	**.003**
Psychology					
Between Groups	4.303	2	2.152	2.562	**.007**

Pearson Correlation * significant correlations at the level 1% ($p < 0.01$).

Considering holistic approach, the relationship between nurse and patient requires particular attention not only to the patient's verbal expressions, but also non-verbal and creation of a professional attitude in communication, which can to affect the nursing process on both sites – the nurse and the patient himself. In a similar intentions we can understand Plato's and Nightingale's holistic, scientific, and humanistic approach to living, and to care practice in all its dimensions, grounds the discipline of nursing in a liberal arts and critical thinking matrix, elevating nursing to higher ethical, safe, and professional levels of standards [7]. Hypothesis H_4 has been only partially confirmed, because of significant differences between both groups considering all two study subjects – Professional Ethics and Law and Legislation in Health Care. A result only underlines the importance of good preparation in behavioural subjects for health care students. See Table 4.

Table 4. Approach of students to holistic approach.

ANOVA					
	Sum of Squares	df	Mean Square	F	Sig.
Professional Ethics					
Between Groups	4.952	2	2.476	2.769	**.008**
Law and Legislation in Health Care					
Between Groups	4.504	2	2.252	2.904	**.009**
Psychology					
Between Groups	2.994	2	1.497	1.311	**.596**

Pearson Correlation * significant correlations at the level 1% ($p < 0.01$).

It is even though adherence of ethical principles is not subject to punishment. Observance of the legislative norms was also comfirmed in all hypotheses.

4 Conclusions

The transition from deep-rooted paternalism to individual responsibility of the patients created a gap which is needed to be filled in education. Behavioural teaching subjects such as: ethics, law and legislation in health care as well as psychology are in preparation for the health care professions needed, not only recommended.

References

1. E. Moraučíková, M. Kopáčiková, *Sexuálna výchova v rómskej komunite. Aktuálne dianie, perspektívy a vývojové trendy v sociálnej, edukačnej a zdravotníckej práci s marginalizovanými skupinami,* 223–228, Verbum (2012)
2. M. Lehotská, *Aplikácia prvkov transkultúrneho ošetrovateľstva u znevýhodnených skupín. Interdisciplinárna kooperácia v ošetrovateľstve, pôrodnej asistencii a sociálnej práci,* 102–109, Verbum (2012)
3. K. Zrubáková, M. Magerčiaková, *Les spécificités des soins professionnels et laïques auprès de la personne âgée à domicile.* Rev. Int. Sci. Hum. Nat. **2**, 25–38 (2014)
4. M. Walker, A.R. Mann, *Exploration of mindfulness in relation to compassion, empathy and reflection within nursing education,* Nurs Edu. Tod. **40**, 188–190 (2016)
5. C. Petrucci, C. LaCerra, F. Aloisio, *Empathy in health professional students:A compatrative cross-sectional study,* Nurs Edu. Tod. **41**, 1–5 (2016)
6. I. Kahriman, N. Nural, U. Arslan et al., *The effect of empathy training on the emphatic skills of nurses, Iran.* Red. Cress. Med. J. **18**, e24847 (2015)
7. J.M. Arnone, V. Fitzsimons, *Plato, Nightingale, and nursing: Can you hear me now?* Int. J. of Nurs. Know. **26**, 156–162 (2015)

Spiritual needs of dying and emergency health care

I. Andrasi

Catholic University in Ruzomberok, Slovakia; St. Elizabeth University, Bratislava, Slovakia

Abstract. Dying process can take different forms, and each death is related to leave-taking of a human with what he liked, with those whom he loved and who were close to him. Daily contact with the dying is very difficult, very demanding not only in physical terms, but especially in terms of psychological burden. The essence of care for dying patients is to ensure adequate quality of life that is highly affected by understanding of human existence and the meaning of human life. Paramedics must realize that quality of life is associated with satisfying the needs that in each patient is different, so the quality of life is a very individual concept. Coping with dying is significantly affected by the ability of their very faith in God's existence, length of service and communication skills.
Key words: spiritual needs, dying, paramedics, emergency health care.

1 Introduction

Dying process can take different forms, and each death is related to leave-taking of a human with what he liked, with those whom he loved and who were close to him. The World Health Organization (WHO) defines health as a state of complete physical, mental and social wellbeing and not merely the absence of disease or malfunction. In the modern era health is increasingly complement from the perspective of the needs and the spiritual well-being of the patient, which in many cases dominates in dying patients. Health care workers in daily contact with the dying are not just health professionals, for whom the care of these patients is very demanding in terms not only physical, but especially in terms of psychological burden. Paramedics involve salvaging not only sick part of the human body, but they work with sick/dying person and subsequently with relatives/survivors. They act by their professional interventions, as well as all their personality, relationship with them, degree of professional qualities, level of competence and professional adaptation [1]. The essence of care for dying patients is to ensure adequate quality of life that is highly affected by understanding of human existence and the meaning of human life. A paramedic must realize that quality of life is associated with satisfying the needs that in each patient is different, so the quality of life is very individual concept. Through interview with the patient a paramedic has the opportunity to find out their spiritual needs and then set a goal of satisfying these needs. Satisfying spiritual needs helps to cope with suffering more tolerable. Satisfying spiritual needs is of great importance for the sick person. A paramedic should particularly respect the faith of the sick, to obtain information on their spiritual needs, to inform the patient and relatives [2]. The concept of spiritual literacy is introduced as the nurse's and paramedic's ability to read the spiritual signs of the human experience.

Main concern is to assist the patient to alleviation. The participants resolved this by discerning the healing path, which comprises three stages: tuning in on spirituality, uncovering deep concerns, and facilitating the healing process. these three stages are accompanied all the way by the participants' willingness to overcome own comfort zone and building a trusting relationship [3].

Our aim was to find out the attitude of emergency workers to religion and spirituality as well as the impact of traumatic events at work on their attitude to spirituality of patients; to find out how paramedics perceive spiritual needs of patients and how often they meet the requirements of the prayer, the presence of a priest.

2 Survey methods

In our survey, we used the online standardized questionnaire – The Spiritual Needs Assessment for Patient (SNAP) concerning psychosocial and spiritual needs. The sample consisted of 88 respondents: 69.4% males and 30.4% females. Age of the respondents was 35.4 ± 9.7 years. Length of practice: 5 years 42.9%, 5–15 years 37.6%, more than 15 years 19.5%. Religion and the existence of God: 74% believe in the existence of God. We analyzed questionnaire by ANOVA statistics with $p < 0.01$. Survey was held in 2014 and 2015. Reliability of a psychometric test – Cronbach's alpha = 0.921.

3 Results

The results were analyzed in relation between the perception of spiritual needs in dying patient and the length of practice. We found that perception of spiritual needs of the patient depends on the length of service and spiritual needs of emergency workers. There is a significant difference between all three groups (according to length of practice) as is shown in Table 1.

Table 1. Perception of spiritual needs in dying patients.

ANOVA					
	Sum of Squares	df	Mean Square	F	Sig.
Lenght of practice < 5 years					
Between Groups	7.753	2	3.385	6.344	**.000**
Lenght of practice 5 – 15 years					
Between Groups	9.703	2	4.847	4.804	**.003**
Lenght of practice > 15 years					
Between Groups	4.992	2	2.477	2.585	**.008**

The spiritual needs of a dying person can sometimes be overlooked in the busyness of physical care. However for those experiencing it, spiritual distress is very real. The time, when the patient faces their own death is the time to start thinking about ways of meeting the spiritual needs [4]. 94.73% of respondents with up to 5 years' experience states that they are focused primarily on the medical aspects; 63.6% of emergency workers with 5–15 years' experience stated that they perceive the needs of patients and meet them as much as possible; 82.35% with experience longer than 15 years stated that suspension of transport would request the presence of a priest prior to transport to the hospital and act without any delays. The results are shown in Fig. 1.

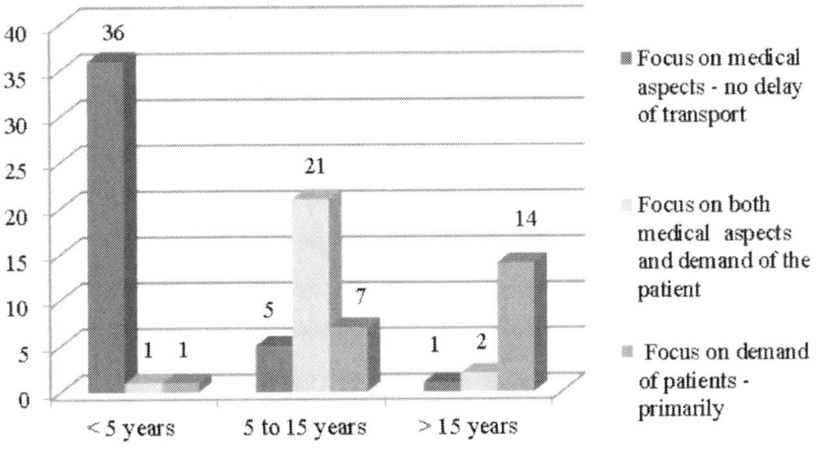

Fig. 1. Focus influenced by length of practice.

How often paramedics meet the requirements of the prayer, the presence of the priest and how to implement it? Most often it is the ratio between cities and countryside 17%:83%. The problem is the lack of preparedness for the situation. Similar research was held also in different care organizations in a rural community in Sweden with participation of n = 33 paramedics. The results showed that less than 50% lacked education in the spiritual/existential areas [5].

Often we meet medical staff's fear of dying. Often they ask themselves how it will affect them personally and how it affects their career attitudes. At the beginning of work in their profession, or while preparing for a career, medical staff is expected to experience a greater fear than after several years of practice. A study conducted in Hungary, however, shows other results. Cross-sectional comparative study conducted to evaluate and compare the current and future ambulance personnel's fear of death and its influencing factors including 106 participants (active rescue ambulance personnel, n = 45 persons; paramedic students, n = 61) discovered that there is no significant difference between the levels of fear of death of the present and future ambulance personnel [6].

An important fact is the perception of the quality of dying. This factor is associated with the staff's perception of high-quality dying and death. Positive death experiences are associated with absence of patient pain, dyspnoea, anxiety, or agitation and the presence of spiritual, psychological, and culturally appropriate care for the patient and family [7]. This topic is becoming more and more relevant, as the incidence of patients requiring this type of symptom care is on the increase. The results of research held in Slovakia in oncology patients had shown that the symptom occurrence of patients in palliative care directly correlates with the quality of their lives. Among the most common symptoms reported were: loss of appetite (n = 102 = 99%), weight loss (n = 101 = 98%) and pain (n = 97 = 94.2%). The symptoms that were least reported were: dizziness (n = 31 = 30.1%), flatulency (n = 27 = 26.2%) and itching (pruritus) (n = 16 = 15.5%) [8].

4 Conclusion

Paramedics are interested and want to talk about the spiritual life of the patient and satisfy their spiritual needs. It is important that professionals be properly prepared when patients need this communication. An evidence-based training intervention could provide such preparation [9]. But it is significantly affected by their faith in God's existence, length of service and communication skills. In severe traumatic events and raised doubts about the strength of faith and ability of God to do miracles it does not affect their spiritual life and faith. Paramedics with active spiritual life are happier in their profession, they easier overcome traumatic events, are more active in approach to meeting the spiritual needs of the patient and are more balanced in medical assistance.

References

1. E. Moraučíková, *Multidisciplinary care about remainder*, Log. Pol. **2**, 82–87 (2015)
2. K. Zrubáková, A. Herinková, *Význam spirituality u onkologicky chorých. Posudzovanie symptómov a spirituálnych potrieb pacientov v paliatívnej starostlivosti.* SAP 94–114 (2015)
3. S. Hlinková, E. Moraučíková, *Spirituality and Spiritual Needs in the Palliative Care. Satisfying spiritual needs of dying.* Tribun, 31–52 (2014)
4. T. Giske, P.H. Cone, *Discerning the healing path – how nurses assist patient spirituality in diverse health care settings.* J. of Clin. Nurs. **24**, 2926–2935 (2015)
5. G. Eriksson, T.W. Bergstedt, Ch. Melin-Johansson, *The need for palliative care education, support, and reflection among rural nurses and other staff: A quantitative study.* Pal.& Sup. Carre **13**, 265–274 (2015)
6. E. Pek, Z. Szekely-Benke, B. Radnay et al. *Am, balance personnel and paramedic student's attitudes regarding death and dying.* Crit. Care Med. **43**, 342 (2015)
7. R. Bennett, J. Proudfoot, *What does the staff think? Factors associated with clinical staff perceptions of what constitutes high-quality dying and death.* J. of Hosp.& Pal. Nurs. **18**, 470–476 (2016)
8. V. Littva, M. Šichman, *Troublesome Symptoms in Palliative Care. 7th International Conference of Hospice and Palliative Care.* Solen, 6–13 (2015)
9. I. Henoch, S. Strang, M. Browall et al. *Development of an existential support training program for health care professionals.* Pal. & Sup. Carre **13**, 1701–1709 (2015)

Social support as the role of social worker in work with dialysed clients

N. Kovalcikova and J. Zakova

Trnava University in Trnava, Trnava, Slovakia

Abstract. The quality of life of individuals with chronic illness is strongly influenced by anticipated social support. In this regard we wanted to know how social support is reflected (perceived) in dialysis patients. The aim of the study was to measure the level of perceived social support in patients undergoing dialysis treatment. We utilized a quantitative research strategy using a standardized social support questionnaire – MOS. The respondents of the survey were clients of dialysis centres in Slovakia undergoing dialysis or peritoneal dialysis and willing to participate in our research. The results showed statistically significant difference in the perception of social support depending on the status of cohabitation. We did not find any significant difference relating to the type of dialysis treatment, its duration, age of respondents, involvement in self-help groups and the use of social services and compensations. The results show that the family acts as the key factor of providing social support and that family support is of the utmost importance for clients who undergo dialysis treatment. The study also contains recommendations for social work.

1 Introduction

Chronic kidney diseases are now a global problem that has an increasing tendency in developed countries. It is estimated that one of ten people in our population suffers from a type of kidney disease. Kidney diseases are becoming not only medical, but also socio-economic problem [1]. When kidneys fail to function properly, a complex nephrology treatment must start, in cases of more serious malfunction dialysis is prescribed and the patient undergoes dialysis treatment. Dialysis is a method of purifying blood by removing waist products of nitrogen metabolism, especially urea and creatinine. Kidney dysfunction is treated by dialysis – either haemodialysis or peritoneal dialysis [2].

Along with dialysis, patients must also follow a strict regime and diet as this greatly influences quality of life for both patients and their families. Šafránková and Nejedlá [2] indicate that dialysis also correlates with somatic symptoms in patients – cardiovascular complications, blood-related conditions, digestive disorders and pain. This treatment is accompanied by complications and patients also deal with associated diseases. There are also certain psychiatric issues such as fear, stress, feeling of being dependent on a machine, prospect. Economy-related issues relate to leaving the labour market and receiving disability pension and of course, higher treatment cost. Regarding social aspects, along with this treatment come changes in the life style, hobbies, often with a negative impact on family relations and similar. Znojová and Sulková [3] state that among the greatest stress factors in these patients, occur such factors as reduction of physical activities, restricted ability, impairment of body functions, fatigue and fluid restrictions. The total stress score did not significantly vary in sexes and was statistically greater in older dialysis patients (over 50 years old). Kidney malfunction is therefore a burden for people, it can change how people feel, behave and think. Psychosocial influence can have an impact on comfort in the patients and the disease course. One of the possible approaches in dealing with the burden brought by the serious disease is the strategy to gain social support. A sick person seeks for somebody to share his problems with, somebody who could understand. When this social support has been gained, the person usually feels better [4, 5]. Mares [6] defines social support as well-doing that is offered to someone who is close and this activity has either an immediate or later coming positive effect. It can also be defined as help that is offered to people who find themselves in a difficult situation. Social help is important especially when a person falls sick or is affected by an injury, when he or she find themselves in a crisis and has to make a serious decision [7]. Znojová's [8] study of people with serious diagnosis showed that those patients who are given reasonable social support from their family and close people and friends achieve better treatment, cooperate better with healthcare professionals, live longer despite the disease and also the overall quality of their life is higher. Andrášiová [9] also emphasizes that social support is an important protective factor when coping with stress brought by the disease and treatment, especially during the period when the patient is not surrounded by a vast spectrum of personal protective factors. Dobríková [10] adds that lonely people, who only enjoy low social supports in comparison to the control sample, showed a weakening of the immune system.

Social support can be achieved from various sources. We can turn to non-professionals, such as family, friends, colleagues or we can seek help among professionals, this can then be given either individually or in a form of group sessions [11]. Sources of social support are [12, 13] either natural, usually represented by close relatives, such as parents, husband, wife, brothers or sisters, friends, etc., or institutions, e.g., non-government organisations, state organisations, support groups that are organised by specialists, self-help groups, religious organisations and helping professions, among which is also social work.

2 Participants and methodology

The main goal of the study was to find the degree of anticipated social support in clients undergoing a dialysis treatment. Krpoun [14] points out that anticipated social support can be understood as a feeling of being accepted and belief that there is somebody you can turn to, when it is needed. On the other hand, there is also acquired social support that reflects a specific pattern in interaction between two and more people in a specific situation. As for partial goals, we wanted to determine degree of social support in relation to the type and duration of dialysis treatment, age, cohabitation status, meeting in self-help groups, and receiving disadvantage compensation and social services.

The objects of our study were dialysis clients. Only clients whose health condition was stabilised, who agreed to participate and who were in the process of active dialysis treatment – either haemodialysis or peritoneal – were accepted in the study. The sample group consisted of 182 respondents, with 55% men and 45% women. The average age of the study respondents was 64.27. The respondents were not younger than 15 or older than 90. Of them, 51% were married, 26% – over a quarter of the group – were widows or widowers and 17.05% of the respondents were single. 4% were unmarried partners and 1% was divorced. 12% of the respondents lived on their own, with majority of them being older age and having lost their husband or wife. 37% of the respondents lived with their husband/wife or partner. 47.73% of the respondents lived with their family members and stated that they lived with their children, son in law or daughter in law. One respondent was a resident of a nursing home. In our sample haemodialysis prevailed, with 86% of the respondents receiving this type of treatment, 14% of the respondents received peritoneal dialysis treatment. Most of the respondents, 47%, received dialysis treatment for 1–4 years, shorter than 1 year – 13% of the respondents, 5–9 years – 25% and more than 10 years – 13% of the respondents. 86% of the respondents did not attend meetings of self-help groups, while 13% of the respondents were active in self-help groups.

We used a questionnaire to collect the data. This questionnaire consisted of two parts; the first part dealt with basic questions, demographic information and specification of the dialysis treatment. The second part of the questionnaire consisted of standardised questions for social support MOS. The method Social Support Survey – MOS, used in this study, was developed by Rand and Medical Outcomes Study. This method, designed to be used in general population, measures degree of anticipated social support, reflects subjective interpretation of the individual's social inclusion. The original 50-item group of questions was reduced to 19 questions that covered five dimensions of social support, namely emotional, information, material assistance, social interactions and affective dimension of social support [15]. To have a Slovak version of the questionnaire, we edited a Czech version of the questionnaire that uses three dimensions: *Practical Intervention* and *Emotional Intimacy* are equivalents of Tangible Support and Affectionate Support in the American version, items Positive Social Interaction and Emotional/Informational Support are represented in the Czech version as *Supportive authority* – see the questionnaire for social support – MOS: Vnitřní struktura nástroje [15].

3 Results and discussion

Dialysis treatment is necessary when kidneys fail to function. It is a life-saving therapy that allows patients to keep on living, but also brings many restrictions and changes in life. Patients who receive dialysis treatment live in stress, this is brought by both progress of the illness and also the dialysis treatment itself. Unless transplantation is performed, patients with malfunctioning kidneys are dependent on the dialysis treatment for the rest of their lives. Regarding quality of their life and overall satisfaction with life it is important that clients and their families continue as normal life as possible so that they are not completely paralysed by thinking about the illness and treatment procedures. It is social support that can help here; we take it as an important component that helps patients overcome problems and burden that come with their disease and its treatment. By general influence of social support we understand the positive effects that good interpersonal relationship brings in the social network of the person, in everyday life, these can positively influence mental and physical health. Specific influence relates to concrete life situations (stress, distress) [16]. We specifically aimed our research at finding out the level of social support in people receiving treatment in form of haemodialysis or peritoneal dialysis. We used 19 questions in a standardised questionnaire. The patients' answers reflected high rankings in all of the questions, the lowest average rank was 4.27 and highest average rank was 4.69. The most frequently used ranking was 5. Question number 13, "How often is somebody, who can help you distract from worries, available?" got the lowest average ranking (4.27). This question reflects the supportive authority dimension. For a dialysis patient it is somebody who can help distract his or her worries and is available most of the time. Question number 19 deals with emotional closeness, i.e. how often the person who you like and who needs you is available for you. The top average rating for this question was 4.69. We can say that the person a dialysis patient likes and who needs him/her is always available. The questions dealing with *practical intervention* dimension got ranking between 4.63 and 4.67. The answers testify that somebody who takes the dialysis patients to the doctor's, who can cook for them when they are not able to do it and who helps them with house chores is commonly available. The questions dealing with *emotional closeness* also got very high ranking (4.35–4.49). Somebody who can hug the dialysis patient when he or she needs it is available most of the time. There is always somebody who shows affection, compassion and makes them feel well, somebody who cares about them and who needs them. Dimension *supportive authority*, studied through 11 questions, got just a little lower ranking when compared to the previous two dimensions (4.33–4.49). For dialysis patients, there is usually somebody available, somebody who listens to them, understands their problems, helps them in critical situations, distracts negative thoughts and shares their fears and worries, offers information that is necessary to understand certain situation, somebody they can confess their worries and problems to and who helps them relax, do something that pleases them, somebody whose advice they need and who is ready to help them with personal problems.

In the study, we were strongly interested whether there was a connection between the level of perceived social support and type of dialysis treatment. We started with the assumption that peritoneal dialysis allows more freedom in personal life, but also requires certain social background and often also other people's cooperation during dialysis treatment or care and maintenance of the dialysis catheter. We expected that the level of social support would be appraised considerably higher in the clients undergoing a peritoneal treatment and that the average rankings in the assessment of social support would be higher. We acquired results using the Mann–Whitney U test for two independent variables. This test did not testify a significant difference based on the type of dialysis treatment ($Z = -621$, $p = 0.535$). Basing on the test results, we can say there was not any statistically important difference in degree of social support based on type of dialysis.

Duration of a dialysis programme has a great influence on patient's mental health. Mortality rate during dialysis treatment is still high despite improvement in care of such patients. 80% of patients survive in the first year of treatment, but only half of the patients live longer than 5 years. Apart from kidney disease, patients also suffer from co-morbidities (subsidiary illness, especially diabetes mellitus and arterial hypertension). Unfortunately, a large group of ill people do not know about their disease, have never been checked up and suddenly have to start a dialysis treatment [1, 17]. It is also a sort of intermediate station, where patients who are given dialysis treatment wait for kidney transplantation that greatly improves quality of their life. The length of the waiting time is individual and the success or failure of transplantation is determined by various factors that the patient cannot influence. During this time patients experience states of ambivalence that are brought not only by dialysis itself, but also by feeling hope/despair when they wait for the transplantation. Basing on these arguments, we assumed that there would be statistically significant difference in assessing degree of social support that relate to the duration of dialysis treatment ($H(3) = 5.680$, $p = 0.128$). The lowest ranking of social support was found in patients who had received dialysis treatment for between 1–4 years (83.40) and the highest ranking was in patients that had been in dialysis programme for 5–9 years.

Basing on Šolcová and Kebza's [18] findings, age was taken as one of the predictors for social support. Therefore we wanted to see in our study whether there were any age-related differences in anticipated social support ranking. However, the Kruskal Wallis test did not testify this ($H(4) = 2.704$, $p = 0.608$). In practical intervention, the lowest value of 17.83 was found in dialysis patients between 30 and 44 years old and highest values of 20.00 were seen in 60–74 years old patients. In the dimension of emotional closeness, the lowest value of 17.86 was seen in 45–59 years old patients and the highest value of 18.75 was in 60–74 years old patients. The lowest result of 47.91 in the dimension of supportive authority was in the 75–89 age group and the highest, 50.50, in the 15–29 age group. This was probably related to the fact that the youngest respondents did not live on their own, and in contrast, in older age social interactions are somewhat limited, as Špatenková [19] and Stuart-Hamilton [20] claim.

We agree with Vágnerová [5] that a serious disease does not affect only the sick person, but brings its consequences also to the whole family. For a sick person, it is the family that is his or her social group, the social group that helps with social integration and thus makes it accessible and approachable, the social group that also offers emotional safety and support. Kebza [21] sees family and family relations as a crucial factor here; family is often used as index of social support. Ondrušová et al. [12] understands family as the most important tool of social help. We understand the fact that this was proven to be the most important outcome. Basing on the claims mentioned above, we wanted to find out whether there was a significant difference in the scope of declared social support in relation to the different status of mutual cohabitation. It was confirmed that there were differences in perceiving the scope of social support in dialysis patients with different form of cohabitation ($H(3) = 10.725$, $p = 0.013$). There was a difference between respondents that lived alone and respondents that lived with their wife/husband, partner or with their family. Those dialysed patients who lived alone appraised the overall social support almost 50% lower than clients who lived with their spouse or partner. The acquired results showed that social connection with partners, families or relatives positively influences the social support appraisal. In connection to this, Andrášiová [9] points out that sick people who enjoy support from their family are commonly more active and better motivated to fight the disease. The Mann–Whitney *U* test testified that there are differences in social support appraisal between those who live alone and those who live with their family. The significance value in overall social support $p = 0.003$, practical intervention $p = 0.007$, emotional closeness $p = 0.002$ and the dimension of supportive authority $p = 0.009$. Research study performed by Dobríková [10], which can be implemented and used also for work with dialysis patients and their relatives, revealed that people with tumours need their relatives to use natural approach in communication, body language that agrees with words they say, support and adequate help, encouragement, respect and tolerance. The need of physical touch is individual, some patients do not like more frequent physical touches, they need understanding especially in the period when they are temporarily emotionally unstable, they appreciate positive news about their surrounding and need others to respect a new daily regime and another changes that are triggered by the illness. However, Mátel and Hardy [22] argue that families differ in their ability to function properly even under pressure. Some families can cope with problems and need support during only a more stressful period; other families need permanent help from organisations and authorities.

For another assumption, we started with arguments presented by Schavel et al. [23] who, basing on research conducted abroad, points out that members of self-help groups are more independent, have greater possibilities for social contact than other people in similar situation and are also less prone to depressions. In self-help groups that consist of parents with chronically ill children, where the parents help each other, was found a very positive influence brought by social support provided by the self-help group [4]. We therefore wanted to know whether there was a relation between degree of anticipated social support and meeting in self-help groups. Surprisingly, the result value determined by the Mann–Whitney *U* test showed there was no difference in degree of declared social support among dialysed patients who met in self-help groups and those who did not ($Z = 412.500$, $p = 0.593$). Similarly, there were no statistically significant differences testified in individual dimensions of social support, practical intervention $p = 0.968$, emotional closeness $p = 0.734$, supportive authority $p = 0.538$. We believe that this result could have been influenced by the small number of participants in our sample group of people who met in self-help groups. Yet, Křivohlavý [4] notices that meeting in self-help groups is an increasing practice. An individual person looks for somebody who could help. Self-help groups are thus becoming an important addition to the system of health care. The majority of respondents who meet in self-help groups were at least 45 years old. We had expected that younger respondents meet in self-help groups in a larger number. It may have been influenced also by the fact that there were fewer respondents of this age group (44 years old and younger) in our sample group, but also because the average age of dialysed clients in our sample group was 64.

We would like to speak about a broader character of social support regarding its levels (macro-level, mezzo-level, micro-level) that generally distinguish those who provide social support in regard to their distance from receivers of social support. Křivohlavý [16], Šolcová and Kebza [24] speak about macro-level, which means help form the state and which is also recognised in legislative. As an example, we can take the instrumental assistance (practical assistance or material assistance) and information assistance (providing guidance or advice) that is offered by social workers. We were therefore interested in the extent people used social assistance, social services, disability benefits for seriously disabled people receiving dialysis treatment. Regarding Slovak legislation provisions that are applied in social services, benefits to the seriously disabled, health care (Act No. 447/2008 regarding disability benefits, amendments and modifications to certain laws, Act No. 448/2008 regarding social services, as amended, Act No. 577/2004 on the scope of health care, as amended), we asked in question 13 whether dialysed patients used social services or benefits to compensate the impact of the serious health condition. 21% of the respondents stated that they used social services and 79% answered "no". Question 14 – "Do you receive financial compensation?" was answered "yes" by 70% of the respondents, while 30% of the respondents answered "no". The answers indicate that chronic kidney disease and dialysis treatment influence negatively the economical perspective of life in dialysis patients and compensations can be understood as an effort for a micro-level social support that can be given by the state. We tested statistically also whether there was a co-relation between the scope of social support and extent to which people apply for social services or compensation benefits for their serious health condition. Our assumption was not confirmed here.

Chronic renal insufficiency and following dialysis treatment are health conditions that mean long-time burden for the sick people and their close but also wider social surrounding. Social assistance in the form of emotional, instrumental and material support is a significant factor for coping chronic stress. The results of our study show that family is the primary source of social support in people who undergo a dialysis treatment. In our sample group, the extent of social support was dependent neither on age of the patients, type and duration of a dialysis treatment, nor on macro-level help – using social assistance and receiving benefits to compensate a serious health condition.

This paper was prepared as a part of the project The Identity of Social Work in the Context of Slovakia [APVV-0524-12] funded by the Slovak Research and Development Agency.

References

1. R. Ryšavá. Chronické onemocnění ledvin- role nefrologa a praktického lékaře. Via Practica. ISSN 1336-4790, 2015 , issue 2, **12**, p. 87–89. (2015)
2. A. Šafránková, M. Nejedlá. Interní ošetrovatelství II. 212 p. ISBN 80-247-1777-8. (Praha: Grada Publishing, a. s.2006)
3. M. Znojová, S. Sulková. Hodnocení stresu u dialyzovaných pacientú. Časopis lékařu českých [online]. 2001, issue 4, p. 108–111. Available on the Internet at

http://www.prolekare.cz/casopis-lekaru-ceskych-clanek/hodnoceni-stresu-u-dialyzovanych-pacientu-25786. (2001)

4. J. Křivohlavý. Psychologie nemoci., 200 p. ISBN 80-247-0179-0. (Praha: Grada Publishing, 2002)
5. M. Vágnerová. Současná psychopatologie pro pomáhajíci profese. 816 p. ISBN 978-80-262-0696-5. (Praha: Portál, s. r. o, 2014)
6. J. Mares. Nevyužívání či odmítaní sociální opory. Psychológia a patopsychológia dieťaťa. ISSN: 055-5574, 2001, **36**, 3, p.214–224. (2001)
7. J. Křivohlavý. Psychologie zdraví. 280 p. ISBN 80-7178-551-2. (Praha: Portál, 2001)
8. M. Znojová. Kvalita života peritoneálne dialyzovaných pacientu. Stežeň [online]. 2005. 16, 2. Available on the Internet at http://stezen.xf.cz/html/stezen/2005/0502.html
9. M. Andrášiová. Psychologické minimum o psychickom prežívaní onkologického pacienta. Via practica. Solen. [online]. 2007, 4, p. 42–46 Available on the Internet at http://www.solen.sk/pdf/7f83efdd5580e66f37f104ac3c86bf9a.pdf
10. P. Dobríková. Zvládanie záťažových situácií. Ako si zachovať duševné zdravie. 126 p. ISBN 978-80-89271-20-7 (Bratislava: SAP – Slovak Academic Press, 2007)
11. V. Hanzalíková, K. Levická. Sociálna opora v procese smútenia. Solen [online]. 2012, **5**, issue 2, p. 43–45 Available on the Internet at http://www.solen.sk/pdf/b905064bdede79a4a2e8ea896e4aeb7a.pdf (2012)
12. Z. Ondrušová. Základy sociálnej práce. 144 p. ISBN 978-807392-109-5. (MSD Brno, 2009)
13. J. Levická. Sociálna práca I., 168 p. ISBN 978-80-969454-2-9. (Trnava: OLIVA, 2007)
14. Z. Krpoun. Sociální opora: Shrnutí výzkumné činnosti v české a slovenské oblasti a inspirace do budoucnosti- nadané děti. E-Psychologie, Elektronický časopis ČMPS [online], 2012, year **6**, issue 1, p. 42–50. Available on the Internet at: http://e-psycholog.eu/pdf/krpoun.pdf
15. J. Kožený, L. Tišanská. Dotazník sociální opory – MOS: Vnitřní struktura nástroje. Československá psychologie. ISSN 0009-062X, 2003, **47**, 2, p. 135–143. (2003)
16. J. Křivohlavý. Psychologie zdraví. 280 p. ISBN 978-80-7367-568-4. 78 (Praha: Portál,s.r.o., 2009)
17. A. OKŠA. Diabetická nefropatia- diagnostika, prevencia a liečba: Odporúčané postupy Slovenskej nefrologickej spoločnosti a Slovenskej diabetologickej spoločnosti. Forum diabetologicum. ISSN 1805-3807, 2013, **2**, 2, p. 51–56. (2013)
18. I. Šolcová, V. Kebza. Prediktory sociální opory u české populace. Československá psychologie. ISSN 0009-062X, **XLVII**, 3, p. 220–226. (2003)
19. N. Špatenková. Krize. Psychologický a sociologický fenomén. 129 p. ISBN 80-247-0888-4 (Praha: Grada, 2004)
20. I. Stuart Hamilton. Psychologie stárnutí. 319 p. ISBN 80-7178-274-2 (Praha: Portál, 1999)
21. V. Kebza, Psychosociální determinanty zdraví. p. 296. ISBN 80-200-1307-5 (Praha: Academia, 2005)
22. A. Mátel, M. Hardy. Vybrané kapitoly z metód sociálnej práce II. 507 p. ISBN 978-808131-074-3. (Bratislava: Vysoká škola zdravotníctva a sociálnej práce sv. Alžbety, 2013)

23. M. Schavel. Sociálna prevencia. 267 p. ISBN: 978-80-89271-22-1. 80 (Bratislava: Vysoká škola zdravotníctva a sociálnej práce sv. Alžbety, 2010)
24. I. Šolcová, V. Kebza. Sociální opora jako významný protektivní faktor. Československá psychologie. ISSN 0009-062X, 1999, **43**, 1, p. 19–38. (1999)

Impact of different physical activities on executive functioning

K. Sneidere[1], J. Harlamova[1], V. Arnis[1], Z. Ulmane[1], I. Mintale[2], K. Kupcs[2], and A. Stepens[1]

[1]Rīga Stradiņš University, Riga, Latvia
[2]Pauls Stradins Clinical University Hospital, Riga, Latvia

Abstract. The aging of the Western society has become one of the key issues in research. Lately research has shown relationship between improvement in cognitive functioning and involvement in physical activities. The aim of the study was to examine the impact of different types of physical activities on executive functioning. Overall 30 seniors aged from 65 participated in a two-day study. They were divided into three groups per their life-style – aerobic activity, strength training and sedentary. Results indicated differences between the groups in cognitive efficiency ($F = 6.791, p < 0.01$) and working memory ($F = 6.179, p < 0.01$) as well as the inhibition of attention ($F = 4.311, p < 0.01$). The mean scores were higher in the aerobic group. The results indicate that involvement in aerobic physical activity might lessen the decline in executive functioning in seniors.

1 Introduction

WHO prognosis is that by year 2050 population aged over 60 will almost double, which means that approx 22% of population will meet difficulties associated with aging, one of them being decline in cognitive functioning [1]. Research has shown that with aging there is a decrease in the brain volume, especially, in frontal, parietal and temporal cortices [2], which leads to decline in cognitive functioning, for example, there are indications of decline in executive functions [3, 4].

Executive function is broad construct of attention thought to be involved in higher level cognitive tasks performed in the service of a goal [5]. Executive functions have been researched extensively and there are several theories on the division of the functions. One of the theories involves three core executive functions – inhibition – which involves being able to control your attention, behaviour, thoughts and emotions to override a dominant reaction, for example, inhibitory control of attention allows us to focus on our task at hand. Working memory is another core function and involves ability to work with information mentally holding it in mind, for example, doing math. Cognitive flexibility refers to ability to adjust, change perspectives, it is involved in problem solving (for example, thinking outside the box) [6].

More and more research results indicate an increase in grey and white matter volume in prefrontal cortex as well as increased functioning of key nodes in the executive control network in people who are involved in aerobic exercises [7]. Evidence of greater brain volume in temporal brain regions in physical active older adults in comparison with less active older adults can be found as well [8]. Increasing amount of research suggests increase of volume in hippocampus and reduced brain atrophy due to physical exercise [7, 9].

2 The present study

Aim of the present study is to examine the impact of different types of physical activities on executive functioning, focusing on inhibition of attention and working memory. We hypothesised that participants involved in aerobic physical exercise will show better results than participants involved in strength training with aerobic activity elements in lifestyle or participants with sedentary lifestyle. This study was funded/partly funded by the Latvian National Research Programme Biomedicine for Public Health (BIOMEDICINE) 2014–2017.

3 Materials and methods

3.1 Participants

To assess the impact of different types of physical activities on executive function, 30 participants aged 65–80 (M = 72.27, SD = 4.55) were included in the data analysis. 30% of the participants were male. Most of the participants had higher education (53.3%), secondary/secondary professional education (33.2%) or incomplete higher education (10%). Only one participant had less than secondary education (3.3%). The mean years spent in education were 15, longest being 22 years (see Table 1). All of the participants were native Latvian speakers, who use this language daily.

Table 1. Participant characteristics.

Characteristic	Mean	Range	SD
Age	72.27	65–80	4.55
Education (y)	15	7–22	3.69

Note. SD – standard deviation.

Seniors were divided into three groups – seniors involved in regular aerobic activities (e.g. orienting, running, swimming, Nordic Walking, etc.), seniors involved in regular strength training (e.g. long jumping, hammer throw, etc.) with irregular aerobic activities, such as walking daily or running, and seniors with sedentary lifestyle. Both physically active groups included participants involved in physical activities for leisure and participants involved in physical activities as competitive athletes.

We indicated several inclusion/exclusion criteria regarding language and medical history that might impact the results of cognitive tests, such as known history of cardio-vascular diseases, metabolic disorders, pulmonary and respiratory diseases, ongoing oncologic diseases, rheumatologic diseases and mental diseases (for detailed description see Table 2).

3.2 Instruments

For overall screening, *Montreal Cognitive Assessment Scale* (*MoCA*) [10] was used. This test consists of twelve tasks, assessing visual-spatial perception, executive functions, short and long-term memory as well as working memory, attention, thinking processes, language skills and orientation in time and place. The maximal count of points is 30, the accepted norm is 26 points.

Table 2. Inclusion/exclusion criteria.

Inclusion/exclusion criteria	Description
Language	native Latvian speaker
No history of:	
cardio-vascular diseases as	uncontrolled arterial hypertension [>140/>80mmHg], angina pectoris, cardiac failure, stroke
metabolic disorders as	diabetes, obesity [BMI > 30]
pulmonary and respiratory diseases that	require regular inhaler use
ongoing oncologic disease	
rheumatologic diseases that	require regular painkiller use
mental diseases as	depression, dementia, ongoing schizophrenia that require medication

To assess working memory, *The Numbers Reversed test* [11] was used. In this test, participants are presented with a series of digits which must be repeated in reverse order. The norms for this test depend on the age of the participant.

To assess inhibition of attention 3 instruments was used:

- *Visual Matching test* [11]. This test requires crossing out similar digits from a sequence of digits with time limit.
- *Map Test of Everyday Attention* [12]. This task requires finding and circling an icon of a gas station on a city map within the time limit of two minutes. For the first minute, participant is asked to circle the icons with one colour (e.g. green), and for the second minute, participant is asked to circle the icons with different colour (e.g. red). There are a total of 80 gas station icons. The aim of this test is to measure the selective attention.
- *Reaction time test* [13]. This is a computer-based test, used to measure simple reaction time and the complex reaction time and consists of two tasks. In the first task, the simple reaction time was measured. The participants were presented with a picture of an athlete. In a randomized period of time, a black ball appeared by the athlete's left hand. The participants had to press a "STOP" button as soon as the ball appeared. The second task measured complex reaction time. In this task, participants were presented with the same picture of an athlete; however, this time the black ball would appear in either of two squares next to athlete's left hand. Depending from the square, participants had to react by pressing "E" or "U" on the keyboard.

To determine the physical activity of each participant, two questionnaires were used – International Physical Activity Questionnaire for Elderly (*IPAQ-E*) [14] and a modified version of *Social Determinants of Health Behaviours* [15]. IPAQ-E questionnaire consists of four questions regarding the physical activity in the last seven days – time spent sitting, time spent walking, time spent doing moderate physical activities and time spent doing vigorous physical activities. The data were measured in minutes and multiplied by a fixed coefficient. To identify the overall tendencies in each of the participant's lifestyle, a modified version of *Social Determinants of Health Behaviours* questionnaire was used.

This questionnaire includes questions regarding their physical activity over their course of life (e.g., what kind of aerobic exercises have you been involved during your life?), the regularity of the exercises and activities for the past years, months and weeks. The questionnaire also includes questions that reflect participants' habits of alcohol use, smoking, and diet.

3.3 Procedure

Data collection for each participant lasted for two days. On the first day, the assessment of executive functioning was conducted, on the second day the data on each participant's lifestyle was obtained. Ethical approval has been obtained from the Riga Stradiņš University Ethics Committee. Participation was voluntary and written consent was obtained from each of the participants prior to data collection.

3.4 Data analysis

SPSS version 21. (Statistical Package for the Social Sciences) was used to process the data. The data was analysed with Spearman correlation coefficient and ANOVA analysis with Post Hoc test (LSD).

4 Results

The mean results of the tests assessing executive functioning were calculated.

Table 3. Mean results of the cognitive assessment tests.

Test	Group	Mean	SD	ND
MoCA		25.3	2.48	.065
	Aerobic group	26.08	2.18	.114
	Strength training group	25	1.85	.769
	Sedentary group	24.56	3.24	.679
The Map Search test (Minute 1)		30.13	11.22	1
	Aerobic group	34.23	9.97	.999
	Strength training group	31.75	13.89	.907
	Sedentary group	22.78	6.92	.928
The Map Search Test (Minute 2)		23.73	7.63	.653
	Aerobic group	26.85	6.59	.777
	Strength training group	21.75	7.44	.873
	Sedentary group	21	8.32	.879
The Map Search Test (total)		53.87	15.06	.076
	Aerobic group	61.08	9.31	.192
	Strength training group	53.5	18.15	.361
	Sedentary group	43.78	14.32	.596
Simple reaction time		.417	.268	.011
	Aerobic group	.349	.05	.878
	Strength training group	.345	.09	.369
	Sedentary group	.580	.455	.209

Complex reaction time		.693	.280	.021
	Aerobic group	.609	.09	.874
	Strength training group	.692	.16	.363
	Sedentary group	.816	.471	.424
Cognitive efficiency		102.5	13.25	.767
	Aerobic group	110.85	9.45	.990
	Strength training group	98.75	15.68	.387
	Sedentary group	93.78	8.60	.880
Visual Matching test		101.47	12.71	.859
	Aerobic group	107	10.48	.885
	Strength training group	99	15.02	.893
	Sedentary group	95.67	11.38	.931
The Numbers Reversed test		102.10	11.50	.997
	Aerobic group	109.23	8.97	.938
	Strength training group	98.25	11.17	.956
	Sedentary group	95.22	9.96	.998

Note: SD = Standard Deviation, ND – Normal Distribution.

In all tests participants, involved in a life style with aerobic physical activities, showed higher results than participants with strength training or sedentary lifestyle. When compared with the means from the sedentary life style group, the group with strength training indicated higher results (see Table 3).

To determine whether there are significant differences across the three groups ANOVA with LSD correction was used. There were significant differences between the groups in cognitive efficiency (F = 6.791, p < 0.01) and working memory (F = 6.179, p < 0.01). Differences between groups were also found in the total score of the Map Search Test which was used to measure the inhibition of attention (F = 4.311, p < 0.01) (see Table 4).

Table 4. Differences across three groups of different physical activities.

Test	*F*	Sig.
MoCA	1.105	.346
The Map Search Test (total score)	**4.311**	**.024**
Simple reaction time	2.648	.089
Complex raction time	1.484	.245
Cognitive efficiency	**6.791**	**.004**
Visual Matching test	2.572	.095
The Numbers Reversed test	**6.179**	**.008**

To examine the relationship between the physical activity and executive function, MET minutes for walking, moderate activity and vigorous activity were collated with the results from tests measuring executive function, Spearman correlation coefficient was used. There was a statistically significant correlation only between the complex reaction time and MET minutes for walking (r_s = –.396, p < 0.05) and complex reaction time and MET minutes for vigorous physical activity (r_s = –.436, p < 0.05).

5 Discussion

The aim of this study was to examine the impact of different types of physical activities on executive functioning, focusing on inhibition of attention and working memory. The results supported our hypothesis that participants involved in aerobic physical exercise would show higher results than participants involved in strength training with aerobic activity elements in lifestyle, or participants with sedentary lifestyle. Participants involved in aerobic physical activities (such as walking, Nordic Walking, swimming, cycling etc.) did show higher results in cognitive testing, which complies with the evidence given by Enzo Iuliano [16] or Geoffroy Boucard [3].

As predicted, there were differences in working memory scores across all three groups. Working memory is a memory system that involves holding information and mentally manipulating it. It can be either verbal or visual-spatial and is essential in seeing connections, making sense of written or spoken language, doing maths, translating instructions into action plans, etc. [6], thus being one of the essential parts of executive functions. Working memory is associated with the grey matter volume in frontal-parietal circuit [17], which is one of the areas strongly impacted by ageing [2] and is also the area of brain, where an increase in the volume of grey and white brain matter is observed after an aerobic physical activity intervention [7, 8]. This indicates that the higher results in the aerobic activity group were consequential rather than circumstantial.

The concept of cognitive efficiency here is defined as the capacity of the cognitive system to process information automatically. This process includes processing speed and short-term memory [11]. Though there has been research on the impact of cognitive training on cognitive efficiency (e.g. [18]. However, research on processing speed indicates a relationship between physical activity and improved processing speed. For example, research indicates that higher fitness improves processing speed in memory tasks [7].

This study has several limitations. Firstly, the sample of the participants is still very small and the results are not statistically relevant and might change with the increase in sample. Secondly, we did not check for education, which might impact the results. Thirdly, we did not check for the longevity of the physical activities, as well as for the previous experience, focusing mainly on the current physical activity status.

6 Conclusions

The topic of impact of different physical activities on the executive functions has yet to be researched more thoroughly. However, the results of this study indicate that there are differences across the groups in such executive functions as working memory and processing speed, allowing us to conclude that the involvement in aerobic physical activities could in fact lessen the decline in executive functioning in the aging population. Results also indicated relationship between walking and processing speed and processing speed and regular involvement in vigorous aerobic activities, which also indicates benefits not only from regular vigorous aerobic activities, but also from as simple activity as a daily walk.

This study was *funded/partly funded by the Latvian National Research Programme Biomedicine for Public Health (BIOMEDICINE) 2014–2017.*

References

1 World Health Organization, "World report on Ageing And Health," 2015

2 S.J. Colcombe, K.I. Erickson, N. Raz, A.G. Webb, N.J. Cohen, E. Mcauley, and A.F. Kramer, "Aerobic Fitness Reduces Brain Tissue Loss in Aging Humans," **58**,

no. 2, pp. 176–180, 2003

3 G.K. Boucard, C.T. Albinet, A. Bugaiska, C.A. Bouquet, D. Clarys, and M. Audiffren, "Impact of Physical Activity on Executive Functions in Aging: A Selective Effect on Inhibition Among Old Adults," pp. 808–827, 2012

4 M.W. Voss, S. Heo, R.S. Prakash, K.I. Erickson, H. Alves, L. Chaddock, A.N. Szabo, E.L. Mailey, S.M. White, N. Gothe, T.R. Wo, E. Mcauley, B.P. Sutton, and A.F. Kramer, "The Influence of Aerobic Fitness on Cerebral White Matter Integrity and Cognitive Function in Older Adults: Results of a One-Year Exercise Intervention," **2985**, no. February 2012, pp. 2972–2985, 2013

5 L. Vaughan and K. Giovanello, "Executive function in daily life: Age-related influences of executive processes on instrumental activities of daily living.," *Psychol. Aging*, **25**, no. 2, pp. 343–355, Jun. 2010

6 A. Diamond, "Executive Functions," *Annu. Rev. Psychol.*, **64**, no. 1, pp. 135–168, Jan. 2013

7 K.I. Erickson, M.W. Voss, R. Shaurya, C. Basak, and A. Szabo, "Exercise training increases size of hippocampus and improves memory," **108**, no. 7, pp. 3017–3022, 2011

8 C. Benedict, S.J. Brooks, J. Kullberg, R. Nordenskjöld, J. Burgos, M. Le, L. Kilander, E. Larsson, L. Johansson, H. Ahlström, L. Lind, and H.B. Schiöth, "Association between physical activity and brain health in older adults," *NBA*, **34**, no. 1, pp. 83–90, 2013

9 V.R. Varma, Y.-F. Chuang, G.C. Harris, E.J. Tan, and M.C. Carlson, "Low-intensity daily walking activity is associated with hippocampal volume in older adults.," *Hippocampus*, **25**, no. 5, pp. 605–15, May 2015

10 Z.S. Nasreddine, N.A. Phillips, V. BÃ©dirian, S. Charbonneau, V. Whitehead, I. Collin, J.L. Cummings, and H. Chertkow, "The Montreal Cognitive Assessment, MoCA: A Brief Screening Tool For Mild Cognitive Impairment," *J. Am. Geriatr. Soc.*, **53**, no. 4, pp. 695–699, Apr. 2005

11 N. Woodcock, R.W. McGrew, K.S. Mather, *Woodcock-Johnson III Test Manual*, III. Itasca, IL: Riverside Publishing Company, 2001

12 I.A.N.H. Robertson, T. Ward, V. Ridgeway, and I.A.N. Nimmo-smith, "The structure of normal human attention: The Test of Everyday Attention," pp. 525–534, 1996

13 A. Molotanovs, "Sacensību darbības optimizēšana handbola vārtsargiem (uz HK LSPA komandas piemēra)," Latvijas Sporta pedagoģijas akadēmija, 2009

14 A. Hurtig-Wennlöf, M. Hagströmer, and L.A. Olsson, "The International Physical Activity Questionnaire modified for the elderly: aspects of validity and feasibility," *Public Health Nutr.*, **13**, no. 11, pp. 1847–1854, Nov. 2010

15 FINBALT, "Heath Behaviour among Latvian Adult Population, 2008," 2008

16 E. Iuliano, A. di Cagno, G. Aquino, G. Fiorilli, P. Mignogna, G. Calcagno, and A. Di Costanzo, "Effects of different types of physical activity on the cognitive functions and attention in older people: A randomized controlled study," *Exp. Gerontol.*, **70**, pp. 105–110, 2015

17 K.H. Karlsgodt, "Working Memory," in *Brain Mapping: An Encyclopedic Reference*, **3**, Elsevier Inc., 2015, pp. 319–326

18 V.B. Ignjatović, S. Kalabić, S. Batić, and M. Žikić, "Improvement of Cognitive Efficiency Through Cognitive Training in Healthy Subjects," *Acta Clin. Croat.*, **54**, no. 2, pp. 169–78, Jun. 2015

SHS Web of Conferences **40**, 03004 (2018) https://doi.org/10.1051/shsconf/20184003004
Int. Conf. SOCIETY. HEALTH. WELFARE. 2016

Accessibility problems in home environment for seniors in Latvia: Experience from the research project *innovAge* (WP2)

S. Tomsone, L. Saukuma, Z. Liepina, and J. Zalkalns

Rīga Stradiņš University, Riga, Latvia

Abstract. Studies on relations between health and environment indicate that accessibility positively influence healthy life expectancy, as accessible home environment and neighbourhood support mobility, daily activities and social participation in old age. One of the tasks in the EC 7th framework research project innovAge (2012–2015) aimed to develop an IT tool to support seniors' involvement in decisions about their housing options and help them to become active partners in choice and supply of housing.

This study aims to explore seniors' opinions about accessibility in their physical home environment based on analysis of data – interview records and notes – from two study stages in the research project innovAge (WP2). Participants, at the first stage – 8 seniors and at the second stage – 10 seniors, represented variety of living conditions, family situations and functional abilities. Participation in the study was voluntary and the participants had right to withdraw from the study at any stage.

Usually, barriers in home environment seniors consider late, when functional abilities decreases substantially. The most common places where environmental barriers appear are sanitary spaces, kitchen and entrance of the home. Despite the environmental barriers and functional limitations, seniors remain in their homes, mainly because in Latvia do not exist traditions to seek options for relocation, as well as there are no policies and services to support this. In few municipalities options exist to adapt home environment, but it refers only to cases of severe disability for persons using wheelchair. Usually, the need for environment adaptations for seniors is replaced by care services, thus meeting also the needs for seniors' socialization. Regarding possibilities to adapt home environment, strong prejudices exist (bath vs. shower, disadvantages of living on the first floor, etc.). Seniors admitted that they lack knowledge and experience on housing adaptations, but emphasized that the environmental accessibility in a wider context, including the neighbourhood and services, is essential.

The close cooperation with participants facilitated the researchers' understanding of the seniors' complex views and needs regarding their home environment and related issues, which are important for developing housing policies for senior citizens in Latvia. Despite the specified accessibility problems in home environment, the seniors evaluated usability of their homes generally high.

Key words: project innovAge, seniors, accessibility, housing.

1 Introduction

Populations are rapidly aging and more people are expected to live longer with disabilities and reduced functional capacity [1]. Such trends involve major societal challenges, e.g. increased demands for accessible housing that supports active and healthy ageing.

Social innovations are considered as one of solutions for global challenges risen by the rapidly ageing population. Social innovations defined as new ideas (products, services and models) that strive to address pressing current needs; they simultaneously meet social needs (more effectively than alternatives) and create new social relationships or collaborations [2]. The EC 7th framework research project *innovAge* (2012–2015) was dedicated to developing and testing, as well as surveying and cataloguing, social innovations that will have a solid impact on improving the quality of life and well-being of older people [3]. One of the tasks aimed to develop an ICT tool to support seniors' involvement in decisions about their housing options and help them to become active partners in choice and supply of housing.

The home is the most important living space of older people where most things in daily life happen [4]. Research on housing and health among very old people in Europe has shown that home and neighbourhood environments affect older people's daily activities and participation, and adequately designed housing affects independence positively [5, 6]. It is widely known that participation in daily activities gives meaning to people's lives and is considered a determinant of well-being and health [7]. Older people themselves stress the importance of having a functional and secure home in order to be as independent and autonomous as possible in daily activities and maintain participation [8].

Studies on relations between health and environment indicate that accessibility positively influence healthy life expectancy, as accessible home environment and neighbourhood support mobility, daily activities and social participation in old age. Accessibility can be defined as the relationship between a person's functional capacity and the prevalence of physical environmental barriers in the environment. According to this definition, accessibility is an aspect of person-environment fit (P-E fit) [9].

One well-established research-based instrument, available for professional assessments of housing accessibility, is the Housing Enabler (HE) [10, 11]. It is useful not only for research and education but also to support practitioners in producing reliable and valid assessments as a basis for interventions targeting housing accessibility problems, at individual and group levels [11]. The HE consists of a three-step assessment and analysis based on one checklist of functional capacity in the individual (personal component) and one checklist of environmental barriers (environmental component), followed by an analysis of P-E fit resulting in a quantitative measurement of accessibility problems [10]. Research using the HE has shown that environmental barriers are common in housing across Europe [12].

In Latvia there are not much studies conducted to explore the older people's needs regarding environment, especially their home environment. According to results from Project ENABLE-AGE (2002–2004) participants in Latvia had lower housing standard but presence of environmental barriers was quite similar with situation in Germany and Sweden [12]. Eurostat statistics on living conditions showed that in Latvia about 65% of the population lived in apartments in multi-family housing [13]. According to EU-SILC data [14], Latvia had Europe's highest overcrowding rate for housing at 58% of the population (the average for European countries is 18%). As to the situation of senior citizens in society, in the Active Ageing Index (AAI) across EU member states Latvia is among the countries at the bottom [15]. The aim of the present study was to explore seniors' opinions on accessibility in their physical home environment.

2 Materials and methods

The present study is based on Latvian results from one part (WP2) of the *innovAge* project, financed by the European Commission (2013–2015). The WP2 targeted people aging with functional limitations to raise their awareness about accessibility issues in their current dwellings and in available rental apartments within the ordinary housing stock. The ambition was to develop the ICT tool based on methodology of the Housing Enabler (HE) that would support decision-making and to make it possible for users, older people, to get an overview and make predictions about housing accessibility in relation to their present or projected functional capacity. Participants from Latvia were involved in research circles [16] in order to generate knowledge of how senior citizens express their own needs and expectations regarding housing options, and to establish the users' requirements for a housing accessibility ICT tool [17] as well as in usability testing of ICT tool prototype.

The common sampling procedures and strategy for data collection were used in all countries (Sweden, Germany, Italy and Latvia) participating in WP2 of project *innovAge* and they are described elsewhere [17, 18] and Table 1 shows details for both data collection stages in Latvia. The first step of the project in Latvia involved 8 senior participants, and the second stage – 10 seniors, they represented variety of living conditions, family situations and functional abilities [17, 18]. Participation in the study was voluntary and the participants had right to withdraw from the study at any stage.

Interview records and notes from both study stages in Latvia were analyzed for present study. Primary data were reviewed by authors separately and information units related to accessibility aspects were extracted. Extracted information was discussed by the authors and organized into three main topics – information, recognition of problems and solutions.

Table 1. Information on two stages of data collection in Latvia (research project *innovAge*, WP2).

	Research circle	**IT tool test**
Research sites	in Sweden, Latvia, Germany and Italy	in Sweden and Latvia
Participants in Latvia (*N*)	8	10
Research team	Researcher and assistant	Researcher and assistant
Other participants	Experts	–
Structure and organization	Guidelines	Guidelines
Data	Audio records, transcripts, researchers notes	Video/audio records, questionnaires, researchers notes

3 Results and discussion

For seniors in Latvia accessibility problems in home environment were related to information about accessibility, recognition of problems and possibilities to solve them.

3.1 Accessible housing-information

Participants stressed lack of information on accessibility problems (potential hazards, long-term consequences of more or less accessible environment) and their potential impact on everyday life. They considered that more knowledge of impact that different environmental barriers can have on individuals is important to raise awareness about one's own environment and to facilitate ability to initiate changes in order to make housing more accessible. Seniors perceived that it was difficult to find suitable information on accessible environments. They emphasized that the environmental accessibility is essential in a wider context – including the neighbourhood and services (e.g. public transport).

3.2 Accessible housing – recognition of problems

The general belief was that individuals do not become aware of environmental influences until something happens to them affecting their day-to-day life, when functional abilities decrease substantially and supportive environment becomes more important. The specific features of home design were well recognized; e.g. Soviet time standards for housing typically means small living space, narrow doors and passages which are difficult to adapt. Most often accessibility problems in home environment were recognised in hygiene facilities (only bath instead of shower, lack of handles for support and lack of other assistive devices useful for bathing), kitchen (narrow spaces for moving around, high shelves) and entrance in house (steps in front of doors, poor condition of walking surfaces, lack of handrails near stairs). Technical problems in communal areas of houses also were mentioned, e.g. elevators do not work properly or lightning is poor.

3.3 Accessible housing – solutions

Despite the environmental barriers and increasing functional limitations, seniors remain in their homes, mainly because there do not exist traditions to seek options for relocation in Latvia as well as there are no policies and services to support such relocation. Seniors are dealing with very limited private financial resources and therefore issue of high costs was brought up associated with refurbishment, housing adaptations, or rents in newly built accessible housing. The participants viewed themselves as badly informed on existing laws and regulations regarding housing adaptation or on whom to contact in order to receive information on accessible housing. In few municipalities in Latvia exist options to support adaptations at home but it implies only in case of severe disability for persons using wheelchair (only three out of the 118 municipalities in the country provide housing adaptation grants). Usually, the need for environment adaptations for seniors is replaced by care services (e.g. care assistant), thus meeting also the needs for senior's socialization. Regarding possibilities to adapt home environment exist strong prejudices (bath vs. shower – use less water and save expenses on water, disadvantages of living on the first floor, etc.). Personal creativity was mentioned as a possible recourse to solve problems arising from inappropriate environment, e.g. refurnishing in order to get more space or to get better support to move around.

4 Conclusions

The close cooperation with older people facilitated the researchers' understanding of the seniors' complex views and needs regarding their home environment and related issues, which are important for developing housing policies for senior citizens in Latvia. Despite the specified accessibility problems in home environment, the seniors generally evaluated usability of their homes as high.

The *innovAge* project was part of Social Innovations Promoting Active and Healthy Ageing, HEALTH.2012.3.2-3 No. 306058 within the 7th EU framework programme. The authors would like to thank all project participants, consortium and team members for their contribution. The authors are specifically grateful to WP 2 leading team from Lund University (Sweden) – Prof. Susanne Iwarsson, Prof. Torbjörn Svensson, Björn Slaug, Maria Haak, Thomas Ladö, Steven M. Schmidt, Oskar Jonsson and Knut Mårtensson.

References

1. A.B. Newman, J.A. Cauley, (Eds.) The epidemiology of ageing. Dordrecht, Germany: Springer, 2012
2. R. Murray, J. Caulier-Grice, G. Mulgan. The Open book of Social Innovations. The Young foundation and NESTA: London, UK, 2010
3. About project *innovAge*. Retrieved May 2016 from `http://www.innovage.group.shef.ac.uk/about-innovage.html`
4. R. Chilvers, S. Corr, H. Singlehurst. Investigation into the occupational lives of healthy older people through their use of time. Australian Occupational Therapy Journal, 2010, (57) 24–33
5. F. Oswald, H.-W. Wahl, O. Schilling, C. Nygren, A. Fänge, A. Sixsmith, J. Sixsmith, Z. Széman, S. Tomsone, S. Iwarsson. Relationships between housing and healthy ageing in very old age. Gerontologist, 2007, **47**(1), 96–107
6. H.-W. Wahl, A. Fänge, F. Oswald, L.N. Gitlin, S. Iwarsson. The home environment and disability-related outcomes in ageing individuals: What is the empirical evidence? Gerontologist, 2009, **49**(3), 355–367
7. E. Townsend, H. Polatajko (Eds.). Enabling occupation II: Advancing and occupational therapy vision for health, well-being & justice through occupation. Ottawa: Canadian Association of Occupational Therapists, 2013
8. M. Haak, A. Malmgren Fänge, S. Iwarsson, S. Dahlin-Ivanoff. The importance of successful place integration for perceived health in very old age: a qualitative meta-synthesis. Int J Public Health, 2011, **56**(6), 589–595
9. S. Iwarsson, A. Ståhl. Accessibility, usability and universal design – Positioning and definition of concepts describing person-environment relationships. Disabil Rehabil 2003; **25**, 57–66
10. S. Iwarsson, B. Slaug. Housing Enabler – A method for rating/screening and analysing accessibility problems in housing. Manual for the the complete instrument and screening tool, 2nd rev ed.; Veten & Skapen HB och Slaug EnablingDevelopment:Lund & Staffanstorp, Sweden, 2010
11. S. Iwarsson, M. Haak, B. Slaug. Current developments of the Housing Enabler methodology. Br. J. Occup. Ther., 2012, **75**(11), 517–521
12. S. Iwarsson, C. Nygren, F. Oswald, H.-W. Wahl, S. Tomsone. Environmental barriers and housing accessibility problems over a one-year period in later life in three European countries. J. Hous. Elderly, 2006, **20**(3), 23–43
13. Living conditions in Europe (2014). Eurostat statistical books. Retrieved May 2016 from `http://ec.europa.eu/eurostat/documents/3217494/6303711/KS-DZ-14-001-EN-N.pdf/d867b24b-da98-427d-bca2-d8bc212ff7a8`
14. EU-SILC data (2012). Retrieved May 2016 from `http://ec.europa.eu/eurostat/web/income-and-living-conditions/data/main-tables`
15. A. Zaidi. Active ageing index: A legacy of the European year 2012 for active ageing and solidarity between generations. 2015. Retrieved May 2016 from `http://www.euro.centre.org/detail.php?xml_id=2489`
16. M. Kylberg, M. Haak, A. Ståhl, E. Skogh, S. Iwarsson. Research brief #5: User participation. Research with and about user participation. Stockholm, Sweden: Forte [Swedish Research Council for Health, Working Life and Welfare], 2015
17. M. Haak, B. Slaug, F. Oswald, S. Schmidt, J.M. Rimland, S. Tomsone, T. Ladö, T. Svensson, S. Iwarsson. Cross-national user priorities for housing provision and

accessibility – Findings from the European innovAge project. International Journal of Environmental Research and Public Health, 2014, **12**(3), 2670–2686

18. O. Jonsson, M. Haak, S.Tomsone, S. Iwarsson, S.M. Schmidt, K. Mårtensson, T. Svensson, B. Slaug. Cross-national Usability Testing of a Housing Accessibility App: Findings from the European innovAge project. Journal of Usability Studies, 2016, **12**(1), 26–49

Comparison of satisfaction with happiness in families with children between the Czech–Latvian people

L. Stasova[1] and L. Vilka[2]

[1]University of Hradec Kralove, Czech Republic
[2]Riga Stradiņs University, Latvia

Abstract. It is an assumed opinion that the family is the main unit of society. The society and the family are in reciprocal influence: the healthier the families are, the healthier the society and vice versa. Subjective wellbeing and satisfaction with life are important factors in family stability. Feeling of happiness and the image of a happy family in contemporary Czech and Latvian society are the focus of the research. In the research the data from an international survey on the lifestyle of contemporary families that have been collected during 2013–2014 in the Czech Republic and Latvia were used. A self-administered questionnaire data were collected in both countries as well. A mixed approach with emphasis on the quantitative approach to research was used in data analysis. The results showed a significant similarity, although the "feeling of happiness" differed and varied in each specific family. So it was possible to use the same categories of "a happy family" that had been found in both samples of data (Czech and Latvian) material, and they were: financial prosperity, good relationships and success in the family, cohesion within the family, doing things and spending leisure time together and a healthy family.

The research results allow us to conclude that the family has a high value in people's minds despite the rapid changes in the family structure and de facto high rate of divorce in both countries.

Key words: family, happiness, happy family, lifestyle.

1 Introduction

Family is a unique entity in a society and a basic unit with the reciprocal impact on many processes in a societal development dynamics. Due to the reciprocity effect family studies are important for economic and welfare policy strategies in the country.

In this research we use the concept of family as a relatively stable social group with adults dealing with household and caring for children.

Advantage of having families can be seen not only on the societal level but also in everyday life of its members. Everyday life we understand as consisting of provision of daily needs, routine work activities and leisure time, inner family self-organization and in social relationships within the environment.

Family capacity to deal with routine, solve problems and difficult situations indicates the stability of the family as the social institution.

One of the important and strengthening moments in family life is a *feeling of happiness* and *satisfaction* which brings a feedback to family members and improves their cohesion and activity.

Happiness is a multi-sided subjective category. Feeling of happiness depends on the idea about the source of happiness and different contexts: material wellbeing, emotional wellbeing, self-fulfillment and career, connectedness and ties with friends, healthy family, etc.

In particular, happiness questions are increasingly used in population surveys and it is suggested that a new science of happiness is emerging which aims to measure happiness, identify the main factors affecting it and quantify their relative importance [1].

In the Czech Republic, research related to the life satisfaction has been focused mainly on the psychological aspects of happiness (e. g. Kožený, Csémy, Tišanská, 2007 [2], Sobotková, Reiterová, Hurníková, 2011 [3]. and links between the family status and family environment and happiness (e.g. Hamplová, 2006 [4]). In the analysis of 21 European countries Hamplová tested assumption regarding family status and happiness in life. Different previous surveys had declared that married people live longer and have a lower probability of death, suffer less from mental diseases, they are more healthy, happy and satisfied in their lives. Hamplová confirmed these hypothesis by the European data [5][1] positive influence of marriage and partnership can be explained not only by a better health status and financial situation of families in comparison with people living alone, but with high probability, it could be supported by better understanding, support, help and stimulation in the family environment.

Researcher's interest on subjective wellbeing of the inhabitants has raised also in Latvia during the last ten years. Studies on life satisfaction and feeling of happiness mainly have been done in the context of welfare studies from different perspectives. But two studies deserve special attention as they are focused on this specific field: *The structure of Latgale inhabitants feeling of happiness and content of psychological well-being.* The thesis for obtaining Doctoral degree in Psychology (Eriks Kalvāns (2013)) and *Comparative analysis of subjective well-being of Latvia's inhabitants in the context of economic development of the Baltic States* (Jelena Lonska (2015)).

To be happy in a family brings for the family members strength to overcome troubles in their jobs, in unemployment, in illness, and different types of losses.

What prerequisites people need to be happy? What does it mean to be happy in a family? How the idea of happiness in people's mind is constructed? What components create a satisfaction in family life? How happy are families with children today?

These questions we try to follow in this paper. We used the data from an international survey on the life style of contemporary families that have been collected during 2013–2014 in Czech Republic and Latvia.

A self-administered questionnaire was translated from Czech to Latvian and data were collected in both countries. Our questionnaires contained items related to specific areas of life style. This text is only concerned with data regarding perceived happiness and satisfaction within families. The data came from what the respondents declared, and are therefore as reliable as the honesty of their answers. It should be also declared that the main idea of our study was not focused on the welfare and happiness in family but followed a life style of family as a whole concept. From this reason our questions related to happiness could create only a small part of our questionnaire and could not be studied in their broader context.

---

[1]Unfortunatelly Latvia has not been included in this study.

Despite complications related to the methodology used, and the sample obtained in both countries, we try to compare basic trends and show aspects of the life styles of contemporary families in a new light.

2 Family in the Czech Republic and Latvia

For starting this comparative research we find it useful to get a general view on the happiness as the study object in both countries. As the relative point of reference for measuring the feeling of happiness can be used the ranking of happiness in The World Happiness Report (2016) [6].

Ranking of Happiness 2013–2015 (Part 1) shows quite big interval between Czech Republic and the Baltic States: placing Czech Republic (6.596) in the 27th position following Lithuania (5.813) in the 60th position, ***Latvia (5.560) in the 68th position*** and Estonia (5.517) in 72nd position [6].

The feeling of happiness is not static. During the period from 2005–2007 to 2013–2015, more rapid dynamics in changes of happiness was observed in Latvia (0.872) placing it in the 5th position into the ranking of the changes. While in Czech Republic dynamics in the changes of happiness were slowlier – Czech Republic (0.126), 55th position in the ranking of changes.

The differences can be explained by the transformations in the context of life. During this period Latvian people felt themselves particularly inspiring and enthusiastic, because Latvia in 2004 became the member state of EU.

European families differ in many ways, although some trends overcome boarders and could be found in every country at the different level.

Current trends among Czech families are demonstrated by a number of specific features [7]. *Firstly*, Czech families tend to have lower stability, which is expressed by a high divorce rate. Statistical data support this long-lasting trend. In recent years, the divorce rate has oscillated at about 50%, which means that half of all marriages fail.

Contemporary Czech spouses can finish their relationships relatively easily, especially when there are no dependent children in the family, and tend to solve interpersonal problems by leaving the partnership. Moreover, the lower stability of partnerships and families has a big impact on children.

More than half of them experience the divorce of their parents (in 56.9% of divorces there were dependent children), and the majority live with a new partner of their mother/father.

Secondly, Czech families are less and less based on formal marriage. Young people perceive that marriage is not necessary for a happy partnership, and that marriage can ever destroy a healthy relationship. These trends are documented in statistical evidence which shows a decrease in the marriage rate and an increase in the number of cohabitating couples. According to the current level of first marriage rates, only slightly more than 50% of people enter marriage during their life. In relation to both trends mentioned above, a growing number of Czech children are born out of marriage (47.3%), but this does not mean that they do not live in a two-parent family.

The age of parenthood in the Czech Republic is increasing, and now most mothers give birth at the age of 30, the average is 29.9.

Generally, contemporary parents wait for a stable professional position and a convenient economic level before having children. Young people declare that good housing and salaries are important preconditions for creating a family. However, family/ work competition often continues much further, because Czech families need two working adults who secure the economic stability of the family. This pushes many mothers to spend a lot of time at work even if they have small children.

Part-time jobs in Czech are unusual and the labour market does not offer enough possibilities for contemporary parents to balance their families and careers. Generally, this means less time for family activities, children and partners. It is possible to observe growing difference among families related to their financial situation, material wealth and lifestyle.

Czech families are also based on liberal relationships, primarily at the parents/child level. In relation to the decreasing number of children in families (the average is 1.4 children per mother), children's position in the family has increased, giving them more opportunities for discussion and negotiation, which can influence what the family does.

Latvian families were represented by the number 588 000 according to the census in 2011. 54% of them consisted of married couples with children, 29% were single mother families, 4% are single father families and 13% are unmarried couple families. The most represented group of families – married couple families – had one child 40%, 2 children in 22% and 3+ children – 6%. The cohabited families had in 35% one child, 2 children in 18% and 3+ children in 7%; Single mother families had a structure – family with 1 child – 75%, 2 children – 20%, 3+ children – 5%; Single father families have 1 child – 82%, 2 children – 16%, 3+ children – 2% [8].

In 2011 in Latvia were fixed 10 760 marriages and 8302 marriages were divorced. Crude marriage and divorce rates were 5.2 and 4.0 (per 1000 inhabitants).

In 2011 in Latvia 18 825 babies were born: it is 9.1 per 1000 inhabitants. The highest birth rates are for women at the age 25–34 – 58%. 434 babies were twins in the families [9].

Family situation in Latvia has strongly been influenced by financial–economic crisis.

It should be noted that at the beginning of 2015 the population declined to 1.99 million [CSP, Demography 2015, 21] and, accordingly, calculations were made, with the assumption that the average family size remained the same as in the 2011 Population Census, it was concluded that at the beginning of the year 2015 the number of families has decreased to 565 thousand, or 10% compared to 2000 [10].

3 Life style of family and happiness in family life

3.1 Methodology of a survey

Joint research – life style of the contemporary family

Our questions were aimed at:

- What is an image of happiness in family?
- What creates a feeling of being happy with family life?
- Satisfaction in family life related to the free time activities?
- What are the differences between Czech and Latvian families?

We used a mixed approach with the stress on the quantitative one. We prepared a questionnaire with some open questions for a qualitative analysis. Our samples in both countries were created by families with children in the age 0–17.

The Czech sample was composed of 1308 families. More than half of them (51.8%) had two children, about 28% had one child, 11.6% had three children, and slightly more than 3% had four or more children or they did not have children. The age of parents ranged from 20 to 56, and the age differences between spouses were not enormous. Adult occupations varied from unskilled, low-paid positions to highly specialized, well-paid positions. 24.4–25.7% of our respondents (male and female) had a university degree.

The Latvian sample was composed of 383 families from different regions and local communities. 32% of them have 1 child, 42% 2 children, 15% 3 children, 7% 4 children,

3% 5 children, 1% 6 children. The age of family members who represented family ranged from 28 to 74.

The old age of respondents is explained with the high number of emigrated families in hope to find a better paid job abroad, who leave their children with their grandparents or other relatives.

There are families who after being abroad take their children abroad as well.

Adult occupation varied from unskilled, low-paid positions to highly specialized ones. 16.5% of our male respondents and 15.5% of females had a university degree. Data were collected from 21 region centers and towns in Latvia, excluding Riga, the capital of Latvia. The women education level in Latvia comparing with men education level is higher.

Table 1. The data collected in 2014.

	Czech Republic	**Latvia**
Number of interviewed families	1308	383
No. of children – 1	28.9%	32.5%
No. of children – 2	55.0%	41.8%
No. of children – 3	12.3%	15.2%
No. of children – 4+	3.8%	10.5%

3.1.1 What is your image of a happy family?

In *the first group of answers* in our questionnaire, there were some open items used that were completed by the respondents in their own expressions.

One of these items was a description – *what is your image of a happy family?* Afterwards, we used the qualitative analysis and identified four groups of answers. The analysis was made separately in the Czech Republic and Latvia.

Subsequently, we put our results together and realized that our results showed a significant similarity, in spite of the *"feeling of happiness"* differs and vary in concrete family. Because of that we could use the same categories that had been found in both samples of data (Czech and Latvian)

Categories of "a happy family"

1. ***material and financial prosperity***
2. ***good relationships and success in the family***
3. ***family cohesion, doing things and spending of leisure time together***
4. ***healthy family***

These categories in many correspond also to some authors' viewpoint on prerequisites' of happiness:

"A happy life involves prerequisites, particularly external prosperity. Aristotle mentions, for example, health and food (McKeon, 1947; Ross, 1925). We would like to add that fundamental to health, in modern Western living, is access to medical care, if and when needed, sanitation, transportation, and education, among other factors."; Money is not an end as such, but rather a tool with which a happy life can be enhanced (McKeon, 1947; Ross, 1925; Tabensky, 1998) Sharing pleasures can be regarded as an initial path to happiness; stable relations seem to contribute, in general, to feelings of happiness in spouses/co-habiters and children (Hetherington & Koch, 2002; Paul, 2002) [11].

In the answers of respondents in both research samples we could find a strong category that was fed by ***such expressions like – happy family needs to have*** enough money, good housing, good job of parents, assurance of employment, no debts, etc.;

Respondents declared in their statements, that for them it is important to have money, good housing, get a good job and have a stability of employment to feel the happiness in family life.

These conditions, mostly from the external environment, showed how important is for family members the family policy supporting basic needs of different family types. These aspects can be guaranteed by the state and its policy.

Second group of answer surrounded **family relationships** and included feeling of love, friendship, tolerance, reliability, respect, good communication, trust and responsibility. These internal aspects show that processes within the family are very important to be happy in the family. People expect to find love, friendship and reliability in the family environment and to have a **safe space** for their everyday life that they can rely on.

Third category of answers has been saturated with expressions of **cohesion, doing things together and spending leisure time of family together.** Being together as a group, feeling of collective spirit and belonging to it has inspired some respondents to find their happiness in these issues. This category could be understood as an active dimension of family life and happiness in it.

Our respondents declared that they feel happy in the family when they can do things together with other family members. This "active group belonging" supports positive relationships and creates a cohesive climate in families.

The last category in our data has showed that **health of family** brings another dimension to happiness. Not to feel sickness, not to face sickness of any member of the family means an important determinant in possibility to feel happy in family. This factors lies mostly out of the decision of family members as an external factor. Social and health policy can only support a development of different preventive activities to avoid the most common diseases and civilization illnesses.

What we could add to this categorization is, that in some cases respondents did not answer this question or declared that happy family was their family.

After this qualitative analysis in both samples that has revealed most important pillars of happiness and satisfaction in family we came to the quantitative analysis of our data. From this point of view, for Czech respondents, happiness was mostly linked to a health of family (51%), harmonic relationship (36%) and financial prosperity (43%).

Table 2. Happiness and satisfaction.

	Czech Republic (%)	**Latvia (%)**
Health (category 4)	51.2	45.6
Good living (category 1)	7.2	3.2
Job, employment (category 1)	11.6	15.6
Financial prosperity, without debts (category 1)	43.4	21.2
Leisure time spent together (category 3)	22.7	13.8
Tolerance, help, support (category 2)	18.2	9.1
Harmonic relationships, emotional satisfaction, no conflicts and tensions (category 3)	36.5	56.8
Succesful children, education of children (category 1, 2)	6.4	4.8
Complete family – both parents and children (category 2, 3)	8.1	16.2

Like my family	4.2	0.6
No image of a happy family	1.6	0.3
Other answers	1.4	0.8

3.1.2 Satisfaction with some aspects of spending of leisure time

We have also asked our respondents, how much they are satisfied with their leisure time and family activities in leisure time.

Our data in both countries have showed that families were mostly satisfied with the quality of their leisure time during holidays and free time. Quantity of free time has been sufficient during weekends.

Table 3. Satisfaction of spending of leisure time (*1 – unsatisfied, 6 – very satisfied*).

	Czech Republic		**Latvia**	
	Mean	**Std. deviaton**	**Mean**	**Std. deviation**
With the quantity of free time after work	4.03	1.514	4.35	1.428
With the quantity of free time during the weekend	4.84	1.393	4.92	1.346
With the length of yearly leave	4.64	1.587	4.92	1.475
With the quality of rest which I get during my holiday	5.13	1.313	4.88	1.300
With the quality of rest I have in my free time	4.78	1.276	4.99	1.188
With the quality of free time which I have for my hobbies	4.06	1.534	4.80	1.449
With the quantity of free time which I can be with my loved ones	4.40	1.451	4.72	1.581
With diversity of my free time	4.40	1.456	4.74	1.372

We tried to ask about differencies between both countries in the leisure time happines and satisfaction. We used a parametric and non-parametric test (*t*-test, Mann–Whitney test) to test the significance of differencies between both countries. Data of these tests showed a significant differencies between both populations in their satisfaction related to free time.

Furthermore, we could observe a tendency of people in Latvia to be more satisfied with their leisure time in almost all items we asked about at the significant level (chi square sig. lower than 0.05).

4 Conclusions

We tried to study happines in family from different points of view. *Firstly* we focused on what is an image of a happy family for people living in families with children. *Secondly* we analysed questions related to the happines with activities in leisure time and spending of leisure time. We used a qualitative approach for *the first* topic and discovered four main categories which contributed to the feeling of being happy in family. In relation to the social and family policy, we tend to be persuaded about possibilites to support happy family life by different social means and tools.

Table 4. Differencies in satisfaction with free time in Czech and Latvian families.

	T-test	Mann–Whitney
	Sig.	**Sig.**
With the quantity of free time after work	0.000	0.000
With the length of yearly leave	0.003	0.002
With the quality of rest which I get during my holiday	0.013	0.013
With the quality of rest which I get during my free time	0.001	0.001
With the quantity of free time which I have for my hobbies	0.000	0.000
With the quantity of free time which I can be with my loved ones	0.000	0.000
With diversity of my free time	0.000	0.000

Second part, devoted to satisfaction with leisure time, showed a slight difference between both countries. Latvian respondents and families declared a higher satisfaction with their leisure time activities, but we have not found any important variable that could explain these results. It could be an inspiring topic for the next research.

As Latvians indicated more satisfaction with quantity and quality of free time in comparison with Czech people, after a time distance we compared our results with the results from the research done by the "Baltic International Bank Latvian Barometer" in October 2016.

Within the framework of the research respondents were asked to evaluate their satisfaction with life, to name things that give them a sense of happiness, and to describe their approach to setting goals of life. According to the results of the research, respondents ***more often evaluate their lives positively than critically***. By describing aspects that could make them more satisfied with life, the most commonly reported participants were ***better health and more opportunity to earn***. In between, things that are more enjoyable in life are often called ***spending time with the family** (67%)*. Also, in the answer to the question what they would most like to achieve, ***a happy family was most often mentioned*** [12].

From the research results it can be concluded, that *family* deserves a high value in people's mind despite the rapid changes in the family structure and high rate of divorce in *de facto* in both countries.

References

1. Dimitris Ballas1 and Mark Tranmer2. (2012). Happy People or Happy Places? A Multilevel Modeling Approach to the Analysis of Happiness and Well-Being. International Regional Science Review **35**(1) 70–102 http://journals.sagepub.com.db.rsu.lv/doi/pdf/10.1177/0160017611403737
2. Kožený, J., Csémy, L., Tišanská, L. Strukturální analýza modelu životní spokojenosti adolescentů. Československá psychologie, 2007, **51**, 3, p. 224–237

3. Sobotková, I., Reiterová, E., Hurníková, K. Rozdíly mezi otci a matkami v pohledu na fungování rodiny, v životní spokojenosti a rovnováze mezi prací a rodinou. *Československá psychologie*, 2011, **55**, 2, p. 139–151
4. Hamplová, D. Životní spokojenost, štěstí a rodinný stav v 21 evropských zemích (Life satisfaction, happiness and family status in the 21 European countries). *Czech Sociological Review*, 2006, **42**, 1, p. 35–55
5. Hamplová, D. Životní spokojenost, štěstí a rodinný stav v 21 evropských zemích (Life satisfaction, happiness and family status in the 21 European countries). *Czech Sociological Review*, 2006, **42**, 1, p. 43–52
6. World Happiness Report 2016. http://worldhappiness.report/wp-content/uploads/sites/2/2016/03/HR-V1Ch2_web.pdf
7. Czech Demographic Handbook 2015. https://www.czso.cz/csu/czso/czech-demographic-handbook-2015
8. Share of couples living in consensual union is growing. http://www.csb.gov.lv/notikumi/pieaug-neregistreta-kopdzive-dzivojosu-paru-gimenu-ipatsvars-39126.html#_ftn1
9. Marriages and divorces. http://www.csb.gov.lv/statistikas-temas/metodologija/noslegtas-un-skirtas-laulibas-36799.html?order=name&sort=asc
10. Līga Āboliņa Ģimene un tās atbalsta politikas attīstība Latvijā (1990–2015) Promocijas darbs., Rīga, 2016.42.lpp.
11. Cited from Lars-Göran Sund, Kosmas X. Smyrnios. Striving for Happiness and Its Impact on Family Stability: An Exploration of the Aristotelian Conception of Happiness. http://journals.sagepub.com.db.rsu.lv/doi/pdf/10.1111/j.1741-6248.2005.00038.x
12. Laime un dzīves mērķi. Baltic International Bank Latvijas barometrs (2016). https://www.bib.eu/uploads/2017/02/2016.11.Laime_dzives_merki_Baltic_International_Bank_Latvijas_barometrs.pdf

Dynamics of psychological safety in mothers raising children with special needs

I. Atamanova*, N. Kozlova, I. Glebova, E. Meshcheryakova, T. Levitskaya, and D. Zueva

National Research Tomsk State University, Tomsk, Russia

Abstract. There is a considerable decrease in psychological well-being among mothers raising children with special needs. Addressing the issues of providing psychological support for them is of particular importance for both researchers and practitioners. The paper presents a study aimed at exploring the dynamics of psychological safety in mothers raising children with special needs. 32 mothers whose children underwent a rehabilitation programme in a rehabilitation centre participated in the study. The key element of the psychological support programme developed for these mothers was a specific psychotherapeutic space combining hippotherapy with environmental, social and personal factors. Compared to the control group, the study participants' parameters of psychological safety showed statistically significant positive dynamics. The results obtained contribute to studying factors of psychological well-being in mothers raising children with special needs and suggest an effective way of enhancing their sense of psychological safety.

1 Introduction

The World Health Organization (WHO) reports that the number of featured children (those with autism, cerebral palsy, Down syndrome, cognitive impairments, etc.) is increasing every year globally [1]. The same tendency has been observed in Russia; e.g., statistics (2015) on cerebral palsy in children reported an increment in the number of new cerebral palsy cases diagnosed annually from 6854 in 2013 to 7734 in 2014 [2]. In turn, children with special needs have a transformative effect on their families as a whole and each family member, especially mothers [3–6]. Recent studies have shown that there is a considerable decrease in psychological well-being among mothers raising children with special needs [7–11]. The researchers point out that mothers of children with special needs, regardless of the nature of abnormal development, have some similar experience. They are often depressed, feeling guilty due to the birth of a disabled child and experiencing social exclusion, and their excessive involvement in raising such a child mostly leads to frustration of their basic needs. Depression, intrapersonal conflicts and neurotic reactions, distortion of parental attitudes and mother-child relationships – all these testify to a violation of adaptation to their life event connected with the birth of a child with special needs. The vast majority of these mothers are in a dire need for qualified psychological support. Therefore, addressing the issues of providing psychological support for them to

* Corresponding author: iatamanova@yandex.ru

enhance their psychological well-being is of particular importance for both researchers and practitioners.

The paper presents a study aimed at exploring the dynamics of psychological safety in mothers raising children with special needs. Being a complex phenomenon, psychological safety plays a crucial role in helping people overcome difficulties in their lives and can be viewed as an indicative factor of one's psychological well-being [12–14]. According to Eksakusto and Lyz [12], Baeva [13] and others, psychological safety of a person is interpreted as an ability to keep his/her own stability in the environment with certain parameters, including stressful events and traumatic impacts, and to resist various destructive internal and external influences. This manifests itself in a feeling of inner security or insecurity in a particular situation and an ability to recognize, anticipate and overcome life difficulties. In other words, psychological safety can be regarded not only as that of being threatened under stress, but also as a personal resource enabling people to cope with stressful events.

Based on Bogomaz and co-workers' viewing of psychological safety (see, for example, [14]) and theoretical considerations by Lyz [12], it is supposed to be an integrated personal characteristic including a number of parameters. In the case of mothers bringing up children with special needs, evaluating their psychological safety requires taking into account their basic beliefs, meaningfulness in life and hardiness as well as their level of personal and situational anxiety and their emotional state.

The present study was organized as follows. Mothers whose children with special needs underwent a rehabilitation programme in a rehabilitation centre agreed to participate in a programme of psychological support. The study participants' psychological safety parameters were evaluated before and after implementing the programme to reveal their dynamics. The key element of the programme developed for such mothers was a specific psychotherapeutic space combining hippotherapy with environmental, social and personal factors. It should be noted that the idea of psychotherapeutic space is connected with Alekseičik's and Kočiūnas' views [15] on interaction between the therapist and the client in a specifically created space contributing to solving psychotherapeutic goals and objectives. A combination of such factors as being outdoors, horse riding, taking care of horses and feeding them, the participants describing their actual experience, individual and group work with riding instructors – all these were included in the programme to create this specific psychotherapeutic space.

2 Research materials and methods

The main objective of the study presented was to examine the dynamics of psychological safety in mothers raising children with special needs as an effect of their participation in a specifically developed programme of psychological support for them.

2.1 Study design

The study was designed as follows. Mothers whose children with special needs underwent a rehabilitation programme in a rehabilitation centre were invited to participate in a programme of psychological support. The study participants' psychological safety parameters were evaluated at the beginning and at the end of the programme to reveal their dynamics. The key element of the programme developed for such mothers was a specific psychotherapeutic space combining hippotherapy with environmental, social and personal factors, namely being outdoors, horse riding, taking care of horses and feeding them, as well as individual and group work with riding instructors and giving feedback about actual experience. The study conducted involved four steps: (1) *psychodiagnostic I*, i.e. evaluating the study participants' psychological safety parameters and collecting their demographic data; (2) *practical*, i.e. implementing the programme with one-and-a-half hour sessions once a week for three months; (3) *psychodiagnostic II*, i.e. re-evaluating the study participants' psychological safety parameters; (4) *reflective*, i.e. providing feedback by the study participants six months after their participation in the programme.

2.2 Study sample

32 mothers, aged from 25 to 45, participated in the study. Middle-aged women from 30 to 40 years old comprised 75% of the study sample. Women under 30 and over 40 were presented in the sample in equal amounts of 12.5%. The mean age of the study participants was 35.3.

Their children with special needs, aged from 3 to 11, underwent a rehabilitation programme in a rehabilitation centre. The most common diagnosis in the children whose mothers took part in the study was autism spectrum disorder (47%), the second one was cerebral palsy (28%). Among others were the following: speech delay and mental developmental delay (10%), motor alalia (9%), Down syndrome (3%) and mental retardation or intellectual disability (3%).

The total sample was divided into two groups: experimental and control. The experimental group participated in the developed programme of psychological support for mothers raising children with special needs.

2.3 Research methods

The psychodiagnostic package aimed at evaluating the study participants' psychological safety parameters involved the following scales and research tools:

1) The World Assumptions Scale by Janoff-Bulman [16] modified and psychometrically checked by Padun and Kotel'nikova [17]. This scale is applied to evaluate one's basic beliefs. The modified version (in Russian) is a 37-item scale to identify one's assumptions of benevolence of the world, justice, luckiness, self-worth and controllability.

2) The Noetic Orientations Test [18] based on the Purpose-in-Life Test by Crumbaugh and Maholick [19]. This personality test (in Russian) is a self-report attitude scale to measure how purposeful and meaningful one perceives one's life to be. There are five subscales: purpose in life, process of life, result of life, locus-control (self) and locus-control (life).

3) Maddi's Hardiness Survey [20] translated into Russian and psychometrically checked by Leontiev and Rasskazova [21]. Hardiness is a personality construct that includes three relatively autonomous components: commitment, control, and challenge. The instrument used consists of 45 items to evaluate the three elements mentioned and hardiness in general as the sum of its components.

4) The Test Anxiety Inventory [22] modified by Hanin [23]. This 40-item tool is applied to measure both personal anxiety and situational anxiety.

2.4 Data analysis

The data collected were statistically treated using the SPSS Statistics package (descriptive statistics, analysis of variance and Student's paired t-test).

3 Results and discussion

The study results indicated a low or transient level of psychological safety among the study participants before their involvement in the programme of psychological support. This was manifested in their poor mental state and reduced emotional background, a high level of anxiety and a distorted image of their future. They also lacked in self-confidence and distrusted others. It should be noted that the results obtained are in line with the literature [7–10]. Table 1 presents the study participants' initial psychological characteristics in the context of their psychological safety. It is seen that there is a number of parameters that are statistically different from the normative ones for the corresponding age group. Parameters characterising the study participants' basic beliefs (the WAS subscales) were not included in the table because there was no statistically significant difference from the normative values.

Table 1. Initial parameters of psychological safety in the total sample ($N = 32$).

Parameters	Descriptive statistics				ANOVA	
	Obtained values		Normative values		*F*	*p*
	Mean	SD	Mean	SD		
Maddi: Hardiness	72.16	2.68	80.72	18.53	2.17	0.15
Maddi: Commitment	33.03	1.58	23.23	4.64	8.19	**0.01**
Maddi: Control	23.34	0.63	16.43	3.47	25.83	**0.00**
Maddi: Challenge	15.41	0.76	12.43	2.98	3.28	0.07
NOT: Purpose in life	30.22	1.08	38.91	3.20	13.84	**0.00**
NOT: Process of life	23.66	1.07	35.95	4.06	27.85	**0.00**
NOT: Result of life	21.50	0.61	29.83	3.00	39.42	**0.00**
NOT: Locus of control – Self	19.34	0.92	24.65	2.39	7.11	**0.01**
NOT: Locus of control – Life	23.41	1.11	34.59	4.44	21.75	**0.00**
TAI: Situational anxiety	38.31	1.51	37.50	7.00	0.06	0.80
TAI: Personal anxiety	47.50	1.18	37.50	7.00	15.33	**0.00**

Comparison of the parameters under study in the experimental and control groups before implementing the programme of psychological support did not reveal any statistically significant difference between them.

Participating in the programme of psychological support helped mothers enhance their sense of psychological safety. Student's paired *t*-test revealed statistically significant differences in all the indicators under study in the experimental group (see Table 2). Accordingly, negative values of *t*-criterion in the table illustrate an increase in the parameter in question after participation in the programme of psychological support. Conversely, positive values of *t*-criterion indicate a decrease in the parameter of interest after participation in the programme.

In other words, participating in the programme developed helped mothers raising children with special needs become more relaxed and more determined about their time perspectives. Their level of situational and personal anxiety significantly decreased. Many of the programme participants realized their life goals more clearly, as well as their image of the future. Their level of control in turn increased, indicating that they became more confident in their ability to influence their life events and situations. Table 2 also illustrates that these mothers became less irritated by other people and the environment and more steady to negative impacts of others, thus being more aware that the world around appeared to be much better and they could trust others. Women who participated in the programme showed greater involvement in their life process and became more adaptive to stressful situations, overcoming some of their fears and feeling more freedom.

Table 2. Paired differences in the parameters of psychological safety in the experimental group before and after participation in the programme of psychological support ($N = 16$).

Parameters	Paired differences			t	p-value
	Mean	SD	Std. Error Mean		
Maddi: Hardiness	–14.313	10.512	2.628	–5.446	0.000
Maddi: Commitment	–9.188	8.199	2.050	–4.482	0.000
Maddi: Control	–3.625	3.008	0.752	–4.820	0.000
Maddi: Challenge	–1.813	2.762	0.691	–2.625	0.019
NOT: Purpose in life	–1.625	3.008	0.752	–2.161	0.047
NOT: Process of life	–3.750	3.109	0.777	–4.825	0.000
NOT: Result of life	–3.313	2.243	0.561	–5.908	0.000
NOT: Locus of control – Self	–3.188	2.762	0.691	–4.616	0.000
NOT: Locus of control – Life	–2.750	3.890	0.973	–2.828	0.013
TAI: Situational anxiety	6.875	4.209	1.052	6.533	0.000
TAI: Personal anxiety	3.688	3.198	0.800	4.612	0.000
WAS: World benevolence	–2.313	1.815	0.454	–5.095	0.000
WAS: Justice	–1.375	1.025	0.256	–5.367	0.000
WAS: Self worth	–1.813	1.328	0.332	–5.461	0.000
WAS: Luckiness	–1.188	1.223	0.306	–3.884	0.001
WAS: Controllability	–2.313	1.138	0.285	–8.126	0.000

Compared to the control group, the programme participants' parameters of psychological safety also showed statistically significant positive dynamics ($p < 0.02$). In addition, Student's paired t-test did not reveal any statistically significant differences in the parameters of psychological safety in the control group.

4 Conclusions

Thus, the results obtained contribute to studying factors of psychological well-being in mothers raising children with special needs and suggest an effective way of enhancing their sense of psychological safety via participating in a programme of psychological support based on a specific psychotherapeutic space combining hippotherapy with environmental, social and personal factors, namely being outdoors, horse riding, taking care of horses and feeding them, as well as individual and group work with riding instructors and giving feedback about actual experience. One of the study limitations seems to be a mixed sample in the context of diagnoses in the children with special needs whose mothers were involved in it. We realize that mothers raising children with autism, cerebral palsy or Down syndrome may differ in their psychological deficiencies and hope that our work and further research will allow us to collect more data to provide targeted psychological support.

References

1. World Health Organization. Fact Sheets, (2016)
2. Statisticheskaya informatsiya za 2014 god [Statistical data for 2014], **5** (2015)
3. O.L. Romanova, Korsakov Journal of Neurology and Psychiatry, **9** (1988)
4. N.Ye. Karpova, T.A. Strekalova, *Proceedings of the II All-Russian scientific conference «Psychological problems of a modern Russian family»* (MSU, Moscow, 2005)
5. J.A. Resch, M.R. Benz, T.R. Elliott, Rehabilitation Psychology, **57** (2012)
6. A. Harper, T. Taylor Dyches, J. Harper, S. Olsen Roper, M. South, J. Autism Dev. Disord., **43** (2013)
7. Yu.M. Milanich, Vestnik of Pushkin Leningrad State University. Psychology, **3** (2010)
8. E.Yu. Shebanets, Bulletin of the Adyghe State University. Pedagogy and Psychology, **1** (2011)
9. E.Yu. Shebanets, T.K. Khozyainova, Bulletin of the Adyghe State University. Pedagogy and Psychology, **4** (2011)
10. E. Yu. Shebanets, F.R. Khatit, Bulletin of the Adyghe State University. Pedagogy and Psychology, **4** (2015)
11. N.V. Kozlova, D. Ye. Zueva, I.A. Glebova, Med. Psihol. Ross., **3** (2016)
12. T.V. Eksakusto, N.A. Lyz, Siberian Journal of Psychology, **37** (2010)
13. I.A. Baeva, Herzen University Journal of Humanities & Science, **145** (2012)
14. N.V. Kozlova, S.A. Bogomaz, *Proceedings of All-Russian scientific-practical conference* (Southern Federal University, Taganrog, 2014)
15. A.E. Alekseičik, R. Kočiūnas, *Psychotherapy by life: Intensive Therapeutic Life by Alexander Alexeichik* (HEPI, Vilnius, 2008)
16. R. Janoff-Bulman, Social Cognition, **7** (1989)
17. M.A. Padun, A.V. Kotel'nikova, Psychological Journal, **4** (2008)
18. D.A. Leont'ev, *Test smyslozhisnennykh orientatsiy [The Noetic Orientations Test]* (Smysl, Moscow, 2000)
19. J.S. Crumbaugh, J. of Individual Psychology, **24** (1968)
20. S.R. Maddi, S.C. Kobasa, *The hardy executive: Health under stress* (Dow Jones-Irvin, Homewood, 1984)
21. D.A. Leont'ev, E.I. Rasskazova, *Test zhiznestoykosti [The hardiness test]* (Smysl, Moscow, 2006)
22. C.D. Spielberger, R.L. Gorsuch, R.E. Lushene, *The State-Trait Anxiety Inventory: Test manual* (Consulting Psychologist Press, Palo Alto, 1970)
23. C.D. Spielberger, R. Diaz-Guerrero, *Cross-cultural anxiety* (Hemisphere/McGraw Hill, New York, 1983)

The role of cultural involvement of patients for their social welfare

L. Dromantiene[1], K. Staras[2], and M. Uselyte[3]

[1]Klaipėda University, Klaipėda, Lithuania
[2]Vilnius Central Outpatient Clinic, Lithuania; Mykolas Romeris University, Vilnius, Lithuania
[3]Lithuanian Social Research Centre, Vilnius, Lithuania

Abstract. Cultural involvement takes a significant role on quality of life of patients (QoL) as namely culture is an important domain for health and QoL simultaneously. Paper analyzes the possible connection between culture and health, and, attendance of patients the cultural events as one of the important dimensions of QoL. Research is based on social project "Cultural prescription", organized by public cultural institution "Mokytojų namai" (Eng. Teacher's House) in Vilnius and dedicated to Outpatient Clinic "Centro poliklinika" (Vilnius) patients, who were suffering from chronic disease to improve their QoL. The aim of the paper is to explore the possibilities of innovative social measures to increase social cultural engagement of patients, to contribute to their well-being, and to present the approach of health care practitioners towards the expediency and usefulness of inter-sectorial cooperation for the patients' social welfare.
Key words: quality of life, social welfare, health, cultural involvement.

1 Introduction

Putnam identifies such features of contemporary society as growing sense of fear and isolation, the demise of trust, belonging and support [1]. To a large extent health problems lead to these social phenomena. Social networks, inter-sectorial collaboration become more and more important for the wellbeing of the individuals and groups. Consequently, one of the ways to improve QoL is an aspiration to improve health. There are many approaches how the state in general or social network organizations, in particular, can improve residents health. Culture is between those methods. Actually, in modern society cultural usage for the health and QoL purpose increases rapidly [2–5]. Moreover, culture is not only a measure to reach better QoL, but it is also a component of QoL. Culture is acknowledged in many social spheres as a social power to influence the particular issue in a right direction, which had not spared the attention of researchers. Sonnenberg [6] indicates that social cultural formal and informal involvement has a positive effect for the wellbeing of individuals.

The paper concentrates on how non-medical factors influence health, welfare and quality of life of patients and presents research approach from the point of health care practitioners view, as intermediaries between social cultural project organizer's and receivers' (patients') perspective in Lithuania.

The article presents the findings of the social project "Cultural prescription" that was successfully carried out in 2013 in Vilnius when implementing the best practice of social cultural involvement of patients in Turku, Finland (2011). Project was organized by public cultural institution "Mokytojų Namai" (Eng. Teacher's House) in Vilnius, Lithuania and dedicated to outpatient clinic "Centro poliklinika" patients, who were suffering from chronic disease (cancer, cardiovascular disease and mental health disorders (a predisposition to depression)) to improve their QoL. The idea of that project was participants receiving invitation cards – "Cultural Prescription" – to a cultural event or to a cultural institution besides their regular based prescription for medicine. The empiric data was collected by applying a semi-structured interview method. This is an evidence based cross-sectorial cooperation when culture was employed for health care by collaborating cultural and social network enterprises. Precisely, this social project reveals the case when culture enabled for improving patient's health, welfare and QoL simultaneously.

The aim of the paper is to explore the possibilities of innovative social project to increase social cultural involvement of patients seeking to contribute to their health and QoL and present attitudes of health care practitioners on this issue. Methods of the research: analysis of scientific literature/data, qualitative research, semi-structured interview, qualitative content analysis of the data.

1.1 Health related well-being and quality of life

Quality of life (QOL) is a broad multidimensional concept that usually includes subjective evaluations of both positive and negative aspects of life [7]. Researchers of different disciplines have proposed their own definitions of quality of life. Psychologists [8] analyse personal experience based quality of life; economists [9] allocate the biggest focus on objective economical capacity related to living standards; medical scientists [10] analyse QoL from medical perspective and argue with a health-related analysis; sociologists [11, 12] dispute individual's well-being and their reliance to different levels of society. Growing popularity in interdisciplinary studies has led to interdisciplinary of QoL studies. It is important to mention that major studies on QoL has been initiated by such worldwide organizations as UNESCO, World Health Organization (WHO), Organization for Economic Co-operation and Development.

Two historical approaches on QoL based on different scientific fields should be mentioned: Scandinavian and American [13]. American psychological approach is more interested into individual, subjective well-being, strongly related with such concepts as well-being, happiness or life satisfaction and less related with material goods than individual emotional state. While Scandinavian perception developed by sociologists and economists supposed to be more interested in objective living conditions. It considered that happiness and satisfaction may be reached in condition of well-developed welfare state as a state is responsible to create good social conditions and accessible for every resident services. As is noticed by Rapley [14], Scandinavian approach supports believes, that well-established welfare system makes people life easier and allow them to reach and have more in their life, enables them integrated in the society which naturally improves their QoL. Even scholars identify different components of QoL, but some domains are agreed by all, such as physical, mental and social health [15, 16]. Despite objective purpose of health care services, QoL is still improved through medical care, though possible alternatives in more social or cultural activities, some supplementary activities began to emerge.

1.2 Employment of culture for social involvement of patients

Researchers agree that cultural experience and impressions from culture have physical, mental and social effects [17]. Health care practitioners are often skeptical about health related researches if they do not measure solid medical facts. Phillips [18] implies, that non-traditional treatment activities or methods such as cultural or social initiatives frequently does not convince medics as important or having value for objective health status, because, they rather rely on medicine technologies and laboratory tests. Despite this approach of medical practitioners, various activities have been implemented and researches have been still conducted, which proved the great interest in the topic in scientific and social levels. Considering a linkage between QoL and culture the Canadian research [19] outcome may be presented as method of direct correlation between culture and QoL visualization in society, as overwhelming majorities of research participants (province of Ontavio) states, that culture enrich their quality of life and they value culture as important part for it. Augustinsson (2011) added, that all kind of cultural experience helps people to express themselves in daily life regardless of objective living conditions, in this way culture is suggested to be understood as resource of human well-being which directly matters to quality of life. Cuypers et al. [20] notice that even low-level cultural activity may have a protective effect on health during the time. Culture has individual and community levels of influence on QoL. Even, if culture or its attendance is presented as entertainment, culture by nature provides more social and individual values. As culture is an integral part of every society, it can help individual to adapt themselves in society.

At individual level it is stated that cultural experience improves mental health by relieving stress, anxiety and depression and its symptoms, therefore it stimulates positive feelings, experiences and minds. Culture may contribute to the human life and humanity improvement, individual self-esteem and happiness; help for individual to maintain their health, improve daily routine and empower human to use personal potential (capacity). The mechanism of community level influence is presented by Guetzkow [21].

Firstly, culture may have positive effect on building of the social capital, as it is dedicated to community or society. Secondly, culture improves the economy through promoting peoples' capacity, health and QoL which allow them and society in common reach better economical results; traditional attending increase general consumption, in particular cases of social initiative or project, it may have positive economic side for single person or family unit, to attend free of charge cultural events and save money for other purposes. Thirdly, is claimed that the culture is advantageous to community through single individual in holistically approach. Social cultural formal and informal involvement of patients into health care programmes does not only look after health risk factor's prevention but also take into account leisure participation in social, cultural and religious activities as part of their concerns. In this respect cultural engaged governmental or non-governmental social programs and projects traditionally divided into two main categories: active and passive participation, which leads to direct or indirect impact. Active kind of participation guarantees direct impact for a participant, because this kind of programs directly include participants into process of producing art (culture) and belief in the inherent healing power of the creative process of art making. Therefore, this kind of art participation in most cases is called art therapy. Art therapy may be provided as part of regular treatment and as intervention to social issues, but it is always required for a group of active people, who would have determined to participate without being afraid of being supervised by therapist [22]. Individuals directly involved in creating or organizing artistic activity may learn skills that they did not previously have and may demonstrate greater creativity not even in cultural nature, but in daily life and solutions of problems as well [23].

Therefore, it is important to highlight, that art therapies have deeper traditions and history of occurance in societies and has a greater acceptance among medical communities. Passive participation guarantees indirect impact for persons who attending this programs or projects. This kind of initiatives do not include people into culture producing process, rather it invites participants to revel cultural events together with other people despite they are participants of initiative or not, in other words – be part of natural audience. Despite active or passive participation in culture strengthens feeling of belonging to community, stimulates positive experience and feelings [24], any kind of relation between person and culture has a health benefits, despite of participating or attending culture. Cultural engaged social initiatives addressing health and QoL are found throughout the world: in Great Britain prescription of culture (2004) later (2010) recalled "Social Prescription" linked patients to participate in particular social actions (membership in social networks, volunteering clubs, etc.) [25]; in Sweden, Region Skane in 2009–2011 culture was prescribed as supplemented method to long term sick leave patient's rehabilitation [26]; particular Lithuanian case "Cultural prescription" (2013) brought from Turku, Finland, where it was implemented in 2011 while Turku was European Capital of Culture and aimed to uncover the comprehensive role of culture for individuals and communities – well-being providing a valuable perspective. As it was already presented, Lithuanian case Cultural prescription was not linked to art therapy, but it aimed to improve health and QoL through participation in cultural events of those who suffered chronic diseases. A solid number of cultural prescriptions were prescribed (3420 units), which evidence health care practitioner's determination to this social project.

1.3 Empirical research of social project Cultural prescription

The research was conducted on 18–31 October 2013 in four different departments of Outpatient Clinic "Centro Poliklinika" in city Vilnius, Lithuania where prescriptions for the attendance of cultural events were prescribed.

The research was aimed to evaluate health care practitioners 'comprehension about prescribing culture and their experience in familiar practices. Informants provided answers verbally during an interview. The empiric data was collected by applying a semi-structured interview method that induced more relaxed communication atmosphere, better adjustment to the research environment, and provided information contributing to an in-depth discussion of the problem analysis. Five doctors and five nurses were selected by Gatekeeper method of sampling.

Health care practitioners participated in the research were family doctors (n = 5) and nurses (n = 5), average work experience 17 years. The focus during the interview was given to understanding of specific moment of prescribing culture, better understanding of informants 'attitudes and comprehension on actual practise.

The data has been collected until a certain degree of theoretical "satiation" has been achieved. Questionnaire contained twenty questions grouped in five blocks, which covered questions on common information about informants; familiar experience, participation and participants in social project Cultural Prescription; possible influence on health and QoL; evaluation of the project; and health care practitioner's role in the process of prescribing culture.

Research has been conducted following the scientific research ethics: voluntary approach, goodwill, privacy and respect. To ensure confidentiality of respondents, their names have been replaced with number and letter that indicates participant and occupation (D = doctor, N = nurse).

Results were processed using qualitative processing method distinguishing categories and subcategories. Inductive category approach was used. Data derived categories raises attention to the data, which also means information gained from the interviewees. In this case, then voices of health care practitioners are hoping to raise, this approach of categorization is agreed as most suitable to apply, because the actual content of the text is much richer than it could be predicted in advance. The qualitative content analysis presents wider understand to date and more information than only it's content [27].

Analysis of the data obtained during the research revealed the following main categories: ***General perception, Experience and knowledge, Process, Outcomes, Final evaluation*** establishing health care practitioners experience in prescribing culture process, attitudes and related practise. Practitioners identified different perceptions of social project culture prescription.

Table 1. Cultural Prescription perceptions in terms of health care practitioners.

Category	General perception of cultural prescription
Subcategories	Art therapy, Entertainment, Charity

It was established that health care practitioners equate cultural prescription to art therapy, that itself associated to medical terminology and health. Cultural involvement in Lithuania is related to art therapy because of being most known term as a term but not by its meaning.

1 – D: "<...> not unnecessarily various art therapies occur, so <...> Cultural Prescription also exist".

However, this assign deeper understanding of cultural events and health interrelation, because another understanding as entertainment and charity supports an idea, that health care practitioners do not recognise health and cultural event interaction as medically important. Association with entertainment informs us that because of access to the cultural event health care practitioners automatically ascribe it as entertainment without any medical willingness.

4 – D: "<...> it is beneficial for everyone to have fun sometimes".

6 – D: "<...> culture cannot harm people <...>, it is good that this project exist". so exist".

An expression of charity also responds to cultural events and health interrelations acknowledgement.

3 – D: "<...> if social status is high, <...> they do not need for this charity".

2 – N: "<...> those who like <...> this charitable initiative makes them very happy <...>".

Thus, information was found as important due to present how the health care practitioners perceived this project. First category revealed three interpretations of the one phenomenon – culture prescription. Many causal indicators can be behind these different perceptions (Table 2). Obtained direct or indirect experience, scientific knowledge or even absence of interest gained during lifetime may influence presumption of a phenomenon as well as disclose surrounding circumstances of health care practitioner's perception of Cultural Prescription.

Table 2. Health care practitioner's common experience and interest.

Category	Experience and knowledge
Sub-category	Direct/indirect experience, Scientific knowledge, Absence of interest

Participants of the research identified a different status of experience: direct experience – indicate the practices, which had directly included interviewee into the process. The same as social project Cultural Prescription did.

10 – N: "<...> I had an internship at oncology institution <...> there are <...> relaxation therapies, music therapies, painting therapies, light therapy <...>. <...> but it was serious treatments, not projects<...>".

Indirect experience – knowledge about something in relation to culture and health occurs, but no personal link exists.

D – 6: "<...> I have heard, Mental Health Centre had a project about clay <...> something with clay <...> and I participated in their exhibition <...> ".

An absence of interests comprises provision that cultural events and health care are not spheres of family doctors or nurses.

4 – N: "<...> I have nothing to comment <...> me or my patients with me we had never been participated in familiar initiatives".

9 – N: "<...> this is not work of mine <...> ".

6 – D: "No <...>. Never needed. But it seems that maybe somehow it is not my area anyway".

It was established that most often health care practitioners mentioned art therapy practises as direct experience of culture and health most relevant to the prescribed culture. In this order seems more meaningful, that health care practitioners who had direct experience tend to associate Cultural prescription with art therapy. Health care practitioners also indicated Cultural Prescription as a first social project that they were involved as a part of implementers' team. Although they knew about familiar initiatives, they had limited experience of culture applications in their working environment. It was also established, that only a minority (n=2) have encountered with scientific literature.

7 – D: "<...> I remember at the study times I had a course about art therapies for mental disorder patients".

10 – N: "Had to take interest in the scientific literature, because I had to write internship assignments, and there was much to do on this topic".

Interviewees who stated about scientific literature had less than ten years working practice, which also means that studies were graduated later. Time aspect is relevant to be mentioned because in Lithuanian context exist significant differences between the decades due to countries historical development as well as art therapy as a source of cultural engaged social projects. It stresses the importance of the research, to highlight health care practitioner's attitudes and perceptions because health care system is the one that needs to be convinced about the worth to join forces. Moreover, interviewees perception on Cultural prescription are varied as well as their experience and interest are various, which naturally concludes, that current state of knowledge influence the perception of the subject and process of its realization. It was established, that process may vary from fail to succeed and also can be neglected (Table 3).

Table 3. Domains of Culture prescription's process.

Category	Process
Sub-category	Neglect, Fail, Succeed

Prescribing cultural prescription is more complex and liable process than it may seem. Complexity depends on both – patients and health care practitioners sides. Research revealed that health care practitioners neglected cultural prescriptions – conceptually appropriate patient came to owns doctor, not necessarily, one gets a cultural prescription.

8 – D: " <...> I did not prescribe, because last week I had many patients <...> it takes tame to explain, registered into computer <...> difficult to find time<...> ".

1 – N: "<...> they don't say anything about, they don't know. Then you (nurse) on your own are tired or unhappy or slimly a lot of work, <...> and patient stays without prescription".

4 – N: "<...> my patients did not know about this project <....> "

6 – D: "<...> when serious chronic disease patients have already gone I notice that I did not have time or forgot to prescribe this prescription even I could. Our regimes are too high".

Neglect occurs via the complexity of reasons. First, established is a low level of project popularity between patients – it leads to an absence of patients recall about a possibility to get a cultural prescription, which, according to health care practitioners would be helpful. Also when a person is unaware of the project, it becomes more time consuming to answer all questions and explain all information. Second is the requirement of firmly scheduled health care and patient appointment. It was even more problematic if practitioner once had a practise when patient refused or it took too long to prescribe cultural prescription, practitioners enthusiasm about process reduces.

6 – D: "<...> one who refused said, I do not want, will not go, I'm not interested".

10 – N: "<...> happens that people says No, but only because one lives far away, outside Vilnius, but not because one does not want <...> ".

2 – D: "<...> but one has chronic back pain <...> and people refuse to go <...> ".

It was investigated, that absent of interest in cultural life, living location, late evening events also health condition are most popular reasons why patients refuse to take a cultural prescription. However, 3420 units of Cultural prescriptions were prescribed, which evidence health care practitioners determination to this social project and their acceptance of it despite all mentioned encumbrances.

1 – N: "<...> they are very happy about it <...>. We share good emotions".

10 – N: "<...> happens that people come here again to thank <...> ".

Health care practitioner's efforts were necessary to reach this number equally as they had to strive for every patient to get pleasure and benefits from this project. Research revealed that health care practitioners had experienced positive emotions among their workloads and rush that indicate mutual outcomes, for practitioners and patients. Established emotional, social and economic subcategories that reflect findings (Table 4).

Table 4. Distribution of Cultural Prescription outcome.

Category	Outcomes
Sub-category	Emotional, Social, Economic

Was established that prescribed culture play a role in patient's social life and increase their social welfare.

3 – D: "<...> finally meet same-minded people <...> communicate and be in society <...> is needful also for quality of life<...> ".

According to the research, emotional, social and economic factors consist of Prescribed culture outcomes. Successful cases incurred practitioners'good emotions because of being the one who could make patient pleased.

1 – N: "*<...> they are really happy about it*".

Simultaneously patients provided positive emotions when received cultural prescription. Emotions are important to daily life because it allow people feel better in daily routine and easier to access social networks. Moreover, relationships are important with those people or networks that can provide emotional or material support for people: patients, who received Cultural prescription, noticed health care practitioner friendly and favourable themselves and noticed relations as important. Social factor is equally important, because importance of being in society and communicating to other people are agreed indicators of quality of life between scholars [18, 19]. Established relation of Cultural prescription with economic factor: linking possibility to save money for those people, who were generally attending cultural events as well as for those who postponed their attendance due to a shortage of money.

2– D: "<...> appreciate that they can go for free, <...> and save money <...> they are very happy about it".

Considering, that Cultural prescription as socially engaging project has positive influence on improving social relations, activating social participation and so reducing social isolation, made it important to evaluate and generalize this social project in general. To reach this, advantages and disadvantages (Table 5) were aimed to gain information about what is done right and what may be improved in the future due to pursuit the best version of the project. Advantages and disadvantages were asked to identify only for cultural prescription as a social project by nature not its outcomes on patient's health or QoL.

Table 5. Cultural Prescription final evaluations.

Category	Final evaluation
Sub-category	Advantages and disadvantages

It was established, that Cultural prescription for two persons and the possibility to participate in the cultural event as a regular participant are highlighted as significant advantages.

1 – N: "<...> that they can go two".

7 – D: "<...> allow participation for those, who don't actually want participate <...> labels are not stuck there".

Two tickets free of charge allow the patient to take a company together which increases emotional and social ties between close persons. Therefore, enable less determined patients, who do not like to participate in active initiatives also gain social project benefits. That means that this method of providing cultural benefits is more suitable for individuals who are not determined to take part in active cultural initiatives. Limited repertoire, agency between health care institutions and cultural events and timing were established as disadvantages.

1 – N: *"<...> cannot choose those concerts <...> limited choice <...> "*

2 – N: *"<...> that they have to go to pick up their ticket*

9 – N: *"<...> question of time should be considered <...> ".*

List of possible cultural institutions or events did not respond to everyone who gave cultural prescription needs. Informants highlighted that prescribing direct tickets to cultural events could be more productive, especially for those, who received cultural prescription more than once. Thus, opposition to intermediaries is linked to time consuming to patients and health care practitioner simultaneously. Therefore, health care practitioners stressed, that type should be considered in advance.

2 Conclusion

Cultural involvement of patients means an active or passive their participation, as within models of social engagement individuals undertake part and through practical projects provide a rich recourse for development community wellbeing. Study present health care practitioner's attitude toward socially engaged cultural project "Cultural prescription" where chronically ill patients together with medical prescription received tickets to cultural events in order to improve their health, social relations, and quality of life.

Although the current study poses some limitations, to the authors knowledge this is the first research examining the health care practitioner's perspective of such social engaged cultural projects in Lithuania. The findings of study suggest that health care practitioners do not deny benefits of attending culture, but also do not recognize it as capable to have medical significance. The meaning health care practitioners gave to social project Cultural prescription depends on their personal interest in a subject as well as knowledge and experience.

Considering organizational disadvantages and displeasure at occupied time founded, generally health care practitioners have a positive attitude toward this particular initiative for their patients. Following the success of the project took one additional year with corrections based on research results that made project run easier for health care practitioners. The current study and project were presented at the stakeholder's conference where it made a discussion toward cooperation health care and cultural sectors for a common welfare of the residents. It was jointly agreed that Cultural prescription plays a role for improvement of health, social relationships, social participation and QoL in general and makes influence at community level through reducing social isolation, and increasing social connections also promotes interpersonal interactions.

References

1. R. Putnam. Bowling alone. The collapse and revival of American community. NewYork: Simon and Shuster (2000)
2. J. Angus. An Enquiry concerning Possible Methods for evaluation Arts for Health Projects. Bath. UK: Community Health (2002)
3. J. Sauth. Community based arts for health literature review. Leeds Metropolitan University (2004)
4. M. Hyyppa, et al. Leisure participation predicts survival population-based study in Finland. Health Promotion International. Oxford University Press. **21**(1), 5–12 (2011)
5. K. Deane, M. Fitch, M. Carman. An innovative art therapy program for cancer patients. Can Oncol Nurs J. **10**, 147–157 (2012)
6. B. Sonnenberg. Dependencies and Mechanisms of Unemployment and Social Involvement. Findings from the socio-economic Panel Study (SOEP) Berlin. Springer (2013)
7. The World Health Organization Quality of Life Assessment (WHOQOL). Development and psychometric properties. Soc Sci Med **46**, 1569–1585 (1998)
8. E. Diener, R. Biswas-Diener. Will money increase subjective well-being? A literature review and guide to needed research. Social Indicators Research. **57**, 119–169 (2002)
9. V. Starkauskienė. Gyvenimo kokybės veiksniai ir jos kompleksinio vertinimo modelis. Doctoral thesis. Kaunas: VDU (2011)
10. R. Savičiūtė. Sveikatos ir socialinių veiksnių sąsajų tyrimas. Doctoral thesis. Vilnius: Vilniaus Universitetas (2013)
11. D. Raphael. The quality of life project: a health promotion approach to understanding communities. Great Britain, Oxford University Press (1999)
12. R. Veenhoven. The Four qualities of life. Ordering concepts and measures of the qood life. Journal of the happiness studies. United Nations University Press, **1**: 1–39 (2000)
13. H.H. Noll. Social indicators and quality of life research: background, achievements and current trends. N.Genov, Ed.: Advances in Sociological Knowledge over Half a Century. Springer (2004)
14. M. Rapley. Quality of life research: a critical introduction. London: Sage Publications (2003)
15. J. E. Stiglitz, A. Sen, J.P. Fitoussi. Report by the Commission on the Measurement of Economic Performance and Social Progress (2009). Interactive. Retrieved from

http://www.stiglitz-sen-fitoussi.fr/documents/rapport_anglais.pdf

16. D. Bruno, A. Hubley, Z. Hubley. Health and the quality of life. Essays of the quality of life. **19**, 153–183 (2003)
17. S.E. Johansson, B.B. Konlaan, L.O. Bygren. Sustaining habits of attending cultural events as maintenance of health: longitudinal study. Health promotion international. Oxford University Press, **16**(3), 229–234 (2001)
18. D. Phillips. Quality of life. Concept. Policy and practice. London and New York: Routledge (2006)
19. Environics research group. The Arts and the Quality of Life. The attitudes of Ontarians. Canada: Ontario Arts Council (2010)
20. K. Cuypers, et al. Patterns of receptive and creative cultural activities and their association with perceived health, anxiety, depression and satisfaction with life among adults: the HUNT study, Norway. Journal of Epidemiology and Community Health. **66**(8), 698–703 (2011)
21. J. Guetzkow. How the arts impact communities: An introduction to the literature on art impact studies. Priceton: Priceton University (2010)
22. H.L. Liikanen. Art and culture for well-being – proposal for an action programme 2010–2014. Finland: Publications of the Ministry of Education and Culture, Finland (2010)
23. H. Parr. Mental health and social space. Towards inclusionary geographies. Oxford: Blachwell Publishing (2008)
24. A. Svenka, et al. Art therapy improves experienced quality of life among women undergoing treatment for breast cancer: a randomized controlled study. European Journal of Cancer Care. **18**, 69–77 (2009)
25. L. Friedly, C. Jackson, H. Abernethy, J. Stansfield. Five's Cultural Prescriptions Pilot Report, Care Services Improvement Partnership. Great Britain (2012)
26. S. Augustinsson. Prescribed culture. Summary of resulting research. Malmo: Region Skane (2011)
27. M. Hennink, I. Hutter, A. Baley. Qualitative research methods. Los Angeles: Sage (2011)

Adaptive behaviour as a factor of psychological health

E. Kuftyak*

Kostroma State University, Kostroma, Russia

Abstract. The results of the research revealed that psychologically healthy children more often choose the strategy of passive retreat from difficulties, children with mental health problems are more focused on a destructive emotions expression and "risk group" children use defence of "regression" and "replacement". The psychological defence and coping-strategies contributing to the psychological health state have also been revealed during research. We consider that the received results can be used in the development of evidence-based value orientations and practical recommendations for the healthy personality cultivation.

1 Introduction

The study of person's adaptation to challenging life situations is one of the topical tasks of modern psychology. Many personal and social problems (addiction, social deviation, fears, impairments of interrelations with others, etc.) have their roots in the impairments of adaptation processes. Obviously, the ability to overcome difficulties is laid as early as in childhood. The development of ability to resolve difficulties and crisis situations defines both the level of adaptation in childhood and the well-being of personality in adulthood.

Modern psychological approaches to stress and adaptation focus on searching uniform mechanisms of human adaptation involving psychological defenses, coping behavior, behavior control, and anticipatory consistency [1].

The theory of defense mechanisms development, elaborated by Cramer, states that a person processes traumatic experiences at both conscious and unconscious levels, which enables to speak of an adaptation process [2]. Cramer assumes that the adaptation process unites psychological defenses and coping behaviour as mechanisms contributing to satisfying an individual need for adaptation to reality. This viewpoint treats defense mechanisms as mechanisms, which change a true perception of reality and protect a person from excessive anxiety caused either by the perception of a troublesome external event or internal destroying mental state. In this aspect, coping behavior is understood as manifest and covert actions performed to reduce or eliminate psychological distress and stress states.

Nikolskaya offers to unite psychological defense mechanisms and coping behavior strategies in a uniform defense system of personality using the analysis [3]. The author identifies four basic levels: somatic-vegetative (physiological), psychomotor (behavioral), level of psychological defense (unconscious psyche) and level of coping behavior

* Corresponding author: elena.kuftyak@inbox.ru

(consciousness). All these levels are formed sequentially and function simultaneously in adulthood. Nikolskaya argues that a set of defence processes is individual and characterizes the level of personality's adaptation.

Vetrova has conducted the longitude study of a psychological regulation of behaviour in adolescence and early youth, which has revealed that behaviour control, coping behaviour strategies, and psychological defences are closely associated mechanisms of behaviour regulation [4]. It should also be noted that coping behaviour compared to psychological defences is more associated with behaviour control. Psychological defences are more isolated from behaviour control and coping; however, they remain associated with them. The author assumes that coping behaviour strategies and psychological defence mechanisms constitute the continuum of adaptive behaviour based on individual resources of a person. This approach leads the unification of adaptive mechanisms of behaviour (coping-behaviour, psychological defences and behaviour control) in a single space.

The author's earlier research aimed at studying person's uniform defence system in the context of dynamic crises of normative and non-normative family development has revealed that a drawback in the coping repertoire of problem-focused strategies is juxtaposed with a high level of tension of psychological defences [5]. One of the important results of the research is that coping behaviour acts as a stable mechanism of spouses' adaptation in a crisis situation. Psychological defences, in turn, are isolated from behavioural coping strategies; however, they are more associated with coping strategies focused on emotions.

Health characterizes the level of functioning of human organism's systems including the system of mental adaptation, which enables to consider adaptation mechanisms in the spectrum of indices of the psychological health state. The notion "psychological health" introduced in the Russian psychology and practice by Dubrovina enables to reveal a psychological aspect of person's health and relates to personality as a whole [6]. Psychological health is related to the processes, which enable a person to become a mature and an independent subject of activity capable of achieving reference points. Inability to cope with the intensive flow of stresses reduces self-confidence, health resources, and social adaptation. The impairments in the functioning of adaptation processes lead to the development of anxious states, cause neuro-mental and psychosomatic disorders and increase the probability of maladaptive behaviour, especially in children and adolescents.

In clinical psychology, there is a common viewpoint that any somatic impairment is associated with changes in person's mental state. Frequent long-term illnesses in children and adults can be conditioned by a non-constructive system of personality's psychological defence and coping strategies. Insufficient variability and maturity of adaptive mechanisms of behaviour leads to long-term destructive responses to stress and difficulties. Thus, the fixation on infantile defences causes somatic impairments: overstressed immune system increases the level of psychosomatization [7].

We particularly stress that the problems of children's psychological health represent a wider range of manifestations than the sphere of mental disorders in childhood and adolescence. There are some impairments of development, "basic impairments of a pre-nosological level" [7], which cannot be referred to either psychiatric illnesses or clinically apparent disorders. These impairments are diagnosed by the markers of social impairments or causing considerable damage to others, which indicates internal and external problems in children. This considerably reduces their adjustment and prevents from optimal development at cognitive, emotional, or behavioural levels.

2 Materials and methods

Our study aims at evaluating psychological health and the contribution of psychological defences and coping behaviour to its state. To achieve this goal, we have studied a group of pre-school children from 4 to 6 years (mean age is 5.6 years). The study has involved 86 pre-school children (49 boys and 37 girls). The expert assessment of the state of health has enabled to single out three groups of children: "children from the group with impaired psychological health" ($n = 25$), "children from the risk group" ($n = 17$), and "children from the group with preserved psychological health" ($n = 24$).

We have used the following techniques for the expert assessment of the psychological health state.

Assessment of anxiety involving observation (Sear's technique) aimed at defining the level of anxiety in pre-school children [8].

Khukhlaeva's "List of teacher's assessment of child's psychological health" aimed at estimating the state of psychological health indices [8]. We have identified three indices of psychological health in Khukhlaeva's technique (Kuftyak's modification): a personal component, an emotional component, and a social component, which has allowed us to include different meaningful characteristics of health in the study.

The Strengths and Difficulties Questionnaire – SDQ (Goodman) aims at studying common internal and external problems of children from 3 to 17 years [9]. The questionnaire has five scales: emotional symptoms, problems with behaviour, hyperactivity/negligence, problems with peers and pro-social behaviour. The scales of emotional symptoms and problems with peers form a final scale of internalization, while the scales of problems with behaviour and hyperactivity/inattention – a final scale of externalization.

We have estimated psychological defence mechanisms by means of Children Psychological Defence Assessment Map (Peri, Plutchik). The technique enables to evaluate the type of psychological defence. The authors suggest defining eight defence mechanisms (denial, repression, regression, compensation, displacement, projection, reaction formation, and intellectualization) treated as basic in Plutchik's structural theory.

We have studied coping behaviour by means of the Schoolager's Coping Strategies Inventory of Ryan-Wegner adapted by Sirota, Yaltonskiy and modified by Nikolskaya (2000) [3]. We have united 26 coping strategies identified by means of the inventory in 6 basic categories: reflexive withdrawal, passive distraction, discharge; search for spiritual support; destructive emotional expression; active distraction; social contacts (communication) for obtaining support. This technique (unification) has allowed us, first, to make coping strategies used by children closer to the basic styles of personality's coping behaviour and second, to use the methods of mathematical data processing later.

Mathematical-statistical processing of the results of the study is carried out using SPSS v.17 computer packages. It includes a correlation analysis (Spearmen r rank correlation) and the application of non-parameter criteria of evaluating significant differences between non-related samples (Mann–Whitney U-test, Kruskal–Wallis H test).

3 Results and discussion

The expert assessment by the specialists of pre-school establishments (educators, a psychologist) has enabled to single out three groups of children: "children from the group with impaired psychological health" ($n = 25$), "children from the risk group" ($n = 17$), and "children from the group with preserved psychological health" ($n = 24$). Table 1 represents statistical significance in the evaluation of psychological health.

Table 1. Indices of psychological health in the groups of children.

Indices of psychological health	Children from the group with impaired psychological health (n = 25)	Children from the risk group (n=17)	Children from the health group (n = 24)	Kruskal–Wallis H test
	M (SD)	M (SD)	M (SD)	
List of Psychological Health Assessment				
General index of psychological health	42.92 (2.78)	31.88 (2.42)	14.13 (5.28)	57.363***
Personal component	19.24 (1.2)	13.18 (1.28)	6.08 (2.63)	57.653***
Social component	16.84 (1.75)	13.35 (0.78)	5.71 (2.19)	57.675***
Emotional component	6.72 (1.39)	5.18 (1.0)	2.50 (1.21)	48.291***
Sears's scale	8.72(2.42)	5.88(1.05)	1.62(0.77)	52.268***
The Strengths and Difficulties Questionnaire				
Internal problems	8.88 (2.6)	5.76 (4.70)	2.13 (1.19)	37.62***
Emotional symptoms	4.04 (2.28)	2.76 (3.25)	0.08 (0.28)	40.86***
Problems with peers	4.84 (1.14)	3.0 (1.70)	2.04 (1.08)	32.11***
External problems	8.08 (2.14)	3.82 (3.52)	2.25 (1.39)	35.23***
Problems with behavior	3.68 (1.31)	2.0 (1.94)	0.96 (0.99)	29.42***
Hyperactivity	4.4 (1.55)	1.82 (1.78)	1.29 (0.81)	33.12***
Prosocial scale	4.04 (1.48)	7.0 (1.22)	7.12 (1.08)	38.19***
General evaluation of problems	16.96 (3.9)	9.59 (7.99)	4.38 (1.95)	36.44***

The comparison of data represented in Table 1 clearly indicates that the differences in the components of psychological health lie in personal, social and emotional functioning in all the groups of children and that there are differences between all the groups revealed by the indices of internal problems and emotional symptoms and general number of problems. Children with impairments of mental state have a higher level of anxiety, while a high level of manifestation of personal and social components of mental health indicates children's inability to overcome difficulties, to take initiative, disorderliness, lack of self-assurance, and lack of a pursuit of becoming better.

Children from the risk group report a marked level of anxiety, while in a social component of psychological health, regulative capacities are the most impaired – children are incapable of controlling and overcoming difficulties.

Children from the group with impaired health have worse assessments for emotional deviations described as internalized compared to other groups. They experience anxiety, fears, sense of inferiority, helplessness, and reduced mood as well as psychosomatic symptoms. Children from the risk group face emotional problems more rarely; "anxiety" is likely to play an adaptive role in such children and thus provides self-defence behaviour, which enables them to achieve safety in particular social situations.

We have revealed differences in the choice of psychological defences and coping strategies between the groups. Thus, the risk group has been significantly different from the group of impaired health by a weak "regression" defence ($p = 0.02$) and severe "repression" ($p = 0.002$), "compensation" ($p = 0.000$), and "displacement" ($p = 0.001$) defences as well as a general level of tension of psychological defences ($p = 0.05$). The health group has been significantly different from the risk group and the group with impaired health by the severity of "reaction formation" ($p = 0.000$) and "intellectualization" ($p = 0.000$) defences.

It is found that healthy children usually choose the strategies of passive distraction from difficulties ($p = 0.003$), while children with impaired health usually focus on destructive-emotional expression ($p = 0.000$).

The study has also revealed the correlation interrelations between psychological health and its characteristics and indices of psychological defences and coping behaviour strategies in different groups of children.

In our study, the "displacement" ($R = -0.44$, $p = 0.027$) and "regression" ($R = -0.39$, $p = 0.05$) mechanisms in the group of children with impaired psychological health have a negative correlation with the components of psychological health, while a "denial" mechanism is positively associated with an emotional component of psychological health ($R = 0.42$, $p = 0.037$). For this group of children, it is shown that the higher is anxiety ($R = -0.74$, $p = 0.000$; $R = -0.58$, $p = 0.003$; $R = -0.71$, $p = 0.000$), emotional symptoms ($R = -0.50$, $p = 0.011$; $R = -0.72$, $p = 0.000$; $R = -0.47$, $p = 0.018$) and internal problems ($R = -0.52$, $p = 0.007$; $R = -0.61$, $p = 0.001$; $R = -0.49$, $p = 0.013$) in children with impaired psychological health, the more rarely they resort to such defence mechanisms as repression, compensation and displacement; the lower is the level of pro-social behaviour ($R = 0.43$, $p = 0.032$), the more rarely they use such defence mechanism as intellectualization.

The discovered relations to these indices support the assumption that descending to the earlier stage of development in behaviour and transformation of a feeling or an object enable children with impaired health to keep their Ego and reduce anxiety, while rejecting everything unpleasant worsens their emotional state. Note also that the ability to act rationally in an emotionally charged situation is a factor of health and contributes to the manifestation of pro-social behaviour.

For children with impaired psychological health, it is shown that the more severe is anxiety ($R = -0.39$, $p = 0.05$), the lower is the index of choosing emotional-destructive strategies; the higher are the indices of emotional symptoms ($R = -0.453$, $p = 0.023$), internal problem ($R = -0.55$, $p = 0.005$) and general evaluation of problems ($R = -0.45$, $p = 0.024$) of a child, the more rare are the strategies of active distraction. In general, we note that emotional deviation restricts the possibility of resorting to children's behavioural strategies that are socially approved by adults and the most efficient in coping with difficulties.

In the risk group, a defence mechanism "compensation" correlates positively with a social component of health ($R = 0.66$, $p = 0.004$) and hyperactivity ($R = 0.49$, $p = 0.045$), a "denial" mechanism correlates negatively with a social index of health ($R = -0.53$, $p = 0.026$), problems with peers ($R = -0.57$, $p = 0.017$), emotional symptoms ($R = -0.64$, $p = 0.005$) and internal problems ($R = -0.59$, $p = 0.012$), while "intellectualization" correlates negatively with an emotional component of health ($R = -0.55$, $p = 0.021$) and a general index of psychological health ($R = -0.39$, $p = 0.05$). In this group, the index of the "repression" defence mechanism positively correlates with problems with peers ($R = 0.62$, $p = 0.007$), problems in behaviour ($R = 0.53$, $p = 0.029$), hyperactivity ($R = 0.51$, $p = 0.036$)

and internal problems (R = 0.58, p = 0.013). The index of a “projection” mechanism correlates negatively with the problems with peers (R = –0.49, p = 0.04) and internal problems (R = –0.60, p = 0.011), and positively with the index of pro-social behaviour (R = 0.52, p = 0.032).

These correlations indicate that secondary (mature) defences, which do not allow negative information to reach the consciousness and demand an obligatory participation of thinking and waste of energy, reduce energetic background, which leads to child’s limited ability to control his behaviour, anxiety and impulsiveness, as well as growing signs of oppositional behaviour. Projecting responsibility for one’s mistakes on others and transferring unacceptable feelings and strivings to people around reduces “stresses” turned inwards and behavioural deviations and improves pro-social focus of behaviour.

We have revealed positive correlations between coping strategies and the components of psychological health in the risk group. The strategy of destructive emotional expression correlates with the index of anxiety (R = 0.44, p = 0.08), while the strategy of reflexive withdrawal correlates with social (R = 0.48, p = 0.05) and emotional (R = 0.58, p = 0.015) components of health, a general index of psychological health (R = 0.59, p = 0.012), problems in behaviour (R = 0.541, p = 0.025), hyperactivity (R = 0.55, p = 0.021) and external problems (R = 0.546, p = 0.023). For this group, the study has shown that the higher is the level of anxiety the more rarely children resort to the strategy of reflexive withdrawal (R = –0.50, p = 0.04); the higher is the general index of health (which indicates impaired health), there more rarely they choose the strategy of emotional-destructive expression(R = –0.48, p = 0.05); the more severe are the problems in behaviour, the more rare is active distraction (R = –0.541, p = 0.025).

The discharge of emotions, frequent tears, and resorting to confession are known to be a characteristic of anxious, restless and inattentive, naughty children [Nikolskaya, Granovskaya, 2000]. On the one hand, the use of these coping strategies enables to overcome internal tension; on the other hand, it causes impairments in the emotional and social aspects of health in such children.

Children with preserved health reveal negative correlations between the mechanism of “reaction formation” and a social index of health (R = –0.457, p = 0.025), a general index of psychological health (R = –0.454, p = 0.026) and the index of anxiety (R = –0.41, p = 0.04). In this group, it has been shown that the lower is the level of anxiety, the more rarely children resort to a “regression” mechanism (R = 0.41, p = 0.04); the less severe is anxiety, the higher are the values of the index of “repression” mechanism (R = –0.43, p = 0.035). For this group of children, it is shown that the less severe are anxiety (R = 0.41, p = 0.04) and the impairments in the emotional component of health (R = –0.48, p = 0.018), the more rarely children choose the strategy of passive distraction.

Thus, resorting to intellectual resources (developing the feeling of subjective control, learning “higher social values”, searching for ways to correct or displace) to remove emotional experiences allow healthy children to obtain control in various spheres and ability for strong feelings and experiences.

4 Conclusions

1. Children with impaired psychological health use a limited repertoire of psychological defences related to primary (primitive) mechanisms of a psychological defence. Children with preserved mental health are more likely to resort to psychological defences and mostly use defence mechanisms corresponding to a higher level of personal development.
2. Children with impaired psychological health report emotional-destructive coping strategies, which contributes to a negative release of emotions and makes a child more

vulnerable to an illness. The group of healthy children reports the use of socially desirable coping strategies aimed at constructive exit from a difficult situation, which enables them to get rid of accumulated negative experiences fully or partially.

3. Psychological defences and coping behaviour play a resource or a compensatory role depending on the state of psychological health. In children with impaired health, psychological health is mostly conditioned by primary (primitive) psychological defences, in children from the risk and healthy groups – by rational actions in an emotionally charged situation. Children socially approved behavioural coping strategies are a resource of health in children with impaired psychological health. The strategies of reflexive withdrawal and active distraction are a predictor of children's psychological health.

Our results are of great differential-diagnostic importance; besides, they enable to elaborate differentiated methods of psychological prophylaxis and correction of psychological health impairments.

5 Limitations

The study of children's psychological health is based on the expert assessment of specialists who work with children. Therefore, we can get a true picture of health. We investigate the correlation between the indices of psychological health (based on three independent instruments) and psychological defence mechanisms and coping strategies instead of cause and effect relations. Therefore, we can focus on the involvement of adaptation processes into the psychological health.

The reported study was funded by RFBR according to the research project No. 16-36-00038.

References

1. E. Kuftyak, Psihol. issled **22**, 3 (2012)
2. Ph. Cramer, J. of Personality **66**, 919 (1998)
3. I. Nikol'skaja, R. Granovskaja, *Psychological defense of children* (Saint-Petersburg, 2000)
4. I. Vetrova, Tomsk State University J **339**, 135 (2010)
5. E. Kuftyak, Med. psihol. Ross **24**, 11 (2014)
6. I. Dubrovina, Development of Personality **2**, 67 (2015)
7. V. Ananiev, *Health Psychology* (Saint-Petersburg, 2006)
8. O. Khukhlaeva, *Bases of psychological counseling and correction* (Moscow, 2011)
9. R. Goodman, H. Slobodskaya, G. Knyazev, Eur. Child and Adol. Psych **14**, 28 (2005)

Prediction for driving behaviour in connection with socio - demographic characteristics and individual value system

K.J. Lazdins and K. Martinsone

Rīga Stradiņš University, Riga, Latvia

Abstract. The aim of research "prediction for driving behaviour in connection with socio-demographic characteristics and individual value system" was to examine characteristics of individual value system prediction for driving behavior. It raised fundamental question for the research: 1. which of the individual value system characteristics predict driving behavior controlling gender and age.

In the study 108 respondents participated, 40 (37.0%) men and 68 (63.0%) women who filled the questionnaire on the Internet. Two questionnaires were used – "Latvian driving behavior survey" [1], the value and levels of availability relations in different spheres of life" [2, 3].

The results showed that the value system integrity/disintegrity indicator predicts distracted driving, explains 18% of variation and is statistically significantly. Internal vacuum and age statistically significantly negatively predicts risky driving explaining 17% of variation. Age statistically significantly predicts safe and courteous driving, explains 12% of variation. Value system integrity/disintegrity indicator and gender, statistically significantly negatively predicts summary indicator of dangerous driving explain 22% of variation. Age statistically significantly negatively predicts distracted driving, explains 30% of variation.

The results can serve as the basis to create new driving behavior interventions and also applicable to psychologist's professional work, when counseling individuals of this group, as well as can be used in the future development of the field, science and research.

Key words: aggressive driving, distracted driving, driving behavior, individual values, risky driving, safe driving.

1 Introduction

Several countries have made fatal road accident reduction, but the total number of accidents remains unacceptably high – 1.25 million people a year [4]. Although in recent years the number of fatal collision in Latvia has decreased, but in comparison with other European Union countries, it is still quite high. In 2012 there was a significant reduction in road traffic accidents, but unfortunately, in 2013 and 2014, road traffic accidents occur on an upward trend [5].

The number of factors causing road traffic accidents can be divided into: vehicle factors, environmental factors, as well as human factors [6]. Previous research has shown that about

40% of road traffic accidents are directly caused by the human factor and indirectly involved for about 50% of accidents, so then totally approximately 90% of road traffic accidents are directly or indirectly related to the human factor. Environmental factors (roadways conditions, weather conditions) is the possible cause of about 34% of accidents and vehicle factors as causes of accidents provides only for 13% of accidents [7]. By the concept of the human factor are understood human psychological conditions, cognitive processes, personality factors, demographic factors, attitudes [8].

Values are defined as long term belief that a particular form of action or destination of existence is personally, socially preferable and more desirable compared to other form of action or destination of existence [9].

In the doctoral thesis of Latvian researcher Inese Muzikante is shown that individual values can significantly predict risky driving, the value "dominion" is the most important predictor of violation from all values, value "conformism" can be an important predictor of risky driving, value "stimulation" (sensation seeking) does not predict risky driving, value for "safety" in only one part of the study predicts risky driving, which means that it is not stable predictor for risky driving [10].

Currently there is no clear evidence that the individual value system characteristics predict driving behavior, the authors not been able to find such studies, so it is necessary to examine whether an individual value system characteristics predict driving behavior and if so, then how.

The aim of the research was to examine characteristics of individual value system prediction for driving behaviour.

Research question was: which of the individual value system characteristics (internal vacuum, internal conflict, the neutral zone, value system integrity/disintegrity indicator) predict driving behaviour (distracted driving, aggressive behaviour towards other traffic participants, risky driving, safe and courteous driving) controlling by gender and age.

2 Materials and methods

2.1 Participants

The research involved 108 Latvian respondents participated. There were 40 (37.0%) men and 68 (63.0%) women, aged from 19 to 58 years. Based on descriptive statistics, the empirical distribution does not match the normal distribution, which means that the results are attributable to the sample not to the population of Latvia.

2.2 Instruments

2 questionnaires were used in this research:

1) Questionnaire on Latvian driving behaviour [1]. Respondents were asked to answer how often they act while driving like described in the proposed statements, using 5 point Likert scale (1 – never, 5 – very often). The final version of the survey consists of 26 statements that allow to evaluate four driving behaviour types: (1) risky driving (6 statements), (2) aggressive behaviour towards other traffic participants (6 statements), (3) distracted driving (6 statements), (4) safe and courteous driving (6 statements).

2) "The value and levels of availability relations in different spheres of life" (Уровень соотношения «ценности» и «доступности» в различных жизненных сферах) [2, 3]. Respondents were proposed to fill out two tables – at first they needed to create hierarchy from 12 different values, second – they needed to decide which of them is easier to reach from their point of view. The values were: (1) active life, (2) health, (3) interesting job, (4)

beauty of nature and art, also enjoying it, (5) love, (6) material ensured life, (7) reliable and good friends, (8) self confidence, (9) exploration, the opportunity to expand their horizons, education, (10) freedom as the independence of the actions and activities, (11) happy family life, (12) innovation [2]. The method is based on the idea of the two spheres of conscious activity, which by its nature is analogous to these psychological characteristics "significance" and "availability". So, the author values ranked not only by their importance, but also according to their implementation options in reality. In the field of importance includes leading life values, they include personal plans for life goals of future. In the field of availability includes everything that can be reached directly and are associated with specific goals which are located in the zone called "easy to reach zone". According to this method it is possible to calculate whether the value leads to internal conflict, internal vacuum is located in the neutral zone, as well as the value system of integrity/disintegrity indicator.

For both instruments was received permission from authors to use the instruments by an e-mail.

The author includes a demographic survey which includes questions about gender, age, monthly income, education, how long ago acquired a driver's license, how many kilometres they drive on average per month, how much warnings and violations they have received during past 3 years, the brand of personal car.

2.3 Procedure

The survey was published on the internet website (`www.google.docs.com/`) and participants were offered to fill out the questionnaire on the Internet, by placing ads in various social sites, like-minded groups and forums.

Sample was selected using the following criteria – acquired driver's license, gender, age.

Instructions for the participants were given clear and understandable; also an objective of the study was notified. They were informed that the data will be examined as a package and not individually, respondents were asked to provide real and not socially desirable answers. The survey took place during the period from 8 February to 21 April, 2016.

2.4 Data processing methods

Data was verified, entered in Microsoft Office Excel 2008, then imported to the program SPSS Statistics 21.0/22.0, where was carried out all the survey results processing. In the program SPSS 22.0 Statistics were calculated the results of descriptive statistics, Cronbach's alpha scores. To answer to the research question first was calculated the Pearson correlation coefficient, then, based on the results, was calculated hierarchical regression analysis (step method).

3 Results

To answer the research question: which of the individual value system characteristics (internal vacuum, internal conflict, the neutral zone, value system integrity/disintegrity indicator) predict driving behaviour (distracted driving, aggressive behaviour towards other traffic participants, risky driving, safe and courteous driving) controlling by gender and age, based on Pearson correlation coefficient hierarchical regression analysis was calculated.

Table 1. Hierarchical regression analysis for independent variable value system integrity/disintegrity index and dependent variable distracted driving, controlled by gender and age.

	Distracted driving	
	ā	ΔR^2
1. Step		0.14
Gender	–0.12	
Age	–0.39	
2. Step		0.18
Gender	–0.17	
Age	–0.35	
Value system integrity/disintegrity indicator	0.21*	
Sum ΔR^2		0.32

*$p < 0.05$, **$p < 0.01$.

To predict distracted driving, in the first step of regression analysis was included the gender and age. This model explains 14% of the variation, but it is not statistically significant. Based on Pearson correlation coefficient, in the second step in addition to gender and age was included value system of integrity/disintegrity indicator, which additionally explained 18% of the model and is statistically significant (Table 1).

Table 2. Hierarchical regression analysis for independent variable internal vacuum and dependent variable risky driving, controlled by gender and age.

	Risky driving	
	ā	ΔR^2
1. Step		0.09
Gender	–0.18	
Age	–0.31**	
2. Step		0.17
Gender	–0.22*	
Age	–0.26**	
Internal vacuum	–0.28**	
Sum ΔR^2		0.26

*$p < 0.05$, **$p < 0.01$.

To predict risky driving, in the first step of regression analysis included gender and age, this model explains 9% of its variations and age was found as statistically significant negative predictor for risky driving. Based on Pearson correlation coefficient, in the second step, in addition to gender and age was included an internal vacuum, in addition to explaining the 17% of this model and it was found as negative statistically significant predictor for risky driving (Table 2).

To predict safe and courteous driving, in the first step of regression analysis included gender and age, this model explains 12% of variation and age statistically significantly predict safe and courteous driving. Based on Pearson correlation coefficient, in the second step, in addition to gender and age was included in the neutral zone, in addition to explaining the 13% of this model, but it was not statistically significant (Table 3).

Table 3. Hierarchical regression analysis for independent variable neutral zone and dependent variable safe and courteous driving, controlled by gender and age).

	Safe and courteous driving	
	ā	ΔR^2
1. Step		0.12
Gender	–0.17	
Age	0.24*	
2. Step		0.13
Gender	–0.15	
Age	0.22*	
Neutral zone	0.15	
Sum ΔR^2		0.25

$*p < 0.05$, $**p < 0.01$.

Table 4. Hierarchical regression analysis for independent variable neutral zone and dependent variable summary indicator of dangerous driving, controlled by gender and age.

	Summary indicator of dangerous driving	
	ā	ΔR^2
1. Step		0.17
Gender	–0.20*	
Age	–0.42	
2. Step		0.22
Gender	–0.25**	
Age	–0.38	
Value system integrity/disintegrity indicator	0.25**	
Sum ΔR^2		0.39

$*p < 0.05$, $**p < 0.01$.

To predict summary indicator of dangerous driving in the first step of regression analysis included gender and age, this model explains 17% of variation and gender is a statistically significant negative predictor. Based on Pearson correlation coefficient, in the second step, in addition to gender and age was included value system integrity/disintegrity indicator, which additionally explained 22% of the model and was negative statistically significant (Table 4).

4 Discussion

The aim of the research was to examine characteristics of individual value system prediction for driving behaviour.

The study has answered to the fundamental question: which of the individual value system characteristics (internal vacuum, internal conflict, the neutral zone, values system integrity/disintegrity indicator) predict driving behaviour (distracted driving, aggressive behaviour towards other traffic participants, risky driving, safe and courteous driving) controlling by gender and age.

The results show that distracted driving can be predicted by value system integrity/disintegrity indicator which may indicate the fact that if an individual's values do not match with the real possibility of reach the value and as a result, is affected his motivation sphere, which can be expressed with frustration, internal discomfort, internal conflicts, self-realization shortages and unclear own identity, it is more likely that he will not pay full attention to direct driving tasks while driving.

The results show that risky driving can be negatively predicted by age and internal vacuum, it may also indicate that young drivers often becomes involved in risky activities while driving in traffic, as well as the fact that individuals who values are more available than those he has actually necessary and important, it is more likely to drive a car in risky way. Some of these results are in line with previous studies. Summarising up the results of different studies, one can conclude that the best traffic accidents and violations predictors are age and sex [11–15].

The results indicated that a safe and courteous driving can be negatively predicted by age, probably can be attributed to young people's experience and immaturity, as well as the physiological processes that take the place in young person growing up process, many young people sought after self-affirmation, so are more inclined to choose the most risky driving, not safe and courteous driving, as evidenced by the preliminary results and studies. The research highlights that one of the factors that increase the likelihood to drive in risky and aggressive way is age [16].

The results show that the summary indicator of dangerous driving, so the trend in general to drive a dangerous manner, predicts sex, which is in line with the previous studies. Men cars often runs intoxicated and rarely wear seat belts, regardless of their age [17], men tend to show higher rates thoughts of revenge and physical aggression [18]. It also emphasized that one of the factors that increase the likelihood of aggressive driving is simply to be a man [16]. Also summary indicator of dangerous driving predicts value system integrity/disintegrity indicator, which means that if an individual's values do not match with the real possibility of reach value and as a result, is affected his motivation sphere, which can be expressed with frustration, internal discomfort, internal conflicts, self-realization of the lack of and blurred own identity, it is more likely that he was driving a car driving a car in general will drive in aggressive, risky and dangerous way.

5 Conclusions

The aim of the study was achieved. Tasks of research were fulfilled and answers to research question received.

In Latvia until now have not been any studies, which reflect the individual value system, and driving behaviour relationships. Author believes that the choice of the theme is successful and useful for future research and this study although confirms the already existing and proven driving behaviour of forecasters, but also gives an idea about the factors that may incidentally affect driving behaviour.

Limitations of the research were related to the size of the sample, participant alignment, use of new instruments, as well as data collection method. Repeating this research would be desirable to increase the sample size and use of approbated instruments, as well as it would be interesting to find out how the factors of individual value system predict the objective number of accident and violations caused by driver. It would be necessary to obtain objective data on number of the road accidents and violations, as well as to collect data in person, also control the distribution of men and women groups.

The results can serve as the basis to create new driving behaviour interventions and also applicable to psychologist's professional work, when counselling individuals of this group, as well as can be used in the future development of the field, science and research.

References

1. Perepjolkina V., Voita D. Latvijas Autovadītāju uzvedības aptauja. Tehniskā atskaite. Nepublicēts. 2011
2. Фанталова Е.Б. *Диагностика и психотерапия внутреннего конфликта.* Самара: БАХРАХ-М. 2001
3. Jirgena S. Jauniešu pašizjūta un adiktīvas uzvedības izvēle. Promocijas darbs. Rīga: LU. 1999
4. World Health Organisation. Global status report on road safety 2013: supporting a decade of action. 2013
5. Ceļu Satiksmes Drošības Direkcija. 2004–2014. gados notikušo Ceļu satiksmes negadījumu, tajos bojā gājušo un ievainoto skaits. 2015
6. Lazdiņš K.J., Mārtinsone K. Autovadītāju agresijas un braukšanas uzvedības saistība ar personīgās automašīnas jaudu un virsbūves tipu. Nepublicēts bakalaura darbs. 2014
7. Dewar R.E., & Olson P.L. Human factors in traffic safety. Tuscon, AZ: Lawyers & Judges Pub. 2002
8. Muzikante I., Reņģe V. Autovadītāju individuālo vērtību saistība ar riskantu braukšanu. *LU raksti* **768**. 39–50. 2008
9. Rokeach, M. *The nature of human values.* New York: Free Press. 1973
10. Muzikante I. Vērtības, attieksmes un autovadītāju uzvedība uz ceļa. Promocijas darbs. 2011
11. Krahe B., & Fenske I. Predicting aggressive driving behavior: the role of Macho personality, age, and power of car. *Aggressive Behavior* **28**, 21–29. 2002
12. Özkan T., & Lajunen T. Why are there sex differences in risky driving? The relationship between sex and gender-role on aggressive driving. traffic offences, and accident involvement among young Turkish drivers.; *Aggressive Behavior* **31**, 547–558. 2005
13. Rhodes, N., & Pivik, K. Age and gender differences in risky driving: The role of positive affect and risk perception.; *Accident Analysis and Prevention.* 2010
14. Stradling, S.G., & Parker, D. Extending the theory of planned behaviour: The role of personal norm, instrumental beliefs and affective beliefs in predicting driving violations. 1997
15. Underwood G., Chapman P., Wright S., & Crundall D. Estimating accident liability. 1997
16. Tasca L. A review of literature on aggressive driving research. Aggressive Driving Issues Conference. 1–11. 2000
17. Lancaster R., Ward R. The contribution of individual factors to driving behaviour: Implications for managing work-related road safety.; Research Report. HSE Contract Research Report: HSE Books, 3–60. 2002
18. Deffenbacher J.L., Petrilli R.T., Lynch R.S., Oetting E.R., & Swaim R.C. The Driver's Angry Thoughts Questionnaire: A measure of angry cognitions when driving. *Cognitive Therapy and Research* **27**, 383–402. 2003

SHS Web of Conferences **40**, 03010 (2018) https://doi.org/10.1051/shsconf/20184003010
Int. Conf. SOCIETY. HEALTH. WELFARE. 2016

Parenting in low-income families from the perspective of social work

A. Banovcinova*, E. Mydlikova, and M. Vodicková

Trnava University, Trnava, Slovakia

Abstract. The family is the primary socializing factor. Its mission is to create a stable and safe environment for the child and for his comprehensive development. The role of parents is to give to a child the values and norms accepted in the society, which is a prerequisite for successful integration of a child into society. That is why the upbringing function of the family is regarded as one of the main. This feature is, however, influenced by many factors.

The goal of the study was to find how poverty affects parenting. For the data collection was used questionnaire Alabama Parenting Questionnaire (APQ), which measures parenthood through five dimensions (1) positive involvement with children, (2) Supervision and monitoring, (3) use of positive discipline techniques, (4) consistency in Theus of discipline chniques, (5) use of corporal punishment. The sample was divided into two groups, with the first group consisted of 188 parents living in poverty The reference group consisted of parents living in households with income standard (N – 188). Analysis of the results showed differences between parents living in poverty and between parents with a standard rate of income especially in monitoring and supervision, and also in the use of positive disciplinary techniques. On the contrary, there were no significant differences in cooperation between the parents or the use of corporal punishment. Based on the results it is clear that poverty is one of the factors affecting parenting. Therefore, social worker who works with low-income families should focus attention on this area of family functioning.

1 Introduction

Poverty in contemporary society is increasingly widespread problem. The number of low-income families continues to grow. According to statistics from the Central Office of Labour, Social Affairs and Family in December 2016, families with dependent children accounted for 35.38% of the total number of recipients of social assistance benefits [1].

Mydlíková [2] notes that the most common area where there is a disruption of the functioning of the family is the economic function. Although the disorder (dysfunction) manifesting in the disruption of economic functions normally respond to social therapy efficiently and rapidly, the danger is that the disorder typically trigger malfunction in other functions.

* Corresponding author: banovcinova.a@gmail.com

A life of poverty and poor socio-economic situation of the family causes detriment to a child, not only through inadequate meeting his needs due to lack of funds, but also through the influence of parents. Parents play the most important role in the socialization of the child. Through them, the child learns the norms and values preferred within the particular society, accepted patterns of behavior and builds self-regulating mechanisms. Also Backmen and Ferrarini [3] point at the risk of child poverty and the need to tackle this problem. They consider child poverty a more serious problem than adult poverty because children are unable to control it. According to the aforementioned authors, a period of childhood spent in poverty has serious implications for the future development. The risk of social exclusion and poverty in the adulthood to some extent begins to manifest in childhood, when the cognitive abilities are formed. Child poverty is influenced by three main factors, namely the family, the labor market and the state. Main family factors include parents' education, their average age, the number of children in the family, child living with one or both parents.

Barnett, Corwyn and Bradley [4] draw attention to the impact of stress, which is part of everyday life of families with low socio-economic status. Even for adult family members, a bad economic situation is so stressful that affects their mental and health condition, which results in impaired childcare. Parents who tend to make do with whatever funds are often emotionally unprepared to provide necessary support to children. Low family income, poor living conditions, lack of access to services and opportunities, and a lack of common children's activities cause various negative impacts on children's lives [5].

Long-term living in poverty affects all areas of a child's life. Several studies have demonstrated the negative impact of low socioeconomic status families on child's physical development and health. The research carried out by Currie and Stabile [6] showed that children from families that can be classified as poor have on average lower birth weight and higher infant mortality risk. Duncan et al. [7] investigated the effect of economically disadvantaged environments on children's behavior. Their research indicates that these children are more frequently diagnosed with behavioral disorders. Roos, Nair, Deng and Burrell [8] found that children coming from low-income families have a higher likelihood of health problems, which manifest themselves in the first year of age compared to children from wealthy families. Their health condition is caused by insufficient and inadequate diet.

In addition to the physical and health conditions, life in the family at risk of poverty also impact social functioning of the child. Roos, Nair, Deng and Burrell [8] found that long-term poverty increases the probability of nonconformity in children, resulting in problems of dealing with the difficult situation in adolescence. In economically disadvantaged families, parents seldom appreciate developmental achievements and abilities of their children. They communicate a little with children, and there are noticeably large differences in quality content and form of communication. Mothers with low socio-economic status tend to be in education of their children more direct, closer and use punishment more frequently. They rarely spend their free time together with children [4].

Inability of parents to ensure the child enough incentives for all-round development due to insufficient resources is often reflected in the school results of the child.

Family income belongs to the most important factors that influence the success in school. These children suffering from poverty in preschool and school periods tend to have a lower successfulness in school as compared to children and adolescents who have experienced poverty later or never. Economic pressure in the families with adolescents can lead to a deepening division with parents [9]. Research of Hart and Risley [10] has shown that children coming from families who are dependent on social benefits can hear during the first three years of their life one million words from their parents as compared with children coming from families with a standard income. These children can hear during these years up to thirty million words from their parents, which is reflected in building of their vocabulary. Language plays one of the most important roles in the quality of life, future education and personality development. We are therefore inclined to the view that this issue requires sufficient attention if we want to alter thinking in future exempt from the state of poverty.

Lee and Orfield [11] found that students coming from high-income families have in the test better results than students from low-income families. Furthermore, research has shown that parents of students from high-income families have higher education on the basis of which they are able to pass their acquired knowledge to their children. It was also found that students coming from high-income families have more resources. These children have much easier access to modern facilities for education (eg. computers, toys) or go to pre-school which prepare them better for school. On the other hand, students living in low-income families have lower expectations, aspirations and motivation. Parents and teachers of pupils living in poor families do not have high expectations with the result that poor children are insecure with their abilities and are not motivated to perform educational duties at the required level.

A similar conclusion was reached by Havemen and Wolfe [12], who, in addition, found that children from economically disadvantaged backgrounds have worse school results. After completing the education is more difficult for them to integrate into the labor market [13] and they tend to have more health problems [14]. All these factors have ultimately impact on their job prospects and therefore on individual income levels [15].

We agree with the author Kusá [5], who states that parents who live in poverty have to fight daily for survival of the family and must often sacrifice to protect their offspring from the negative effects of poverty. Hoping to find the most suitable deal for themselves and children, they must often prioritize food and clothing for children. Even so, parents are blamed, in most cases, for not paying enough care to children, even when they try to make the most for them, even though they themselves are in bad living conditions. Since parents are responsible for the quality of life of their kids, parents are usually blamed for their problems. Families go into debt, as parents seek to cover family expenses based on the needs of the family.

2 Methodology

It is indisputable that the family is the most important factor contributing to healthy mental and physical development of children. It significantly participates in shaping their personality and is important in building their identities. The quality of relationships between parents and children influences health status and quality of life of the child. If the person copes with adverse life situations a family should create the primary social networks and help him overcome this situation. In pursuit of survival of individuals family environment plays an important role. In order for the family was able to provide its members with optimal conditions for life, it must achieve an adequate degree of functioning in all dimensions and functions. In families living in poverty, there is a failing of economic functions of the family. Research shows, however, that loading of the economic functions of the family affects its functioning in all other areas.

In poor families we are often faced with the presence of stress, which results in depression, anxiety or the wrath of parents, which results in problems in behaviour. Parents who are experiencing such tensions are less sensitive to the needs of their children, which we can observe in the education and care.

The aim of our research was to determine how poverty affects parenting.

2.1 Participants

The sample consisted of 188 respondents living below the poverty line. The condition for inclusion in the survey was being a parent of at least one child younger than 18 years and living with that child in the household (mean age = 37.21 ± 6.824, range = 21–60). The second group of respondents formed a reference group consisting of 188 respondents – parents of children under 18 living in households with standard income (mean age = 40.63 ± 5.840, range = 27–60).

2.2 Measures and procedure

To achieve the objective, we chose a quantitative research strategy. The method of data collection was a questionnaire. To collect data, we used two questionnaires.

The first type was standardized questionnaire Alabama Parenting Questionnaire (APQ), authored by Frick [16]. It consists of 42 closed scaling issues. Our elected questionnaire focuses on measuring five dimensions of parenting: (1) positive involvement with children, (2) supervision and monitoring, (3) use of positive discipline techniques, (4) consistency in the use of such discipline and (5) use of corporal punishment.

The second type was the questionnaire of our own design, which included questions aimed at determining the demographics.

3 Results

The results were analysed using statistical software SPSS.

Table 1. Differences in dimensions of parenting.

Dimension	Group	*N*	Mean rank	*U*	*Z*	*p*
Positive involvement with children	Group 1	188	197.22	16032.500	-1.560	.119
	Group 2	188	179.78			
Supervision and monitoring	Group 1	188	201.53	15222.000	-2.330	**.020**
	Group 2	188	175.47			
Use of positive discipline techniques	Group 1	188	198.24	15840.500	-1.745	.081
	Group 2	188	178.76			
Consistency in the use of such discipline	Group 1	188	199.85	15537.500	-2.037	**.042**
	Group 2	188	177.15			
Use of corporal punishment	Group 1	188	183.30	16694.000	-.936	.349
	Group 2	188	193.70			
Total	Group 1	188	207.87	14031.000	-3.457	**.001**
	Group 2	188	169.13			

*Group 1 – income below the poverty line.
Group 2 – standard income.

https://doi.org/10.1051/shsconf/20184003010

Our goal was to see how poverty affects parenting and its various dimensions in your reports. When comparing the groups of respondents living below the subsistence level, with the group of respondents with standard income, we found differences in several dimensions we monitor.

As presented in Table 1, analysis of the results showed significant differences between groups of respondents living in poverty and the group of respondents with a standard income in the dimension of "Supervision and monitoring" ($Z = -2.330$, $p = 0.020$). Similarly, significant differences between low-income parents and parents with a standard income were shown in the dimension of the "Consistency in the use of dry discipline" ($Z = -2.037$, $p = 0.042$). Mann-Whitney test showed significant differences between the two groups of our interest in the total score ($Z = -3.457$, $p = 0.001$). Analysis of the results did not demonstrate significant differences between the group of respondents living in poverty and the group of respondents with a standard income in the dimension of "Positive involvement with children" ($Z = -1.560$, $p = 0.119$). There were no significant differences in the dimension of "Use of positive discipline techniques" ($Z = -1.745$, $p = 0.081$). The last was monitored dimension was "*Use of corporal punishment*". Even in this dimension, significant differences between the participants living in low-income families and the group of respondents living in families with a standard income were not shown ($Z = -.936$, $p = 0.349$).

4 Discussion and conclusion

Poverty determines the family in many areas and affects its entire functioning. The family is the place where the child is socialized and learns the values and the patterns of behaviour in specific situations. Essential role in this process is played by parents. The aim of our research was to determine how poverty affects parenting and its various dimensions.

The results of the research show us that poverty affects parenting particularly in the dimensions "*Supervision and monitoring*" and "*Consistency in the use of such discipline*". Similar results were reached by several authors dealing with the impact of poverty on the behaviour of the parents and the child. McLeod and Shanahan [17] show a correlation between current family poverty and mother's weak emotional responsiveness and frequent use of physical punishment. Similarly Jocson and McLoyd [18] point to the fact that the disorder, parents' distress, and harsh and inconsistent discipline are important pathways that contribute to the development of socio-emotional problems in low-income children across a wide range of ages. Hashima a Amato [19] associate poverty with unsupportive and punitive behavior of parents. Parental interest in child, manifesting through supervision, has a significant impact on his survival and behaviour. Parental indifference adds stress to the child, which is usually often overlooked. Such overlooked children are unable to concentrate, are often anxious, apathetic, aggressive or enclosed in themselves. Up to 65% of these children had to repeat first grade because the intense stress during this period may result in impairment of the functions of brain centers that affect learning and damage the child's intellect. Children from families where the tension, hatred, and quarrels prevail, may present emotional instability a propensity of jealousy, are often weeping, and also may be hypersensitive and touchy [20]. Lacinova and Škridlikova [21], in turn, mentioned that child from birth responds to the demands of his parents through inborn dispositions and thus provides them with feedback. The parent is able to judge this bond, meaningfully evaluate and change his attitudes and behaviour towards the child. Certainly worth mentioning that, as stated by the authors, in other cases communication between parents and children is usually burdened with stereotypes of certain behaviour. To create a specific process of education, in addition to the child, other influences are also involved, affecting the creation of educational practices, such as parents' previous experience, temperament.

We agree with the Klimáčková [22], that the needs of children and parents are different to everyone, and timely prevention and support creates long-term benefits for society and the family. It contributes to the prevention of child abuse and neglect caused by wrong approach, ignorance or inability to take care of their children. At the same time the conducted research points to the significance of the first years of life for achieving the quality relationships between parents and children. During this period of child's life, excessive stress, caused by neglecting or bullying parents, destroys his brain development. In connection with disciplinary practices such as requiring discipline of children we must mention the arguments of the author Prekopová [23] suggests that currently the most common cons of upbringing is bossiness of children and that it is often expressed through aggressiveness and defiance. They are reluctant to submit to the requirements of adults and unwilling to accept their authority. Such children seek to manipulate the parents, dictate their conditions and they gain them by force. Finally, it is very important to note the negative aspects of the child and be able to recognize causes of such behaviour. Also Morrish [24] considers a negative impact in the upbringing the lack of defined borders, which has a significant impact on children. Without the establishment of adequate border child is usually rebellious, unmanageable and the parents are frustrated with such a style of upbringing.

An important area of parenting is the application of disciplinary techniques. As we mentioned above, our results showed differences between low-income families and families with standard income in the dimension "Consistency in the use of dry discipline". Pinderhughes et al. [25] indicate that the parent discipline practices are integral in theories of children's socialization. Parents' use of physical punishment with their child is of special interest. Numerous theories posit a role for physical punishment in the development of antisocial behavior in children. According to one set of theories, discipline responses are made in the context of multiple influences ranging from more distal factors such as culture, ethnicity, and socioeconomic status to more proximal factors such as available social supports, family structure, and family processes.

According to Matejček [26], a sentence is what the child perceives as punishment and not how his parents have punished him. Such a feeling is humiliating, unpleasant and even embarrassing, and child would like to avoid it in the future. One of disciplinary techniques is the use of corporal punishment. In this dimension, the impact of poverty has not been demonstrated, that the differences between the two families surveyed were not found. Corporal punishment is engaged in many researches. Gershoff and Bitensky [27] point to the importance of corporal punishment which parents most commonly used in response to the behavior of children who shows signs of aggression. On the other hand, several studies confirm the negative effect in the short term but also in the long term. Corporal punishment against children was included in research in 1999. The survey was made by SLONAD on a sample of 5230 children aged 15–16 years which showed that up to one quarter of those surveyed children experienced physical abuse beatings, kicks even cutting from their parents. One-third of abused children answered that their experiences are still traumatizing and they haven't spoken to anyone about them. The most common forms of domestic violence against children include psychological and physical abuse, verbal abuse, particularly threatening [28].

Living conditions of children in families and communities are, as a result of the increasing workload of parents the stress and the collapse of traditional large families, increasingly demanding, so children often suffer from loneliness. As reported by the Research of Institute of Information and Prognosis in Education in 2005, it was found that 47.3% of children and more than 56% of students cannot express their opinion at home due to authoritarian parent. Psychological punishment by parents confirmed 21% of children and physical punishment 15%. Lack of care, bullying, humiliation, insufficient understanding, superior relationship and superficiality were showed, on average, by 6% of children. Children aged 11–16 were more often physically punished [29] (Správa o mládeži, 2010). Research conducted in the Australia in 2006 Griffith University, Brisbane confirms the association between the violence against children in the family and juvenile delinquency. Negative factors affecting the family-juvenile delinquency include not only beating but also the lack of supervision of parents, harsh and temperamental discipline, child abuse, cold attitude and relationship of parents leading to the neglect of their children. This research found that children who were hurt in childhood did not commit more acts of violence than children who were not hurt. Children who were abused also in the adolescence were more prone to delinquency [22].

Use of positive discipline techniques are also among major disciplinary techniques. Nor in this dimension the differences between families living in poverty and families with a standard income were not confirmed. We are inclined to arguments of a number of psychologists who warn that the use of praise in the upbringing of the child shall be controlled. Excessive praising of children or vice versa little praise affects motivation and achievement of their goals. The purpose of our praise is to encourage the child to continue in anything positive and perform well. If a parent praises wrong, the praise will be counterproductive for him [30]. Between years 2001–2003 empirical research was conducted, examining the issue of remuneration and punishment in the family education in a research sample of secondary school students in our country and the Federal Republic of Germany. Samples were selected into research because of the application different styles of education in these countries. The results of the research show that in German and Slovak families negative educational methods are at the forefront. Parents often order children, prohibit, admonish them or threaten them. They evaluate their children more often when they are guilty of something, and hereby reduce their self-esteem and personal pride. On the other hand, the child's good behavior often remains unnoticed and unappreciated, as parents had taken it for granted [31]. According to Kohn [32], the bulk of compliments intended for children are formed by word reward that affect them as well as many other material rewards. Praising does not represent self-serving tactics by parents. They can easily compliment him on the grounds that they are happy to see what he did. If parents rate their child praising them too often, he will rely on them increasingly and will not learn to make choices and proper judgment. If we look at praise from the psychological point of view, Kopriva et al. [33] perceive praise as superior, power and also manipulative use of some sort of positive comments about the behaviour, properties or activities of the other. At the same time it represents an unbalanced relationship in which the individual feels entitled to evaluate others.

The way parents fulfil their parental role is an important factor affecting the bio-psycho-social functioning of the child. The results suggest that parental accompanying and the development of parenting skills should be an area that emphasized by social worker when working with low-income families.

This paper was prepared as a part of the project VEGA 1/0487/14 Key Concepts of Sesected Systemic Theories in Assesing Riskness of the Families within the Purview of Children Social Protection.

References

1. Sociálne dávky [online]. http://www.upsvar.sk/statistiky/socialne-veci-statistiky/2016/2016-socialne-davky.html?page_id=583002 (2016)

2. E. Mydlíková. *Vybrane aspekty sociálnej prace s rodinami dlžníkov nájomného*. In: Zborník príspevkov: Rizikové súvislosti chudoby a rodiny v súčasnej slovenskej spoločnosti. Ružomberok: VERBUM (2011)
3. M. Magadi, Risks Factors for Severe Child Poverty in the UK. Journal of Social Policy **39**, 2,
4. O. Matoušek, H. Pazlarová, *Hodnoceni ohroženeho ditěte a rodiny*. Praha: Portal, (2010)
5. Z. Kusá, *Pre zlepšenie života deti v Europe: vysvetlenie chudoby deti v Europskej unii*. Bratislava: Slovenska sieť proti chudobe a Sociologicky ustav SAV (2013)
6. J. Currie, M. Stabile,. Socioeconomic Status and Child Health: Why Is the Relationship Stronger for Older Children? In The American Economic Review **93**(5), (2003)
7. G.J. Duncan, J. Brooks-Gunn, P.K. Klebonov, Economic Deprivation and Early Childhood Development. In Child Development **65**(2), (1994)
8. M.W. Roosa, S. Deng, R. Nair, G.L. Burrell. Measures for studying poverty in family and child research. Journal of Marriage and Family **67** (2005)
9. J. Brooks-Gunn, G.J. Duncan, The Effect of Poverty on Children. The Future of Children **7**, 2 (1997)
10. B. Hart, T. Risley, Meaningful Differences in the Everyday Experience of Young American Children. Baltimore: Paul H. Brookes Publishing Co, (1995)
11. D. Gerbery, I. Lesay, D. Škobla, *Kniha o chudobe: Spoločenske suvislosti a verejne politiky*. Bratislava: Priatelia Zeme-CEPA (2007)
12. R. Haveman, B. Wolfe, The Determinants of Children's Attainments: A Review of Methods and Findings. In Journal of Economic Literature **33**, 4 (1995)
13. P. Gergg, P. Machin, *The Relationship Between Childhood Experiences, Subsequent Educational Attainment and Adult Labour Market Performance* [online]. `http://personal.lse.ac.uk/machin/pdf/lux2000.pdf` (1999)
14. A. Currie, M.A. Shields, S.W. Price, *Is the Child Health/Family Income Gradient Universal? Evidence from England* . Bon: Forschungsinstitut zur Zukunft der Arbeit Institute for the Study of Labor (2004)
15. S.P. Jenkins, T. Siedler, *The intergenerational Transmission of Poverty in Industrialized Countries* [online]. Berlin: German Institute for Economic Research, (2007)
16. C.A. Essau, S. Sasagawa, P.J. Frick, Psychometric Properties of the Alabama Parenting Questionnaire. Journal of Child and Family Studies **15**, No. 5, October (2006)
17. J.D. McLeod, M.J. Shanahan, Poverty, Parenting, and Children's Mental Health. In American Sociological Review **58**, No. 3 (1993)
18. R.M. Jocson, V.C. McLoyd, Neighborhood and Housing Disorder, Parenting, and Youth Adjustment in Low-Income Urban Families. In American Journal of Community Psychology **55** (2015)
19. P.Y. Hashima, P.R. Amato, P.R. Poverty, Social Support, and Parental Behavior. In Child Development. 1994. **65**, Issue 2 (1994)
20. D. Goleman, *Emočni intelignece*. Bratislava: Columbus (1997)
21. L. Lacinová, P. Škridlíková, *Dost dobří rodiče, aneb, Drobné chyby ve výchově dovoleny*. Praha: Portal (2008)
22. A. Klimáčková, *Podpora rodičovstva – podpora spoločnosti*. Bratislava: Národne centrum pre rovnosť príležitostí (2011)
23. J. Prekopová, *Malý tyran*. Praha: Portál (2007)
24. R. Morrish, *12 klíčů k důsledné výchově*. Praha: Portál (2003)

25. E.E. Pinderhughes, K.A. Dodge, A. Zelli, J.E. Bates, G.S. Pettit, Discipline Responses: Influences of Parents' Socioeconomic Status, Ethnicity, Beliefs About Parenting, Stress, and Cognitive-Emotional Processes. Journal of family psychology: JFP: journal of the Division of Family Psychology of the American Psychological Association (Division 43) **14**(3), (2000)
26. Z. Matějček, *Po dobrém, nebo po zlém?* Praha: Portál (2007)
27. E.T. Gershoff, S.H. Bitensky, The case against corporal punishment of children: Converging evidence from social science research and international human rights law and implications for U.S. public policy. Psychology, Public Policy, and Law **13** (2007)
28. S. Laca, Rodina ako primarne prostredie v spoločnosti, In: Prohuman. [online]. `http://www.prohuman.sk/socialna-praca/rodina-ako-primarneprostredie-v-spolocnosti` (2016)
29. *Správa o mládeži. Národná sprava o štátnej politike vo vzťahu k deťom a mládeži v SR*. Bratislava: Ministerstvo školstva SR (2010)
30. S. Miklovičová, *Chválite svoje deti za všetko? Nerobte to!*. [online]. `https://eduworld.sk/cd/stanislava-miklovicova/1597/chvalite-svojedeti-za-vsetkonerobte-to` (2016)
31. L. Koldeova, *Systém odmien a trestov v kontexte rodinnej výchovy v Slovenskej republike a Nemeckej spolkovej republike*. [online]. `http://www.pedagogika.eu/kiw_03/artykuly/KiW_3_85-97.pdf` (2012)
32. A. Kohn, *Five Reasons to Stop Saying "Good Job!" Young Children*. [online]. Dostupne: `http://www.alfiekohn.org/parenting/gj.htm` (2001)
33. P. Kopřiva a kol, *Respektovat a být respektován*. Kroměříž: Spirála (2008)

SHS Web of Conferences **40**, 03011 (2018) https://doi.org/10.1051/shsconf/20184003011
Int. Conf. SOCIETY. HEALTH. WELFARE. 2016

Why the share of small amount pensions is so substantial in Latvia?

O. Rajevska[1] and F. Rajevska[2]

[1]University of Latvia, Riga, Latvia
[2]Vidzeme University of Applied Sciences, Valmiera, Latvia

Abstract. More than 70% of all old-age pensions in Latvia are smaller than 300 euro, which is close to the monetary value of the at-risk-of-poverty threshold. There is a number of reasons for it: the lack of non-contributory component and inadequately low minimum pensions, the absence of redistribution mechanisms in the mandatory notional defined contribution (pillar I) and funded (pillar II) schemes, an unfair conversion of pre-reform employment record into pension formula, and a high tax burden on pensioners. The authors proposed a package of measures to improve the situation: an introduction of basis pensions, linking minimum pensions to the country average wages, increasing income tax exempt for pensions, restoration of the supplements for pre-reform employment and their regular indexation, removal of the threshold in initial notional pension capital calculation or its reduction from 30 to 20 years.

1 Introduction

Along with many other countries that undertook fundamental pension reforms in 1990s, Latvia went through an essential reform of its old-age pension system in 1995 by introducing the notional defined contribution scheme, later complemented by mandatory and voluntary private pension funds.

It was anticipated that the modern multi-pillar system would strengthen the social security of the elderly and bring decent pension benefits to the population. In reality, after twenty years of the reform being effective, the pensioners in Latvia are among the poorest in the EU.

Leading international experts note that "a primary objective of public pension systems is to provide adequate levels of retirement income to ensure that people are not at risk of poverty in old age" [1]. Meanwhile, the Latvian Ministry of Welfare, sees the objectives of the pension system in the provision of long-term financial stability (in the first instance) and in the adequate replacement rate in accordance with individual contributions [2, 3]. Such goal as the poverty alleviation, regretfully, is not articulated by the architects of the Latvian pension system.

2 Occurrence of small amount old-age pensions

The share of old-age pensioners in Latvia who receive small and very small pensions is disastrously high. According to the data of Central Statistical Bureau (CSB) of Latvia, in 2015, 70.5% of all old-age pensioners had pension benefits below 300 euro, 40.3% had pensions below 250 euro and 12.0% – below 200 euro. The monetary value of the at-risk-of-poverty threshold – specified at 60% of the national median equivalent disposable income – was not available for 2015 at the time of writing this paper (January 2017), while the CSB calculations for the previous past years return to 291 euro in 2014 compared to 260 euro in 2013, so one can expect a figure well above 300 euro for 2015. This means that more than 70% of old-age pensions actually were below the poverty threshold. Surely, some of the old-age pensioners are still working and receiving wages in addition to their pensions and some are receiving their pensions from more than one state, but to the majority of Latvian old-age pensioners public pensions are their only source of income (excluding the financial support from their family members).

The distribution of pensions by size is very much skewed: although the average old-age pension in September 2016 was equal to 296.09 euro, ca. 68% of all pensions were lower (see Fig. 1). On the other side of the scale, public pensions in Latvia have no upper limit. 1% of all pensions (not shown on the figure) are higher than 1000 euro, the highest old-age pension in 2016 being as high as 19 thousands euro per month [4]. Polarisation of rich and poor pensioners is only increasing with the course of time. The S80/S20 ratio for old-age pensions is about 3.0 in 2016, while in 2009 it was ca. 1.8 (for comparison – in neighbouring Estonia this ratio was 1.7 in 2016 [5]).

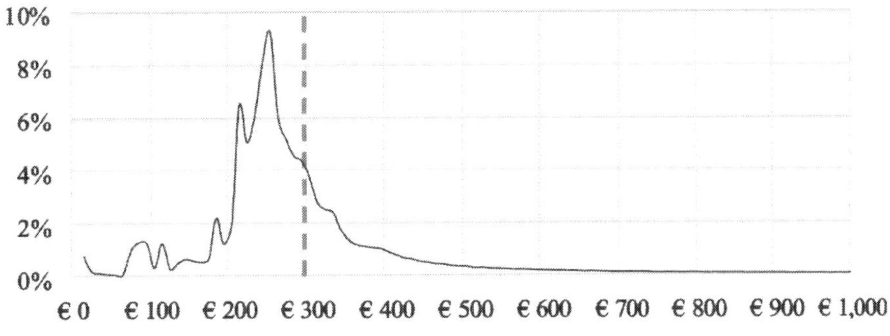

Fig. 1. Distribution of old-age pensions by size compared to the average pension in Latvia in September 2016. *Source: State Social Insurance Agency.*

The next figure (Fig. 2) provides the comparison of the at-risk-of-poverty rates in different age groups in Latvia as opposed to the EU averages during the last eight years. Almost in all age groups (except for the young aged 18–24) Latvians are more exposed to the risk of poverty, but the most critical situation is among those of retirement age, where the 2015 rates were twice higher than the EU averages, which is the second-worst result in the EU after Estonia. The surprisingly good rates in 2010–2012 resulted not from the improvement of financial situation of the pensioners, but were due to the exceptional decline in the incomes of the working age population in the crisis years.

With the population ageing and emigration of the employable people the share of pensioners in Latvia is expected to increase, making the issue of old-age poverty more and more acute. The causes of the existing situation and possible solutions are considered further in this paper.

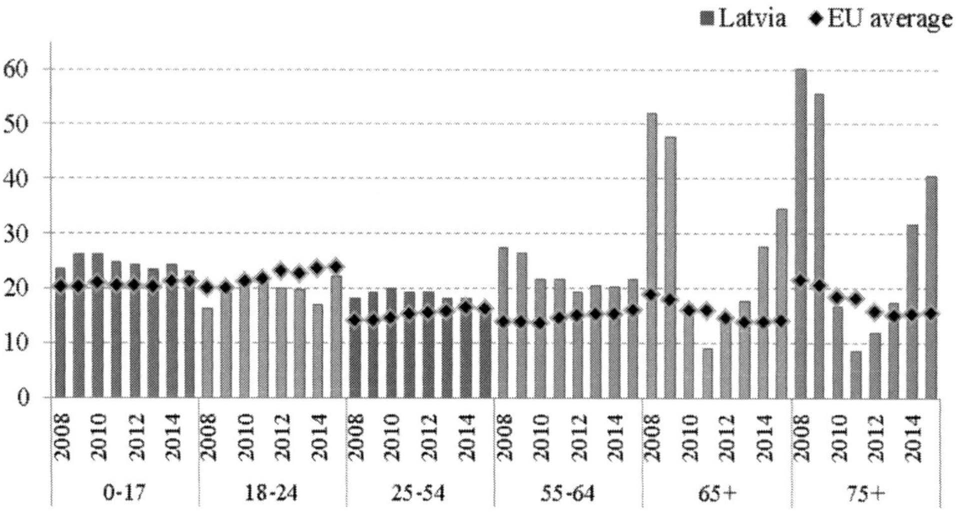

Fig. 2. At-risk-of poverty rates in Latvia and EU in different age groups in 2008–2015 (cut-off point: 60% of median equitable income after social transfers) *Source: Eurostat.*

3 Cause-and-effect analysis

A number of causes contribute to the overweight of small amount pensions in Latvia rooting in the institutional design of the country pension system. Some of them are immanent features of the chosen public Notional Defined Contribution (NDC) pension scheme, while the keys to others lie in ill-considered norms or in unfortunate course of events.

3.1 Lack of non-contributory basic pension

Latvian pension system completely lacks any non-contributory element in the form of basic pension that could be either flat-rate or linked to the length of service record, available to everyone or means-tested.

Guaranteed minimum income provision is not meant for the elderly people. It is not a part of social insurance system and is arranged by local municipalities from general taxes. Indeed, its amount in 2017 is 49.80 euro which is even lower than the minimum old-age pension.

The introduction of minimum income level (MIL) which was supposed to take place in 2016 and improve the situation with the poverty in Latvia has been postponed till 2019 without any political debates.

3.2 Inadequate minimum pension

The eligibility for an old-age social insurance pension is restricted by the minimum mandatory period of work experience – 15 years (will be raised to 20 years in 2025) and the minimum pension amount depends on the length of service (see Table 1):

Table 1. Minimum amounts of old-age pension in Latvia (2016).

Length of service (years)	Amount (EUR)
15–19	70.43
20–29	83.24
30–39	96.05
40+	108.85

These amounts are linked to the amount of the social security state benefit. Although the Cabinet Regulation stipulates that "*the Cabinet shall review the amount of the State social security benefit on the basis of a proposal by the Minister for Welfare in accordance with the possibilities of the State budget, as well as by assessing the economic situation in the State and by taking into account the average actual consumer price index specified by the Central Statistical Bureau*" [6], this wording is non-committal, and the amount of benefit has not been revised since 2006 in fact.

With the shift to individualised pension accounts the role of the minimum pension in poverty protection is increasing [7]. Latvia has not ratified either ILO Social Security (Minimum Standards) Convention, 1952 (No. 102), or ILO Invalidity, Old-age and Survivors' Benefits Convention, 1967 (No. 128). Both of them are setting the lower limits of the old-age pension to the average insured wage (50% of the average insured wage according to the Convention No. 102 or 56.25% of the average insured wage according to the Convention No. 128). When we compare the minimum pension amounts to the average insured wage in Latvia in 2015, we obtain the ratios 10.6–16.4% – which are by far lower than the international standards.

According to the data of the Ombudsman of the Republic of Latvia, 11% of all newly granted pensions are minimum pensions [8].

Within the above mentioned MIL concept, it is planned to link the statutory minimum pension to the minimum income level; however, no legislation drafts are available publicly and the introduction of the MIL has been postponed several times.

3.3 Mandatory pillars I and II: No redistribution

Currently, the design of the Latvian pension system is made up of three pillars, two mandatory and one voluntary.

The first (mandatory) pillar pension benefit is earned by insured individuals by "directing" part of their social insurance contributions to the personalised notional pension capital account. No actual money transfer takes place; this capital exists only as a record in the State Social Insurance Agency database, and the whole scheme is known as NDC: notional (or, in another version, "non-financial") defined contribution. The accrued notional capital is annually valorised in line with increases in the covered wage bill. These annual indices imitate the role of interest rates in funded schemes. The pension amount is the sum of notional capital at retirement divided by the life expectancy at retirement age.

The second (mandatory) pillar benefit is earned by insured individuals by directing part of their social insurance contributions to a private pension fund chosen by the insured person. In this pillar, actual money transfer is made from the state's special budget, accumulating the social insurance contributions, to pension funds, and the insured person acquires a certain amount of pension plan shares with each transfer. Pension funds invest the money into different financial instruments aiming to increase the net asset value per share. At the time of retirement the accumulated capital made of the insured's contributions plus accrued interest is converted into a monthly benefit either by adding it to the person's first-pillar NDC with further transforming into annuity according to the rules described above, or by purchasing a life pension insurance policy from an insurance company. However, the second option is practically not in use yet, only 6–7% of those who retired in 2015 and who were participating in the second pillar did use the option to buy an insurance in exchange of pension accumulations. At the end of 2016 there were four insurance companies offering these policies, three of them require an applicant to have at least 3000 euro accumulated in a second pillar pension fund, and one insurance company sets the minimum capital requirement at 5000 euro. Meanwhile, on 31/12/2015 the average accumulated capital among those with the longest possible record (14 years of participation) was 2878.78 euro on average [9]. Consequently, the majority of pensioners are not eligible for choosing the option to buy a life insurance.

Neither the first (NDC), nor the second (private pension funds) pension pillar in Latvia includes any redistribution from the rich to the poor As John B. Williamson and Matthew Williams have aptly noted, such systems "could actually end up redistributing money from the poor to the rich the rich tend to live longer than the poor, which means that the rich will often end up reaping disproportionately more from the system" [10]. In the funded pillar the reverse redistribution from the poor to the rich arises also because of worse financial literacy of the disadvantaged population who are therefore more exposed to making wrong investment decisions. The gender income disparities in funded and quasi-funded schemes are not evened either and are even amplified in old age [11].

3.4 Treatment of pre-reform service record (employment before 1996)

In 1995 Latvia became one of the first countries in the world to introduce the NDC scheme, and the only one that extended the NDC to all working population irrespective of their age. Therefore a mechanism was needed for incorporating the pre-reform service period into the NDC formula – i.e. the Soviet times when no social insurance contributions did exist at all, and the early 1990s. The so called "initial [notional pension] capital" was introduced for this purpose.

Three other European countries – Italy, Sweden and Poland – have also adopted the NDC in those years and faced similar problems. However, the approaches used by pension architects in relation to the pre-reform service records differ considerably [12]. In the same year of 1995, a pension reform introduced a NDC system in Italy. But in that country the NDC-scheme fully applies only to individuals entering the labour market from 1996 onwards, while people with at least 18 years of contributions in 1995 continued to be the subject to the previous system and people in between (with fewer than 18 years in employment by 1996) have pension calculated by a mix of the old and the new formula: pension rights obtained in the old system are not converted into the new one.

In Sweden, where the NDC concept was originally developed, the pension reform took place in 1999 and the reformed system applied in full to those born 1953 and later. Those born before 1938 were not included into the reformed scheme; those born between 1938 and 1952 received a mix of old defined-benefit and new NDC pension. NDC was "retroactively" applied to individual earnings histories from 1960 for everyone born 1938 and after to calculate their initial capital.

Poland launched NDC in 1999 as well and applied it to those born in 1949 and later, i.e. to those who were under 50 years at the time of implementation. Those aged 50+ continued to acquire their pension rights and have their pensions calculated according to the old defined-benefit formula. For people born after 31 December 1948, their initial capital was calculated according to the old rules reflecting their pension rights acquired before 1 January 1999.

In Latvia, the initial capital amount is based on two variables: (1) the length of the employment and equated periods until 31/12/1995; and (2) the average earnings during the period 01/01/1996–31/12/1999 (reference period) from which social insurance contributions were paid. Non-productive periods when the person did not get any earnings have the same weight as productive ones: the total amount of all earnings is divided by 48 months. To calculate the initial capital, the individual's average earnings in the reference period are extrapolated to their whole pre-reform working career.

This theoretically-derived initial capital is then incorporated into the person's notional capital.

The share of the initial capital in the total notional capital was close to 100% in the first post-reform years and until now the pre-reform service record has made more than a half in the whole employment history for any Latvian pensioner. So the principles of translation into the reformed pension formula are a very important factor influencing adequacy and equity of pensions.

The years taken as the reference period were quite difficult for Latvian economy, when a large share of population suffered from low earnings, long-term unemployment without any benefits, and/or grey "envelope" salaries. In addition, the population was not sufficiently informed of the reformed pension legislation and was unaware of the importance of that period to their future pensions.

It did not take long before the formula started generating a growing amount of very small pensions, so the initial law was amended to improve the situation partially. Should a person have no insured wage in the reference period or should his/her average insured wage in the reference period be lower than the statutory minimum wage, the minimum wage is taken into calculation. Starting from 2002, this provision was amended and split into two separate rules. A special rule of calculating the initial notional capital for people with low wages in 1996–1999 but having 30 or more years of employment record was introduced: Paragraph 33 of the Transitional Provisions to the Law on State Pensions says:

"*33. The calculation of the pension initial capital for an insured person, whose length of period of insurance is not less than 30 years and whose average monthly wage subject to insurance contributions, which has been calculated in accordance with Paragraph 13 of these Transitional Provisions, is less than the average wage subject to insurance contributions in the State in the period from 1 January 1996 to 31 December 1999, consideration shall be made of the 48 month average wage subject to insurance contributions in the State within the previously indicated time period. State pensions, which have been granted to insured persons from 1 January 2000 to 31 December 2001, and to whom the previously referred to conditions are applicable, based upon their request, shall be recalculated from 1 January 2002.*" [13] (Quoted as per wording in force in January 2017)

For the first time this Paragraph appeared in the Law on 20 December 2001 (in force since 01 January 2002), and the first version of the Paragraph included time restrictions: this rule was intended to be in force "up to 1 January 2010". The rule was designed as an ad hoc temporary measure. Later on, it became obvious that the number of such people remains high, and on 19 June 2008, the Saeima (Parliament) extended the validity of the Paragraph for two more years – "up to 31 December 2011". On 8 December 2011, one more two-year extension was granted – "up to 31 December 2013". Further, on 14 June 2012, it was again prolonged for two more years – "up to 31 December 2015". And finally, on 3 April 2014, the time restriction was lifted and the duration of this rule is not limited in time. On 18 June 2015, it was also preconditioned that at least five years of the insured period need to be earned after the reform (valid since 1 July 2017).

For those not having 30 years of employment, another benchmark is applied instead of the statutory minimum wage since 2002: the person's average insured wage in the reference period upgraded by annual valorisation indices is compared to 40% of the average countrywide insured earnings in the year prior to the year preceding the calendar year of retirement (i.e. in 2017 the benchmark year is 2015). Should the benchmark figure be higher, it is used for calculating the initial capital.

The 30-year boundary is arbitrarily dividing pensioners. The authors, therefore, considered worth of comparing two model cases on both sides of the boundary: 30-year record vs 29-year record.

Using the retrospective simulation method [14] we have compared a number of model cases: a person retiring at the end of 2014 and having either 29 or 30 years of employment record and at different wage levels throughout the career: from 50% of the country average wage to 250% of the country average. The official pensionable age in 2014 was 62 years and 3 months, so the maximum possible length of service record was assumed to be 44 years. The probability of a career interruption in each particular year varies in line with the registered unemployment level U in the country in the respective year t (the higher the unemployment – the lower the probability P to have the full year t in personal service record):

$$P_t = \frac{30\ (\text{or } 29)}{44} \times \frac{100 - U_t}{100 - \sum_{1971}^{2014} U_t / 44}, \quad \sum_{1971}^{2014} P_t = 30\ (\text{or } 29) \qquad (1)$$

The historical unemployment rates for the Eq. (1) are taken from LABORSTA (laborsta.ilo.org – Statistical Office of International Labour Organisation) and Eurostat. The earliest available data for Latvia is for 1992 (3.2%), for the period 1989–1991 Estonian figures were extrapolated to Latvia. As well as the rate 0.6% observed in Estonia in 1989 and 1990 was extrapolated to the whole period 1971–1988. Omitting the interim stages of further calculation, the obtained results are shown in Fig. 3 below.

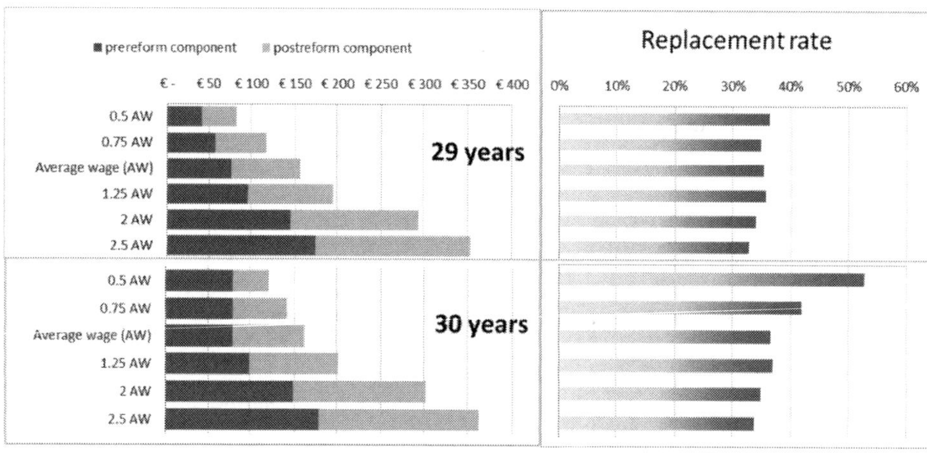

Fig. 3. Theoretical net old-age pension and replacement rate for a person retiring in 2014 depending on the wage and length of employment record. *Source: authors' calculations based on statistical and normative data.*

Due to the arbitrary threshold set at 30 years, the difference in net pension benefit between people with 29- and 30-years' record among low earners in Latvia is inequitably high: one extra year of service would bring 37.99 EUR for people earning 50% of the average wage and 23.94 EUR for people earning 75% of the average wage. Should a person have no employment in some specific "unlucky" years, these differences are pronounced even stronger. Although Paragraph 33 of the Transitional Provisions of the Law on State Pensions considerably improves the adequacy of low-earners with long service record in

Latvia – the replacement rate in the model case "30 years of employment at 50% of the country average wage" is 53% (which is graphically demonstrated above) in contrast to 36% in the model case "29 years of employment at 50% of the country average wage", – this approach is far from being truly equitable.

One more mechanism for compensation for the pre-reform employment was introduced in 2006, namely, the monthly monetary supplements to pensions amounting to one euro for each pre-reform year of service. Initially, they were granted only to people with low pensions assigned both before and after 1996, but starting from 2009 (in the course of pre-elective populist package) they were extended to all pensioners. Since 2012 these supplements have been phased out: those who retired before 2012 are still receiving the supplements, but later pensioners are not. In contrast to pensions, the supplements are not indexed, so their amount has not changed since 2006. Both of these factors: different treatment of pre-reform years for those retired before and after 1 January 2012 and lack of supplement indexation, – run in opposition to the fairness principle and need to be reviewed.

3.5 Taxation of pensions

In Latvia, the pensions assigned before 1996 (i.e. according to the pre-reform pension law) are not taxed for income tax. The pensions, assigned starting from 1996, are taxed for 23% income tax (this is the current rate, historically the rates varied from 23% to 26%). The non-taxable amount is set at the level of 235 euro (this level has not been raised since its introduction in 2009). As was demonstrated above, this threshold is significantly below the average pension level, and ca. 70% of all pensions surpass the tax exempt.

For comparison, the other two Baltic countries: in Lithuania pensions are not taxed at all, while in Estonia both the tax rate is lower (20%) and the tax exempt is higher (395 euro, which is higher than the average pension), making the tax burden on pensioners much less noticeable for the majority of pensioners [15].

4 Conclusions and recommendations

In the absence of redistribution mechanisms in mandatory pension pillars accompanied by high tax burden on pensions, the only systemic element aimed at alleviation of poverty in the old age is the statutory minimum pension, which is not equal to the task.

Therefore, within the existing institutional design of the pension system, the first and foremost way to fight poverty among the elderly population in Latvia is to increase the minimum statutory pension level(s) and to link it to absolute (e.g. consumer basket) or relative (e.g. at-risk-of-poverty threshold) indicators and ensure their annual revision.

The second easy-to-implement measure to address the low income group – is to increase the tax exempt and link it to the absolute (e.g. the statutory minimum wage) or relative (e.g. the average or median pension) indicators and ensure their annual revision.

Our third recommendation relates to the treatment of the pre-reform employment record. Although the very approach of using the short reference period during which serious economic perturbations took place for extrapolation to the major part of working career seems inappropriate, unfair and cannot be recommended to other countries planning their pension reforms, in case of Latvia it would be too costly to re-organise the translation formula fundamentally (like extending the reference period for a longer period; or excluding of non-productive periods from the reference period). Instead, we suggest the measures aimed at the most vulnerable groups: first, the 30-year threshold should be removed and the country average insured wage should be always used when it is higher than the individual's average insured wage. There are not many people with short service record, since employment was mandatory in the USSR. Second, we believe that the supplements for the pre-reform years of employment should be also allocated to pensions granted after 2011, as well as their amount should be increased.

We would also recommend starting a discussion on introducing of a non-contributory element into Latvian pension formula, in the form of a basic pension. In order not to increase the total social insurance contributions rate, this would need a split of the current mandatory pay-as-you-go pillar into two components: some part of the pillar I contributions may be directed to the "common pool" for basic pensions, while the rest still be registered on individual notional accounts in the NDC-scheme. Needless to say, that the exact proportion of this split as well as the form of the basic pension should to be a subject to a thorough analysis.

The paper was supported by the National Research Programme 5.2 EKOSOC-LV.

References

1. B. Shang, Pension Reform and Equity: the impact on poverty of reducing pension benefits. In: *Equitable and Sustainable Pensions*, 87 (IMF, Washington D.C., 2014)
2. Ministry of Welfare of the Latvian Republic, 20 gadi Latvijas pensiju sistēmai, 02/03/2016, http://www.lm.gov.lv/upload/aktualitates/4/20_preseskonference_02032016.pdf
3. I. Alliks, Divdesmit – daudz vai maz?, 02/06/2016, http://www.lm.gov.lv/upload/publikacijas/1/1_ingus_lv_pensiju_sistema_20_konference_02062016.pdf
4. LA.LV News portal, Lielākā pensija Latvijā ir 19 tūkstoši eiro, 25/07/2016, http://www.la.lv/lielaka-pensija-latvija-ir-19-tukstosi-eiro/
5. Postimees.ee News portal, Цахкна хочет поднять пенсионный возраст до 70 лет, 14/09/2016, http://rus.postimees.ee/3836433/
6. Republic of Latvia, Cabinet Regulation No. 1605, http://likumi.lv/doc.php?id=202850
7. A. Chlon-Dominczak, P. Strzelecki, Journal of Pension Economics and Finance, **12**, 326–350 (2013)
8. Ombudsman of the Latvian Republic. Svarīgs paziņojums minimālās algas saņēmējiem! 08/11/2016, http://www.tiesibsargs.lv/sakumlapa/svarigs-pazinojums-minimalas-algas-sanemejiem
9. State Social Insurance Agency. Pārskats par valsts fondēto pensiju sistēmas darbību 2015. gadā (2016), http://www.vsaa.lv/media/uploads/UserFiles/vfps_statistika/parskati/vfps_2015_gada_parskats_v04_kpmg_ar_audita_zinojumu.doc
10. J.B. Williamson, M. Williams. The Notional Defined Contribution Model: an Assessment of the Strength and Limitations of a New Approach to the Provision of Old Age Security. Working Paper 2003–18. Chestnut Hill, MA: Center for Retirement Research at Boston Colledge (2003), http://crr.bc.edu/wp-content/uploads/2003/10/wp_2003-181.pdf
11. O. Rajevska, F. Rajevska. Contributions to Humanities AGH **13/4**, 185 (2014) https://journals.agh.edu.pl/human/article/view/1666
12. E. Palmer, Conversion to NDCs – Issues and Models. In: *Pension Reform*, 182–183 (The World Bank, Washington D.C., 2006), http://siteresources.worldbank.org/INTPENSIONS/Resources/NDC_English.pdf

13. Republic of Latvia, Law on State Pensions, `http://likumi.lv/doc.php?id=38048`
14. O. Rajevska, Economics and Business, **28**, 13–19 (2016), `https://ortus.rtu.lv/science/lv/publications/22095/`
15. O. Rajevska, Economic Science for Rural Development, **43**, 352–357 (2016), `http://www.esaf.llu.lv/getfile.php?id=1588`

AUTHOR INDEX

EDP Sciences – Web of Conferences
17, Avenue du Hoggar
Parc d'Activité de Courtabouf
BP 112
F-91944 Les Ulis Cedex A
France

ISBN 978-1-5108-5727-8

CURRAN ASSOCIATES INC.
proceedings
.com